Bibliographic Atlas of Protein Spectra in the Ultraviolet and Visible Regions

EDITED BY
DONALD M. KIRSCHENBAUM

State University of New York
Downstate Medical Center
Brooklyn, New York

This third volume in a series devoted to the literature on ultraviolet and visible spectra of proteins and protein-containing material presents the location of 2434 different protein spectra. Enough information is given on most of the proteins to enable the user to duplicate the desired spectrum. There are three indexes:

- Alphabetical Name Index—lists all the proteins
- Alphabetical Source Index—lists the literature sources of all the protein spectra
- Additive/Modifier/Modification Index—lists the reagents and conditions that affect the spectra of the proteins listed

Along with Volumes 1 and 2 of the *Atlas of Protein Spectra in the Ultraviolet and Visible Regions,* this bibliographic atlas offers a broad overview of the literature in the protein field during the last ten years. Those working in the fields of biochemistry, analytical chemistry, molecular biology, and biochemical engineering will find the *Bibliographic Atlas of Protein Spectra in the Ultraviolet and Visible Regions* a welcome addition to their libraries.

Design by A. Palagano

Bibliographic Atlas of Protein Spectra in the Ultraviolet and Visible Regions

Bibliographic Atlas of Protein Spectra in the Ultraviolet and Visible Regions

EDITED BY
DONALD M. KIRSCHENBAUM
State University of New York
Downstate Medical Center
Brooklyn, New York

IFI/PLENUM • NEW YORK-WASHINGTON-LONDON

Library of Congress Cataloging in Publication Data

Kirschenbaum, D. (Donald Monroe), 1927–
Atlas of protein spectra in the ultraviolet and visible regions.

Vol. 3 has title: Bibliographic atlas of protein spectra in the ultraviolet and visible regions.
Includes bibliographical references and index.
1. Proteins—Spectra—Charts, diagrams, etc. I. Title. [DNLM: 1. Ultraviolet rays—bibliography. 2. Proteins—Analysis—Bibliography. 3. Spectrum analysis—Bibliography. ZQU 55 K61b]
QC463.P7K5 547.7′5046 77-183566
ISBN 978-1-4684-8838-8 ISBN 978-1-4684-8836-4 (eBook)
DOI 10.1007/978-1-4684-8836-4

Softcover reprint of the hardcover 1st edition 1984

A Division of Plenum Publishing Corporation
233 Spring Street, New York, N.Y. 10013

To Roslyn

He who ascribes a citation unto its author brings redemption unto the world.

—RABBI ELEZAR, citing RABBI HANINA
(Adapted from the Mishnah commentaries on the Megillah of Esther)

Some may think that the same talent and industry would be better devoted to original work: but it must be allowed that to elucidate and render accessible the labors of others may be a service as valuable as the addition of new material to the common store.

Lord Rayleigh [1874]

Cited by D.R. Lide and S.A. Rossmassler, Ann. Rev. Phys. Chem., 24, 135 [1973].

PREFACE

This is the third volume in a series devoted to a compilation of the ultraviolet and visible spectra of proteins and protein-containing materials. In this volume about 2500 citations for the 2434 indexed spectra are given. Of these, 50% are from the period 1971-1975, 45% from the period 1976-1980, 4% from the period 1960-1970, and 1% from the period prior to 1960. Thus, this volume together with its predecessors, Atlas of Protein Spectra in the Ultraviolet and Visible Regions, Volume 1 [1972] and Volume 2 [1974], both published by Plenum, constitutes a major source of spectral data of proteins.

The useful features of this volume are:

1. It furnishes a broad overview of the protein field for the past ten years, revealing the large number of proteins examined spectroscopically.

2. The indices, of which there are three, provide a very useful guide for those seeking a special protein from a special source treated in a special way:

A. The Name Index is arranged alphabetically, using the names given in the references cited.

B. The Source Index allows the user to go directly to a specific source for a specific protein spectrum.

C. The Additive/Modifier/Modification Index allows the user to learn the effect on protein spectra of various substances added to solutions of proteins and of various covalent and noncovalent modifications of proteins. In addition to chemicals which affect the proteins, other environmental modifiers such as pH, temperature, reaction time, and irradiation are listed.

3. At least one citation for each protein is the original data source.

4. Almost all the entries have Solvent and Comment sections which provide the user with the information necessary to repeat the spectral study without consulting the original literature.

Together, the three volumes in this series provide almost complete coverage of the literature concerned with spectral examination of proteins in the ultraviolet and visible regions. While these three volumes may not make all spectroscopists completely happy (e.g., slit widths and band widths are not mentioned), they will provide the information desired by a large number of researchers in various fields. In addition to the spectroscopic data, examination of the literature citations can provide useful marketing information for manufacturers of spectrophotometers.

The library of the Downstate Medical Center was the major source of the necessary journals and books. What was not available in the Library was obtained for me by the very efficient Interlibrary Loan Service. During the summer months of 1973-1981 the Library of the Marine Biological Laboratory, Woods Hole, Massachusetts was my working library. During a sabbatical in 1977 the Biology Library of the University of Sussex, Falmer, England, was used. I thank all the librarians at these libraries for their very necessary assistance.

I thank Ms. Pat Vann of Plenum Press for the helpful answers to my numerous questions, and Ms. Mindy Wolf for the excellent typing of the manuscript.

If any errors remain, they are entirely my responsibility, and I would appreciate having them brought to my attention.

Donald M. Kirschenbaum

CONTENTS

GLOSSARY OF ABBREVIATIONS

AMP	Adenosine 5'-monophosphate
ATP	Adenosine 5'-triphosphate
CoA	Coenzyme A
DNP	Dinitrophenyl
DPN^+/DPNH	Diphosphopyridine nucleotide, oxidized and reduced
DPG	2,3-Diphosphoglycerate
DTNB	5,5'-Dithiobis[2-nitrobenzoic acid]
DTT	Dithiothreitol
EDTA	Ethylenediaminetetraacetic acid
FAD/$FADH_2$	Flavin-adenine dinucleotide, oxidized and reduced
FMN	Flavin mononucleotide
NAD^+/NADH	Nicotine-adenine dinucleotide, oxidized and reduced
NADP, $NADP^+$-NADPH	Nicotine-adenine dinucleotide phosphate, oxidized and reduced
SDS	Sodium dodecylsulfate
TPN^+/TPNH	Triphosphopyridine nucleotide, oxidized and reduced
Tris	Tris[hydroxymethyl]aminomethane
U.V.	Ultraviolet

LISTING OF PROTEIN SPECTRA

1. Accelerator, Thrombotic

   SOURCE: Human Serum
   REFERENCE: R. M. Howell, M. A. Cook, and G. B. D. Scott, Nature, 204, 1088 (1964)

2. Acetohydroxy Synthase

   SOURCE: Escherichia coli
   SOLVENT: 20 mM Potassium phosphate, pH 7.0 - 0.5 mM DTT - 5% glycerol - 1 mM $MgCl_2$ - 10 M thiamine diphosphate - 1 μM FAD
   REFERENCE: H. Grimminger and H. E. Umbarger, J. Bacteriol., 137, 846 (1979)

3. Acetylcholinesterase

   SOURCE: Human erythrocytes
   SOLVENT: 5 mM Phosphate buffer, pH 6.6
   REFERENCE: G. Ciliv and P. T. Ozand, Biochim. Biophys. Acta, 284, 136 (1972)

4. Acetylcholinesterase

   SOURCE: Human erythrocytes
   SOLVENT: a. 0.01 M Tris buffer, pH 9.3; pH 7.3; pH 5.7
   b. 0.005 M Tris buffer, pH 8, urea concentration: 1 M, 2 M, 4 M, and 5 M
   REFERENCE: K. Sihotang, J. Biochem. (Tokyo), 75, 939 (1974)

5. Acetylcholine Receptor

   SOURCE: Electrophorus electricus
   COMMENTS: Tween 80, 1% w/v present
   REFERENCE: R. P. Klett, B. W. Fulpus, D. Cooper, M. Smith, E. Reich and L. D. Possani, J. Biol. Chem., 248, 6841 (1973)

6. endo-β-N-Acetylglucosaminidase H

SOURCE: Streptomyces griseus
SOLVENT: 0.01 M Potassium phosphate buffer, pH 8.45
REFERENCE: A. L. Tarentino and F. Maley, J. Biol. Chem., 249, 811 (1974)

7. endo-N-Acetylmuramidase, β enzyme

SOURCE: Bacillus subtilis YT 25
SOLVENT: 0.01 M Acetate buffer, pH 5.0
REFERENCE: Y. Takahara, E. Michigaki, and S. Murao, Agr. Biol. Chem., 38, 2357 (1974)

8. N-Acetylneuraminate Lyase

SOURCE: Clostridium perfringens
SOLVENT: 0.1 M Phosphate buffer, pH 7
REFERENCE: S. Nees, R. Schauer, R. Mayer, and K. Ehrlich, Hoppe-Seyler's Z. Physiol. Chem., 357, 839 (1976)

9. Acetylpyruvate Hydrolase

SOURCE: Pseudomonas putida 01
REFERENCE: J. F. Davey and D. W. Ribbons, J. Biol. Chem., 250, 3826 (1975)

10. Acetyl-Co A Synthetase

SOURCE: Rabbit heart
COMMENTS: Active and photolyzed enzyme
REFERENCE: E. V. Sharkova, Biokhimiya (Biochemistry), 33, 648 (1968)

11. Acetyl-Co A Synthetase

SOURCE: Yeast
SOLVENT: 0.1 M Potassium phosphate buffer, pH 7.4
REFERENCE: T. Satyanarayana and H. P. Klein, Arch. Biochem. Biophys., 174, 480 (1976)

12. O-Acetylserine Sulfhydralase

SOURCE: Salmonella typhimurium
SOLVENT: 0.1 M Tris-HCl, pH 7.6
REFERENCE: P. F. Cook and R. T. Wedding, J. Biol. Chem., 251, 2023 (1976)

13. O-Acetylserine-O-acetylhomoserine Sulfhydralase

SOURCE: Yeast
SOLVENT: 0.1 M Potassium phosphate buffer, pH 7.8, containing 1 mM EDTA and 100 μM pyridoxal 5-phosphate
COMMENTS: Treated with 10 ml of 0.05 mg of sodium borohydride/ ml 0.05 N sodium hydroxide and spectra taken at various times
REFERENCE: S. Yamagata and K. Takeshima, J. Biochem. (Tokyo), 80, 777 (1976)

14. Aconitase

SOURCE: Pig heart
SOLVENT: 0.015 M Tris-tricarballylate, pH 7.8
REFERENCE: S. C. Kennedy, R. Rauner, and O. Gawron, Biochem. Biophys. Res. Comm., 47, 740 (1972)

15. Aconitase

SOURCE: Yeast
SOLVENT: 0.05 M Tris-HCl buffer, pH 7.4
REFERENCE: T. Suzuki, O. Yamazaki, K. Nara, S. Akiyama, Y. Nakao, and H. Fukuda, J. Biochem. (Tokyo), 77, 367 (1975)

16. Aconitase

SOURCE: Yeast
SOLVENT: 0.05 M Tris-HCl buffer, pH 7.4
COMMENTS: Spectra of holo- and apo-enzymes
REFERENCE: T. Suzuki, S. Akiyama, S. Fujimoto, M. Ishikawa, Y. Nakao, and H. Fukuda, J. Biochem. (Tokyo), 80, 805 (1976)

17. Actin

REFERENCE: W. F. H. M. Mommaerts, J. Biol. Chem., 38, 559 (1951)

18. Actin

SOURCE: Rabbit skeletal muscle
REFERENCE: T. Wieland, J. X. deVries, A. Schafer, and H. Faulstick, FEBS Lett., 54, 73 (1975)

19. Actin, alkylated

SOURCE: Rabbit skeletal muscle
SOLVENT: 10 mM Tris-acetate, pH 7.5
COMMENTS: Change in spectrum upon addition of 1-anilinonaptha-lene-5-sulfonic acid
REFERENCE: K. Mihashi, Biochim. Biophys. Acta, 267, 409 (1972)

20. Actin, Cu^{++}-G-

SOURCE: Rabbit skeletal muscle
SOLVENT: 1 mM N-2-Hydroxyethylpiperazine-N-2-ethanesulfonic acid and 0.5 mM ATP, pH 7.5
COMMENTS: Cu^{++}: actin = 0.8 on a molar basis
REFERENCE: S. S. Lehrer, B. Nagy, and J. Gergely, Arch. Biochem. Biophys. 150, 164 (1972)

21. β-Actinin

SOURCE: Slime mold (Physarum polycephalum)
SOLVENT: 1 mM Sodium carbonate
REFERENCE: K. Maruyama, R. Kamiya, S. Kimura, and S. Hatano, J. Biochem. (Tokyo), 79, 709 (1976)

22. Acyl Carrier Protein (1-74)

SOURCE: Escherichia coli B
SOLVENT: 0.01 M Tris-HCl, pH 7.3
REFERENCE: W. S. Hancock, D. J. Prescott, G. R. Marshall, and P. R. Vagelos, J. Biol. Chem., 247, 6224 (1972)

23. Acyl Carrier Protein

SOURCE: Klebsiella aerogenes
SOLVENT: a. 5 mM Potassium phosphate, pH 7.0
b. 50 mM Ammonium bicarbonate
REFERENCE: a. P. Dimroth, FEBS Lett., 51, 100 (1975)
b. P. Dimroth, Eur. J. Biochem., 64, 269 (1976)

24. Acyl-Coenzyme A Dehydrogenase

SOURCE: Pig kidney
SOLVENT: a. 160 mM Phosphate, pH 7.6
b. 100 mM Phosphate, pH 7.6
COMMENTS: Dithionite reduction and ferricyanide oxidation; anaerobic titration with octanoyl-CoA, with palmitoyl-CoA, and with butyryl-CoA
REFERENCE: C. Thorpe, R. G. Matthews, and C. H. Williams, Jr., Biochemistry, 18, 331 (1979)

25. Adenosine Deaminase

SOURCE: Pseudomonas iodinum IFO 3558
SOLVENT: 0.05 M Tris-HCl buffer, pH 7.05 containing 15% ethyl alcohol
REFERENCE: T. Sakai and H. K. Jun, FEBS Lett., 86, 174 (1978)

26. Adenosine Triphosphatase

SOURCE: Beef heart mitochondria
COMMENTS: Spectra of the protein after reaction with 7-chloro-4-nitro-2,1,3-benzoxadiazole and following addition of dithiothreitol
REFERENCE: S. J. Ferguson, W. J. Lloyd, and G. K. Radda, FEBS Lett., 38, 234 (1974)

27. Adenosine Triphosphatase (ATPase)

SOURCE: Beef heart mitochondria
SOLVENT: 50 mM Triethanolamine-HCl buffer containing 200 mM sucrose, 4 mM ATP, and 2 mM EDTA, pH 7.5 or 8.5
COMMENTS: Spectral changes seen (A) during reaction of ATPase with N-acetyl-O-7-nitrobenzofurazan tyrosine ethyl ester, (B) during reaction of 7-nitro-benzofurazan-modified ATPase, with N-acetylcysteine, and (C) during reaction of 7-nitrobenzofurazan-modified ATPase with sodium dodecylsulfate
REFERENCE: S. J. Ferguson, W. J. Lloyd, M. H. Lyons, and G. K. Radda, Eur. J. Biochem., 54, 117 (1975)

28. Adenosine Triphosphatase

SOURCE: Beef heart mitochondria
SOLVENT: 10 mM Triethanolamine-HCl buffer containing 4 mM ATP, 2 mM EDTA, 200 mM sucrose, pH 9.0
COMMENTS: Spectral change due to shift of 7-nitrobenzofurazan group from O atom to N atom on ATPase
REFERENCE: S. J. Ferguson, W. J. Lloyd, and G. K. Radda, Eur. J. Biochem., 54, 127 (1975)

29. Adenosine Triphosphatase, Inhibitory Subunit

SOURCE: Escherichia coli strain ML 308-225
SOLVENT: 100 mM Phosphate, pH 7.0
REFERENCE: P. C. Sternweis and J. B. Smith, Biochemistry, 19, 526 (1980)

30. Adenosine Triphosphatase

SOURCE: Micrococcus lysodeikticus
SOLVENT: a. Water
b. 6 M Guanidine-HCl
COMMENTS: Spectra of α and β subunits
REFERENCE: J. M. Andreu and E. Munoz, Biochemistry, 18, 1836 (1980)

31. Adenosine Triphosphatase

SOURCE: Micrococcus lysodeikticus
SOLVENT: 0.005 M Tris-HCl buffer, pH 8.0
REFERENCE: E. I. Mileikovskaya, I. A. Kozlov, and G. V. Tikhonova, Biochemistry (Biokhimiya), 40, 846 (1975)

32. Adenosine Triphosphatase

SOURCE: Micrococcus lysodeikticus
SOLVENT: 6 M Guanidinium chloride
REFERENCE: M. Nieto and J. A. Ayala, Biochem. J., 161, 321 (1977)

33. Adenosine Triphosphatase

SOURCE: Yeast (Saccharomyces cerevisiae)
SOLVENT: a. 50 mM Tris-sulfate buffer containing 0.25 M sucrose, 2 mM EDTA, pH 8.0
b. 100 mM Tris-sulfate buffer containing 4 mM EDTA and 50% glycerol, pH 8.0
REFERENCE: K. Takeshige, B. Hess, M. Bohm, and H. Zimmerman-Telschow, Hoppe-Seyler's Z. Physiol. Chem., 357, 1605 (1976)

34. Adenosine Triphosphatase Activator, Ca^{++}-stimulated

SOURCE: Human erythrocytes
SOLVENT: pH 7 to 8
REFERENCE: H. W. Jarrett and J. T. Penniston, J. Biol. Chem., 253, 4676 (1978)

35. S-Adenosylmethionine Decarboxylase

SOURCE: Yeast
SOLVENT: 50 mm Tris-HCl, 1 mM dithiothreitol, 0.1 mM EDTA, 2.5 mM putrescine·2HCl, pH 7.2
REFERENCE: M. S. Cohn, C. W. Tabor, and H. Tabor, J. Biol. Chem., 252, 8212 (1977)

36. S-Adenosylmethionine Decarboxylase

SOURCE: Rat liver
SOLVENT: 160 mM Tris-HCl, 0.1 M EDTA, 1.0 mM dithiothreitol, 2.5 mM putrescine, pH 7.5
REFERENCE: A. A. DeMetriou, M. S. Cohn, C. W. Tabor, and H. Tabor, J. Biol. Chem., 253, 1684 (1978)

37. S-Adenosyl-L-methionine-ε-N-L-lysine Methyltransferase

SOURCE: Neurospora crassa 33933
REFERENCE: P. R. Borumst and H. P. Broquist, J. Biol. Chem. 252, 5651 (1977)

38. Adenylate Kinase

SOURCE: Pig heart muscle
SOLVENT: 50 mM Imidazole-HCl, pH 6.5
COMMENTS: Spectrum of phenylglyoxal-treated material also presented
REFERENCE: J. Berghauser and R. H. Schirmer, Biochim. Biophys. Acta, 537, 428 (1978)

39. Adenylate Kinase, nitro-

SOURCE: Pig muscle
SOLVENT: 0.133 M Buffer
a. pH 10.83 (glycine-NaOH),
b. pH 9.18 (glycine-NaOH),
c. pH 8.06 (Tris-HCl)
d. pH 6.94 [3-(N-morpholino) propanesulfonic acid]
REFERENCE: M. Yazawa and L. H. Noda, J. Biol. Chem., 251, 3021 (1976)

40. Adenylate Kinase

SOURCE: Pig
SOLVENT: 50 mM Potassium N-2-hydroxyethylpiperazine-N'-2-ethanesulfonate, pH 7.9
COMMENTS: Time study of reaction with NBD (7-Chloro-4-nitrobenzo-2-oxa-1,3-diazol)
REFERENCE: N. C. Price, M. Cohn and R. H. Schirmer, J. Biol. Chem., 250, 644 (1975)

41. Adenyl Sulfate Reductase

SOURCE: Desulfovibrio vulgaris
SOLVENT: 0.05 M Phosphate buffer, pH 7.5
REFERENCE: R. N. Bramlett and H. D. Peck, J. Biol. Chem., 250, 2979 (1975)

42. Adrenodoxin

SOURCE: Beef adrenals
SOLVENT: 323 mM Tris buffer in ethanol:water (1:2), pH 8.8
COMMENTS: Changes in spectrum during reaction with triphenylphosphine
REFERENCE: T. Kimura, Y. Nagata, and T. Tsurugi, J. Biol. Chem., 246, 5140 (1971)

43. Adrenodoxin

SOURCE: Beef adrenal cortex
SOLVENT: 50 mM Potassium phosphate buffer, pH 7.4
REFERENCE: K. Suhara, S. Takemori, and M. Katagiri, Biochim. Biophys. Acta, 263, 272 (1972)

44. Adrenodoxin

SOURCE: Beef adrenal glands
SOLVENT: 0.05 M Sodium phosphate buffer, pH 7.4 containing $5x10^{-9}$ M benzylviologen
COMMENTS: Anaerobic titration with $5.2x10^{-3}$ sodium dithionite
REFERENCE: J. J. Huang and T. Kimura, Biochemistry, 12, 406 (1972)

45. Adrenodoxin

SOURCE: Beef adrenal glands
SOLVENT: 0.5 M Phosphate buffer, pH 7.4
COMMENTS: Spectrum after dithionite reduction
REFERENCE: S. S. Mardanyan and R. M. Nalbandyan, Biochemistry (Biokhimiya), 40, 40 (1975)

46. Adrenodoxin, amino-, apo-

SOURCE: Beef adrenal glands
SOLVENT: a. 10 mM Tris buffer, pH 7.5
b. 10 mM Phosphate buffer and 0.50 M potassium chloride, pH 7.4
COMMENTS: a. Apoaminoadrenodoxin, oxidized, reconstituted
b. Apoaminoadrenodoxin, reduced, reconstituted
REFERENCE: T. Taniguchi and T. Kimura, Biochemistry, 14, 5573 (1975)

47. Adrenodoxin, nitro-, apo-

SOURCE: Beef adrenal glands
SOLVENT: a. 10 mM Tris, 0.5 M potassium chloride, pH 7.5
b. 50 mM Tris, 1 M sodium chloride, pH 8.0
c. 50 mM Phosphate buffer, pH 7.4

COMMENTS: a. Nitroadrenodoxin, reconstituted
b. Aponitroadrenodoxin
c. Aponitroadrenodoxin in presence of 4 M urea
d. Apoadrenodoxin in solvent c

REFERENCE: T. Taniguchi and T. Kimura, Biochemistry, 14, 5573 (1975)

48. Adrenodoxin, apo-

SOURCE: Pig adrenal glands

COMMENTS: a. Native adrenodoxin.
b. ^{57}Fe-enriched, reconstituted adrenodoxin.
c. Apoadrenodoxin.

REFERENCE: R. Cammack, K. K. Rao, D. O. Hall, and C. E. Johnson, Biochem J., 125, 849 (1971).

49. Adrenodoxin Reductase

SOURCE: Beef adrenal mitochondria

SOLVENT: 50 mM Potassium phosphate, pH 7.5

COMMENTS: 1. Spectral changes during reoxidation by oxygen with excess NADPH or NADH
2. Spectral changes leading to semiquinone form of enzyme

REFERENCE: T. Sugiyama, R. Miura, and T. Yamano, J. Biochem. (Tokyo), 86, 213 (1979)

50. Adrenodoxin Reductase

SOURCE: Beef adrenal glands

SOLVENT: 0.2 M Phosphate buffer, pH 7.4

REFERENCE: R. P. Foster and L. D. Wilson, Biochemistry, 14, 1477 (1975)

51. Adrenodoxin Reductase, oxidized and reduced

SOURCE: Beef adrenal glands

SOLVENT: 0.01 M Sodium phosphate buffer, pH 7.4

COMMENTS: Anaerobic titration with sodium dithionite

REFERENCE: J. W. Chu and T. Kimura, J. Biol. Chem., 248, 2089 (1973)

52. Adrenodoxin Reductase, oxidized

SOURCE: Beef adrenals

SOLVENT: 10 mM Potassium phosphate, pH 7.76

COMMENTS: a. Anaerobic reduction with NADPH
b. Anaerobic reduction with NADPH, isocitrate and isocitrate dehydrogenase

c. Anerobic reduction with sodium dithionite
d. Reoxidation by air
e. Reaction of reduced enzyme with $NADP^+$
f. Reaction of reduced enzyme with NADPH
g. Reduction with NADPH then reaction with $NADP^+$

REFERENCE: J. O. Lambeth and H. Kamin, J. Biol. Chem., 251, 4299 (1976)

53. Adrenodoxin Reductase, NADPH-

SOURCE: Beef adrenocortical mitochondria
SOLVENT: 0.1 M Phosphate buffer, pH 7.4
REFERENCE: T. Sugiyama and T. Yamana, FEBS Lett., 54, 109 (1975)

54. Adrenodoxin Reductase, NADPH-

SOURCE: Beef adrenal glands
SOLVENT: 0.1 M Potassium phosphate buffer, pH 7.4
COMMENTS: Dithionite-reduced and apo-NADPH-adrenodoxin reductase spectra
REFERENCE: A. Hiwatashi, Y. Ichikawa, N. Maruya, T. Yamano, and K. Aki, Biochemistry, 15, 3082 (1976).

55. Adrenodoxin Reductase, NADPH-

SOURCE: Beef adrenocortical mitochondria
SOLVENT: 50 mM Potassium phosphate buffer, pH 7.4
REFERENCE: K. Suhara, Y. Ikeda, S. Takemori, and M. Katagiri, FEBS Lett., 28, 45 (1972)

56. Adrenodoxin Reductase, NADPH-

SOURCE: Pig adrenal mitochondria
SOLVENT: 0.1 M Potassium phosphate buffer, pH 7.4
COMMENTS: Spectra of oxidized and dithionite-reduced forms
REFERENCE: A. Hiwatashi and Y. Ichikawa, J. Biochem. (Tokyo), 84, 1071 (1978)

57. Adrenodoxin Reductase, NADPH-

SOURCE: Pig adrenocortical mitochondria
SOLVENT: 0.1 M Potassium phosphate buffer, pH 7.4
COMMENTS: Oxidized and dithionite reduced forms
REFERENCE: A. Hiwatashi, Y. Ichikawa, and T. Yamano, FEBS Lett., 82, 201 (1979)

58. Adrenodoxin Reductase-Adrenodoxin Complex

SOURCE: Beef adrenal cortex

SOLVENT: 10 mM Potassium phosphate, pH 7.5
COMMENTS: Complex titrated with NADH and spectra recorded
REFERENCE: J. D. Lambeth, D. R. McCaslin, and H. Kamin, J. Biol. Chem., 251, 7545 (1976)

59. Adrenodoxin Reductase-Adrenodoxin Complex

SOURCE: Beef adrenal cortex
SOLVENT: 10 mM Potassium phosphate, pH 7.0-7.1
COMMENTS: Spectra of reduced and oxidized forms
REFERENCE: J. D. Lambeth and H. Kamin, J. Biol. Chem., 254, 2766 (1979)

60. Agglutinin

SOURCE: Peanut (Arachis hypogaea)
SOLVENT: pH 8.0
REFERENCE: W. W. Fish, L. M. Hamlin, and R. L. Miller, Arch. Biochem. Biophys., 190, 693 (1978)

61. Agglutinin

SOURCE: Wheat germ
SOLVENT: a. Phosphate buffered saline, pH 7.3
b. Phosphate buffered saline, pH 7.35
COMMENTS: Spectra of agglutinin and fluorescein isothiocyanate-labelled agglutinin
REFERENCE: A. P. Cronin, F. Biddle, and F. K. Sanders, Cytobios, 2, 225 (1970)

62. Agglutinin

SOURCE: Wheat germ
SOLVENT: a. 0.1 M Acetic acid, 8 M urea, pH 3.9
b. 0.1 M Citrate buffer, 0.2 M NaCl, pH 6.0
COMMENTS: Spectra of N-bromosuccinimide oxidized and native protein
REFERENCE: J. P. Privat, R. Lotan, P. Bouchard, N. Sharon, and M. Monsigny, Eur. J. Biochem., 68, 563 (1976)

63. Agglutinin

SOURCE: Wheat germ
SOLVENT: 0.15 M NaCl, 0.05 M Tris-HCl, pH 7.2
REFERENCE: M. Monsigny, F. Delmotte, and C. Helene, Proc. Nat. Acad. Sci. 75, 1324 (1978)

64. Alanine Dehydrogenase

SOURCE: Bacillus subtilis
SOLVENT: a. 0.05 M Tris buffer, pH 8.0
b. 0.1 M NaOH
REFERENCE: A. Yoshida and E. Freese, Biochim. Biophys. Acta, 92, 33 (1964)

65. Alanine Racemase

SOURCE: Bacillus subtilis
REFERENCE: R. B. Johnson, J. J. Scholz, W. F. Diven, and S. Shepherd, in Pyridoxal Catalysis: Enzymes and Model Systems (E. E. Snell, A. E. Braunstein, E. S. Severin, and Yu. M. Torchinsky, eds.) 537 (1968), Interscience, New York

66. Alanine Racemase

SOURCE: Escherichia coli
COMMENTS: Enzyme treated with [1,2-^{14}C, 2-^{2}H]-D-fluoroalanine and spectrum recorded
REFERENCE: E. Wang and C. Walsh, Biochemistry, 17, 1313 (1978)

67. Albumin

SOURCE: Human serum
COMMENTS: 30% aqueous solution of twice recrystallized protein
REFERENCE: E. J. Cohn, W. L. Hughes, Jr., and J. H. Weare, J. Am. Chem. Soc., 69, 1753 (1947)

68. Albumin

SOURCE: Human serum
SOLVENT: 67 mM Phosphate buffer, pH 7.4
COMMENTS: Spectra of albumin, urea-albumin, and 2-hydroxy 5-nitrobenzyl modified albumin.
REFERENCE: K. J. Fehske, W. E. Muller, and U. Wollert, Hoppe-Seyler's Z. Physiol. Chem., 359, 709 (1978)

69. Albumin

SOURCE: Human serum
SOLVENT: 0.01 M Glycine buffer, pH 9.0
COMMENTS: Effect of covalently bound salicylic acid on spectrum
REFERENCE: P. Nuhn and M. Strecker, Pharmazie, 29, 10 (1978)

70. Albumin

SOURCE: Human serum
SOLVENT: 0.1 M Ammonium bicarbonate
COMMENTS: Spectrum of dinitrophenylsulfonyl albumin
REFERENCE: M. Wilchek and T. Miron, Biochim. Biophys. Acta, 278, 1 (1972)

71. Albumin

SOURCE: Human serum
COMMENTS: Spectrum of AMP-albumin conjugate
REFERENCE: S. Estrada-Parra and E. Garcio-Ortigoza, Immunochemistry, 9, 799 (1972)

72. Albumin

SOURCE: Human serum
SOLVENT: 0.1 N NaOH
COMMENTS: Spectrum of albumin modified by addition of tryptophan
REFERENCE: G. I. Tesser, R. A. O. M. M. de Hoog-Declerck, and L. W. Westerhuis, Hoppe-Seyler's Z. Physiol. Chem., 356, 1625 (1975)

73. Albumin

SOURCE: Human serum
SOLVENT: a. 0.1 N NaOH
b. 0.05 M Borate-phosphate, pH 6.0
COMMENTS: Spectra of albumin modified with 5-amino-tetrazole, o-aminoazotoluene, 2-chloro-4-nitroaniline, sulfanilic acid, ρ-aminobenzoic acid, α-aminonaphthol, and ρ-methyldiethylammonium aniline
REFERENCE: M. Franek and Z. Pechan, Scripta Fac. Sci. Nat. Brun. Chemia, 2, 67 (1971)

74. Albumin

SOURCE: Human serum
SOLVENT: pH 10.75
COMMENTS: Spectra as a function of iodine content
REFERENCE: W. L. Hughes, Jr. and R. Straessle, J. Am. Chem. Soc., 72, 452 (1950)

75. Albumin

SOURCE: Human serum

COMMENTS: Spectra of haematin albumin and haem albumin.
REFERENCE: J. Keilin, Nature, 154, 120 (1944)

76. Albumin

SOURCE: Beef serum
COMMENTS: Spectra of unirradiated and U.V. irradiated solutions
REFERENCE: I. M. Claesson, Arkiv fur Kemi, 10, 4 (1956)

77. Albumin

SOURCE: Beef serum
COMMENTS: Spectrum with yeast ribonucleic acid also shown
REFERENCE: V. F. Kalb, Jr. and R. W. Bernlohr, Anal. Biochem., 82, 362 (1977)

78. Albumin

SOURCE: Beef serum
REFERENCE: B. Ehresmann, P. Imbault, and J. H. Weil, Arch. Biochem., 54, 454 (1973)

79. Albumin

SOURCE: Beef serum
SOLVENT: 20 mM Tris, pH 8.0
COMMENTS: Spectra of albumin with and without jatrophone
REFERENCE: J. R. Lillehaug, K. Kleppe, C. W. Sigel, and S. M. Kupchan, Biochim. Biophys. Acta, 327, 92 (1973)

80. Albumin

SOURCE: Beef serum
COMMENTS: Spectra of albumin and aquocobalamin and albumin after reduction with formate
REFERENCE: E. L. Lien and J. M. Wood, Biochim. Biophys. Acta, 264, 530 (1972)

81. Albumin

SOURCE: Beef serum
REFERENCE: R. M. Howell, M. A. Cook, and G. B. D. Scott, Nature, 204, 1088 (1964)

82. Albumin

SOURCE: Beef serum

SOLVENT: 0.05 M Potassium phosphate - $1x10^{-4}$ EDTA, pH 7.8
REFERENCE: R. A. Weisiger and I. Fridovich, J. Biol. Chem., 248, 3582 (1973)

83. Albumin

SOURCE: Beef serum
SOLVENT: 0.1 M Tris HCl, pH 8.2
COMMENTS: Spectra of albumin at different persulfuration extent
REFERENCE. A. Finazzi-Agro, G. Federici, C. Giovagnoli, C. Cannelda, and D. Cavallini, Eur. J. Biochem., 28, 89 (1972)

84. Albumin

SOURCE: Beef serum
COMMENTS: Spectra of albumin and N-acetyl-serotonin-albumin conjugate
REFERENCE: L. J. Grota and G. M. Brown, Can. J. Biochem., 52, 196 (1974)

85. Albumin

SOURCE: Beef serum
SOLVENT: a. pH 6.0 - 6 M guanidinium HCl
b. pH 11.7 - 6 M guanidinium HCl
REFERENCE: G. Barth, E. Bunnenberg, and C. Djerassi, Anal. Biochem., 48, 471 (1972)

86. Albumin

SOURCE: Beef serum
SOLVENT: 0.1 M Phosphate buffer, pH 7.0
REFERENCE: T. Ichikawa and H. Terada, Biochim. Biophys. Acta, 580, 120 (1979)

87. Albumin

SOURCE: Beef serum
COMMENTS: Spectra of mixtures of albumin and 8-methoxysporalen stored in the dark
REFERENCE: K. Yoshikawa, N. Mori, S. Sakakibara, N. Mizuno, and P. S. Song, Photochem. Photobiol., 29, 1127 (1979)

88. Albumin

SOURCE: Beef serum
COMMENTS: Spectrum of albumin-glucose derivative
REFERENCE: A. Mohammad, H. Fraenkel-Conrat, and H. S. Olcott, Arch. Biochem., 24, 157 (1949)

89. Albumin - Bilirubin Complex

SOURCE: Human serum
SOLVENT: pH 4.0 - pH 7.3
REFERENCE: G. Blauer, D. Harmatz, and J. Snir, Biochim. Biophys. Acta, 278, 68 (1972)

90. Albumin - Bilirubin Complex

SOURCE: Human serum
SOLVENT: a. pH 5.0
b. pH 4.8
COMMENTS: a. Effect of Na_2SO_4 on spectrum
b. Effect of NaCl on spectrum
REFERENCE: G. Blauer and D. Harmatz, Biochim. Biophys. Acta, 278, 89 (1972)

91. Albumin -Bilirubin Complex

SOLVENT: 0.1 M NaCl, pH 4.2, 9.8, 7.4, 4.0
COMMENTS: Albumin from beef, goat, human, rabbit, pig, and chicken used
REFERENCE: D. Harmatz and G. Blauer, Arch. Biochem. Biophys. 170, 375 (1975)

92. Albumin -Chloranil Complex

SOURCE: Horse serum
SOLVENT: 50% ethanol, pH 8
REFERENCE: J. B. Birks and M. A. Slifkin, Nature, 197, 42 (1963)

93. Albumin, Cr^{III}-

SOURCE: Beef serum
COMMENTS: One chromium cation per six amino acid residues
REFERENCE: A. Danchin, Biochimie, 57, 875 (1975)

94. Albumin, Deoxyribulose-

SOURCE: Human serum
SOLVENT: a. 0.01 N NaOH
b. 0.01 N HCl
COMMENTS: Deoxyribulose derivative made by reaction with ribose
REFERENCE: L. Berrens, Immunochemistry, 4, 37 (1967)

95. Albumin-Iron-X Complexes

SOURCE: Beef serum
SOLVENT: 0.2 M Borate buffer, pH 8.8
COMMENTS: Spectra of complexes with X corresponding to sulfur, selenium, or tellurium
REFERENCE: S. Arakawa and T. Kimura, Biochim. Biophys. Acta, 580, 382 (1979)

96. Albumin, Succinylated

SOURCE: Beef serum
SOLVENT: pH 8.0, pH 7.0
COMMENTS: Spectra of succinylated albumin after 5 min.- 240 min. of desuccinylation
REFERENCE: T. S. Chang and S. F. Sun, Int. J. Prot. and Pep. Res., 11, 65 (1978)

97. Alcalase

SOURCE: Commercial
SOLVENT: 0.01 M Sodium phosphate buffer, pH 8
COMMENTS: Spectra of native and glutaraldehyde treated protein
REFERENCE: J. Boudrant, J. L. Cuq., and C. Cheftel, Biotech. Bioeng., 18, 1719 (1976)

98. Alcohol Dehydrogenase

SOURCE: Acetobacter aceti
SOLVENT: a. 0.1% Triton X-100, 1 mM 2-mercaptoethanol (reduced form)
b. pH 9, 0.1% Triton X-100 (oxidized form)
REFERENCE: O. Adachi, E. Miyagawa, E. Shinagawa, K. Matsushita, and M. Ameyama, Agric. Biol. Chem., 42, 233 (1978)

99. Alcohol Dehydrogenase

SOURCE: Gluconobacter suboxydans
SOLVENT: 0.1% Triton X-100
COMMENTS: Spectra of reduced and oxidized forms
REFERENCE: O. Adachi, K. Tayama, E. Shinagawa, K. Matsushita, and M. Ameyama, Agric. Biol. Chem., 42, 2045 (1978)

100. Alcohol Dehydrogenase

SOURCE: Methylococcus capsulatus
SOLVENT: 20 mM Tris-HCl buffer, pH 8.0
REFERENCE: R. N. Patel, H. R. Bose, W. J. Mandy, and D. S. Hoare, J. Bacteriol., 110, 570 (1972)

101. Alcohol Dehydrogenase, Phenazine Methosulfate-Dependent, Long-Chain Alcohol

SOURCE: Pseudomonas aeruginosa
REFERENCE: J. P. Tassin, C. Celier, and J. P. Vandecasteele, Biochim. Biophys. Acta, 315, 220 (1973)

102. Alcohol Dehydrogenase, Primary Alcohol

SOURCE: Pseudomonas sp. No. 2941
SOLVENT: 50 mM Tris-HCl, pH 8.0
REFERENCE: K. Yamanaka and K. Matsumoto, Agr. Biol. Chem., 41, 467 (1977)

103. Alcohol Dehydrogenase

SOURCE: Rhodopseudomonas acidophila
REFERENCE: C. W. Bamforth and J. R. Quayle, Biochem. J., 181, 517 (1979)

104. Alcohol Dehydrogenase, Zn-, Co-, Cd-

SOURCE: Horse liver
SOLVENT: 0.1 M Tris-acetate, pH 7.0
REFERENCE: D. E. Drum and B. L. Vallee, Biochem. Biophy. Res. Comm., 41, 33 (1970)

105. Alcohol Dehydrogenase

SOURCE: Horse liver
SOLVENT: 0.033 M Sodium phosphate, 25 mM EDTA, pH 8.0
COMMENTS: Spectra in the picolinimidyl, acetimidyl-derivative and native enzyme
REFERENCE: B. V. Plapp, R. L. Brooks, and J. D. Shore, J. Biol. Chem., 248, 3470 (1973)

106. Alcohol Dehydrogenase

SOURCE: Horse liver
SOLVENT: 0.1 M Acetate buffer, pH 4.0
COMMENTS: Spectra of methylated and native enzymes
REFERENCE: C. S. Tsai, Y. H. Tsai, G. Lauzon, and S. T. Cheng, Biochemistry, 13, 440 (1974)

107. Alcohol Dehydrogenase

SOURCE: Horse liver
COMMENTS: Spectrum of reduced alcohol dehydrogenase-pyridoxal-P complex
REFERENCE: J. S. McKinley-McKee and D. L. Morris, Eur. J. Biochem., 28, 1 (1972)

108. Alcohol Dehydrogenase

SOURCE: Horse liver
SOLVENT: a. 0.1 M Sodium phosphate, pH 7.0
b. 0.1 M Sodium phosphate, pH 7.7
c. 0.1 M Sodium pyrophosphate, pH 9.0
COMMENTS: At all pH's: glyoxal derivative
REFERENCE: M. Canella and G. Sodini, Eur. J. Biochem., 59, 119 (1975)

109. Alcohol Dehydrogenase, Co-

SOURCE: Horse liver
COMMENTS: a. Enzyme-NADH and enzyme-NADH-isobutyramide complex spectra
b. Enzyme-NAD^+-pyrazole complex spectrum
c. Enzyme-NAD^+-trifluoroethanol complex spectrum
REFERENCE: J. D. Shore and D. Santiago, J. Biol. Chem., 250, 2008 (1975)

110. Alcohol Dehydrogenase, Co_2Zn_2-

SOURCE: Horse liver
COMMENTS: Enzyme-NADH-isobutyramide complex
REFERENCE: R. A. Harvey and A. Barry, Biochem. Biophys. Res. Comm., 72, 886 (1971)

111. Alcohol Dehydrogenase

SOURCE: Horse liver
SOLVENT: 33 mM Sodium phosphate, pH 8.0
COMMENTS: Diazonium-1H-tetrazole derivatives
REFERENCE: D. C. Sogin and B. V. Plapp, Biochemistry, 15, 1087 (1976)

112. Alcohol Dehydrogenase

SOURCE: Horse liver
SOLVENT: a. 0.2 M Tris-acetate, pH 7.5
b. 0.1 M Sodium phosphate, pH 7.5

COMMENTS: 1. Spectra of zinc and cobalt enzymes
2. Spectra of zinc and cadmium enzymes
3. Spectral changes during carboxymethylation of cysteine-46 in the tetracobalt enzyme (solvent c)

REFERENCE: A. J. Sytkowski and B. L. Vallee, Proc. Nat. Acad. Sci., 73, 344 (1976); A. J. Sytkowski and B. L. Vallee, Biochemistry, 18, 4095 (1979)

113. Alcohol Dehydrogenase, Co_2Zn_2-

SOURCE: Horse liver
SOLVENT: 0.1 M Phosphate buffer, pH 7.0
COMMENTS: This is the Sepharose 4B immobilized material
REFERENCE: H. Schneider-Bernlohr, H. Deitrich, W. Maret, I. Andersson, and M. Zeppezauer, Eur. J. Biochem., 91, 475 (1978)

114. Alcohol Dehydrogenase

SOURCE: Yeast
SOLVENT: 50 mM Phosphate buffer, pH 8.0
COMMENTS: Spectra of reaction with benzylisothiocyanate after different lengths of time
REFERENCE: L. Drobnica and P. Gemeiner, in I.U.B. Symposium on Protein Structure and Evolution (J. L. Fox, Z. Deyl and A. Blazej, eds.), 105 (1975), Marcel Dekker, Inc., New York

115. Alcohol Dehydrogenase

SOURCE: Yeast
SOLVENT: 0.1 M Phosphate buffer, pH 6.5
COMMENTS: Spectrum of enzyme-pyrazole-NAD complex
REFERENCE: D. Karlovic, P. Amiguet, F. J. Bonner, and P. L. Luisi, Eur. J. Biochem., 66, 277 (1976)

116. Alcohol Dehydrogenase, Co_4

SOURCE: Yeast
SOLVENT: 20 mM Tris-HCl, pH 7.5
REFERENCE: A. J. Sytkowski, Arch. Biochem. Biophys., 184, 505 (1977)

117. Alcohol Oxidase

SOURCE: Yeast (Candida boidinii)
Brown rot fungus (Poria contigua)
SOLVENT: 100 mM Sodium phosphate, pH 7.5

COMMENTS: Spectra in presence and absence of sodium azide
REFERENCE: S. Bringer, B. Sprey, and H. Sahm, Eur. J. Biochem., 101, 563 (1979)

118. Aldehyde Dehydrogenase

SOURCE: Methylomonas methylovora
SOLVENT: 50 mM Phosphate buffer, pH 7.0
REFERENCE: R. N. Patel, C. T. Hou, and A. Felix, Arch. Microbiol., 122, 241 (1979)

119. Aldehyde Dehydrogenase

SOURCE: Beef liver
SOLVENT: 0.01 M Ammonium bicarbonate buffer, pH 7
REFERENCE: W. Leicht, F. Heinz, and B. Freimuller, Eur. J. Biochem., 83, 189 (1978)

120. Aldehyde Oxidase

SOURCE: Rabbit liver
SOLVENT: 0.05 M Potassium phosphate, pH 7.8
COMMENTS: Spectra of enzyme at various states of oxidation and reduction
REFERENCE: U. Branzoli and V. Massey, J. Biol. Chem., 249, 4346 (1974)

121. Aldehyde Oxidase

SOURCE: Rabbit liver
SOLVENT: 0.05 M Potassium phosphate, pH 7.8
COMMENTS: Spectra of deflavo enzyme in oxidized and reduced states
REFERENCE: U. Branzoli and V. Massey, J. Biol. Chem., 249, 4339 (1974)

122. Aldolase

SOURCE: Tuna muscle
REFERENCE: T. W. Kwon and H. S. Olcott, Comp. Biochem. Physiol., 15, 7 (1965)

123. Aldolase

SOURCE: Blowfly (Phormia regina [Meigen]) larval and adult muscle
REFERENCE: L. Levenbook, A. C. Bauer, and H. Shigematsu, Arch. Biochem. Biophys., 157, 625 (1973)

124. Aldolase, azo-

SOURCE: Rabbit muscle
SOLVENT: 10 mM Phosphate buffer, pH 7.2
COMMENTS: Spectra of different preparations of aldolase coupled with diazotized p-aminobenzoic acid
REFERENCE: R. Felicioli, L. Nannicini, E. Balestreri, and G. Montagnoli, Eur. J. Biochem., 51, 467 (1975)

125. Aldolase, azo-

SOURCE: Rabbit muscle
SOLVENT: a. 10 mM Phosphate buffer, pH 7.2
b. 0.1 N NaOH
REFERENCE: G. Montagnoli, S. Monti, L. Nannicini, and R. Felicioli, Photochem. Photobiol., 23, 29 (1976)

126. Alhagain

SOURCE: Shrub jawasee (Alhagi pseud-alhogi [M. B.] Desv)
REFERENCE: K. Yeshoda, S. C. Dhar., and M. Santappa, Ital. J. Biochem., 26, 181 (1977)

127. Allergen

SOURCE: Nippostrongylus brasiliensis
REFERENCE: J. Ambler and T. S. C. Orr, Immunochemistry, 9, 263 (1972)

128. Allergen M, Fragments

SOURCE: Cod fish white muscle (Gadus callarius)
SOLVENT: 0.15 M Phosphate buffer, pH 7.2
COMMENTS: Fragments T M-1 and T M-2
REFERENCE: S. Elsayed, K. Aas, K. Sletten, and S. G. O. Johansson, Immunochemistry, 9, 647 (1972)

129. Allergen, Material DS22 and Subcomponents

SOURCE: Cod fish white muscle (Gadus callarius)
COMMENTS: Fractions pI 4.75, pI 4.40, pI 4.85
REFERENCE: S. Elsayed, K. Aas, and T. Christensen, Int. Arch. Allergy, 40, 439 (1971)

130. Allergen

SOURCE: Pollen of Parietaria officinalis
SOLVENT: a. 20 mM Tris-HCl, pH 7.9
b. 0.01 N NaOH
c. 0.01 N HCl

COMMENTS: Allergen ND4 and D4 spectra
REFERENCE: D. Geraci, U. Oreste, and A. Ruffilli, Immunochemistry, 15, 491 (1978)

131. Allergen

SOURCE: a. Purified house dust
b. Feathers
c. Trichophytin
d. Ipecac
e. Human dandruff
SOLVENT: a. 0.01 N NaOH
b. 0.01 N HCl
REFERENCE: L. Berrens, Immunochemistry, 4, 37 (1967)

132. Allergen

SOURCE: a. Dwarf ragweed pollen
b. Great ragweed pollen (Trifidin A)
REFERENCE: a. O. H. Callaghan and A. R. Goldfarb, J. Immunol., 89, 612 (1963)
b. A. R. Goldfarb, A. K. Bhattacharya, and S. K. Koerner, J. Immunol., 81, 302 (1958)

133. Allergen

SOURCE: Kapok linters
SOLVENT: a. 0.01 M Phosphate buffer, pH 7.0
b. 0.01 N HCl, pH 2
c. 0.01 N NaOH, pH 12
REFERENCE: L. Berrens, Int. Arch. Allergy, 29, 575 (1966)

134. Allergen

SOURCE: Caddis fly
REFERENCE: S. Shulman, P. Bronson, and C. E. Arbesman, J. Allergy, 34, 1 (1963)

135. Allergen

SOURCE: a. Tomato
b. Alternaria tenius
c. Trichophytin
d. House dust
e. Human dandruff
f. Cotton linter
REFERENCE: L. Berrens, Monographs in Allergy, 7, 199 (1971)

136. Allergen

SOURCE: Cottonseed
REFERENCE: J. R. Spies and E. J. Umberger, J. Am. Chem. Soc., 64, 1889 (1942)

137. Allophycocyanin I

SOURCE: Mastigocladus laminosus
SOLVENT: 0.02 M Phosphate buffer, pH 5.9
COMMENTS: Spectral changes due to reaction with hydrogen peroxide (0-90 minutes)
REFERENCE: J. Gysi and H. Zuber, FEBS Lett., 68, 49 (1976)

138. Allophycocyanin II, and α- and β-Subunits

SOURCE: Mastigocladus laminosus
SOLVENT: 0.02 M Phosphate buffer, pH 8.0
REFERENCE: J. Gysi and H. Zuber, FEBS Lett., 48, 209 (1974)

139. Allophycocyanin II and α- and β-Subunits

SOURCE: Mastigocladus laminosus
SOLVENT: a. 0.02 M Potassium phosphate, pH 8.0
b. pH 3.0
c. pH 11.0
REFERENCE: J. R. Gysi and H. Zuber, Biochem. J., 181, 577 (1979)

140. Allophycocyanin

SOURCE: Phormidium luridum
SOLVENT: a. 0.05 M Phosphate buffer, pH 7.0
b. 0.2 M Sodium acetate buffer, pH 5
c. 0.05 M Potassium phosphate buffer, pH 7
REFERENCE: A. S. Brown, J. A. Foster, P. V. Voynow, C. Franzblau, and R. F. Troxler, Biochemistry, 14, 3581 (1975)

141. Allophycocyanin

SOURCE: Cyanidium caldarium
SOLVENT: 0.05 M Potassium phosphate, pH 7.0
REFERENCE: A. S. Brown and R. F. Troxler, Biochem. J., 163, 571 (1977)

142. Amidophosphoribosyltransferase

SOURCE: Escherichia coli
SOLVENT: 10 mM Tris-HCl, pH 7.4
REFERENCE: L. J. Messenger and H. Zalkin, J. Biol. Chem., 254, 3382 (1979)

143. Amine Dehydrogenase

SOURCE: Pseudomonas putida
SOLVENT: 50 mM Tris-HCl, pH 7.0
COMMENTS: Spectra of the oxidized and reduced (dithionite) enzyme
REFERENCE: D. R. Durham and J. J. Perry, J. Bacteriol., 134, 837 (1978)

144. Amine Oxidase

SOURCE: Aspergillus niger
SOLVENT: 0.06 M Phosphate buffer, pH 7.0
COMMENTS: a. Reduced with sodium dithionite
b. Reactions with n-butylamine, benzylamine, histamine, putrescine
REFERENCE: H. Yamada, O. Adachi, and K. Ogata in Pyridoxal Catalysis: Enzymes and Model Systems (E. E. Snell, A. E. Braunstein, E. S. Severin, and Yu M. Torchinsky, eds.), 347 (1968), Interscience, New York

145. Amine Oxidase

SOURCE: Aspergillus niger
COMMENTS: Spectral changes on reaction with β-bromoethylamine
REFERENCE: H. Kumagai, H. Uchida, and H. Yamada, J. Biol. Chem., 254, 10913 (1979)

146. Amine Oxidase

SOURCE: Sheep plasma
SOLVENT: Phosphate buffer, pH 7.0
COMMENTS: Reactions with benzylamine and phenylhydrazine
REFERENCE: R. B. Rucker and W. Goettlich-Riemann, Enzymologia, 43, 33 (1972)

147. Amine Oxidase

SOURCE: Pig liver microsomes
SOLVENT: 0.1 M Glycine - 0.02 M pyrophosphate, pH 8.4

COMMENTS: a. Oxidized enzyme
b. Reduced enzyme: dithionite and NADPH used
REFERENCE: D. M. Ziegler and C. H. Mitchell, Arch. Biochem. Biophys., 150, 116 (1972)

148. Amine Oxidase

SOURCE: Beef plasma
SOLVENT: 0.1 M Phosphate buffer, pH 8
COMMENTS: Spectra in the presence of glycine and cycloserine, glycine, phenylhydrazine, and phenylhydrazine and cycloserine
REFERENCE: M. T. Costa, S. Sabatini, P. Turini, and B. Mondovi, FEBS Lett., 94, 112 (1978)

149. D-Amino Acid Oxidase

SOURCE: Pig kidney
SOLVENT: pH 8.3
COMMENTS: Time course of anaerobic reaction with D-norvaline
REFERENCE: K. Yagi and N. Ohishi, J. Biochem. (Tokyo), 84, 1653 (1978)

150. D-Amino Acid Oxidase

SOURCE: Pig kidney
SOLVENT: 0.017 M Pyrophosphate buffer, pH 8.3
COMMENTS: Effect of progesterone on spectrum
REFERENCE: F. Tanaka, M. Bamji, and K. Yagi, Biochim. Biophys. Acta, 522, 43 (1978)

151. D-Amino Acid Oxidase

SOURCE: Pig kidney
SOLVENT: 0.05 M Sodium pyrophosphate buffer, pH 8.3
COMMENTS: Spectra of complexes with benzoate, fluorobenzoates (o-, m-, p-), chlorobenzoates (o-, m-, p-), methylbenzoates (m-, p-), hydroxybenzoates (o-, m-, p-), and nitrobenzoates (o-, m-, p-)
REFERENCE: K. Shiga, K. Horiike, Y. Nishina, A. Isomoto, and T. Yamano, J. Biochem. (Tokyo), 81, 1465 (1977)

152. D-Amino Acid Oxidase

SOURCE: Pig kidney
SOLVENT: 0.017 M Pyrophosphate buffer, pH 8.3
COMMENTS: Spectra of enzyme-benzoate complex and in presence of methanol
REFERENCE: H. Ohama, N. Sugiura, F. Tanaka, and K. Yagi, Biochemistry, 16, 126 (1977)

153. D-Amino Acid Oxidase

SOURCE: Pig kidney
SOLVENT: 0.1 M Sodium pyrophosphate buffer, pH 8.3
COMMENTS: Effect of N-phenylglycine and sarcosine on spectrum
REFERENCE: M. Naoi and K. Yagi, Biochim. Biophys. Acta, 438, 61 (1976)

154. D-Amino Acid Oxidase

SOURCE: Pig kidney
SOLVENT: 0.1 M Pyrophosphate, pH 8.3
COMMENTS: Spectra of chlorinated enzyme in oxidized and reduced states
REFERENCE: D. J. T. Porter and H. J. Bright, J. Biol. Chem., 251, 6150 (1976)

155. D-Amino Acid Oxidase

SOURCE: Pig kidney
SOLVENT: 0.05 M Sodium pyrophosphate buffer, pH 8.3
COMMENTS: Spectra of enzyme inactivated with D-propargylglycine and then reduction with potassium borohydride
REFERENCE: K. Horiike, Y. Nishina, Y. Miyake, and T. Yamano, J. Biochem. (Tokyo), 78, 57 (1975)

156. D-Amino Acid Oxidase, deazaFAD-

SOURCE: Pig kidney
SOLVENT: 0.1 M Sodium pyrophosphate buffer, pH 8.5
COMMENTS: Spectra of complex formation with sodium benzoate and 2-amino benzoate; spectra of time study of reaction with DL-alanine, pyruvate-imine, and D-alanine
REFERENCE: L. B. Hersh and M. S. Jorns, J. Biol. Chem., 250, 8728 (1975)

157. D-Amino Acid Oxidase

SOURCE: Pig kidney
SOLVENT: 0.1 M Sodium pyrophosphate, pH 8.5
COMMENTS: Spectral changes on reaction with β-chloro-α-aminobutyrate
REFERENCE: C. T. Walsh, E. Krodel, V. Massey, and R. H. Abeles, J. Biol. Chem., 248, 1950 (1973)

158. D-Amino Acid Oxidase

SOURCE: Pig kidney
SOLVENT: 0.1 M Sodium pyrophosphate-HCl, pH 8.3
COMMENTS: Spectra of aerobic and anaerobic reaction with β-chloro-D-alanine, and with β-chloropropionate
REFERENCE: Y. Miyake, T. Abe, and T. Yamano, J. Biochem. (Tokyo), 73, 1 (1973)

159. D-Amino Acid Oxidase

SOURCE: Pig kidney
SOLVENT: 0.01 M Sodium pyrophosphate, pH 8.3
COMMENTS: Effect of 0-20% ethanol on spectrum
REFERENCE: H. Ohama, N. Sugiura, and K. Yagi, J. Biochem. (Tokyo), 73, 1123 (1973)

160. D-Amino Acid Oxidase

SOURCE: Pig kidney
SOLVENT: pH 6.3-pH 6.4
COMMENTS: Reaction with singlet oxygen
REFERENCE: M. Yamasaki and T. Yamano, Biochem. Biophys. Res. Comm., 51, 612 (1973)

161. D-Amino Acid Oxidase

SOURCE: Pig kidney
SOLVENT: pH 8.3
COMMENTS: Spectral changes due to addition of D-alanine (anaerobic) and o-aminobenzoate
REFERENCE: K. Yagi, M. Nishikimi, and N. Ohishi, J. Biochem. (Tokyo), 72, 1369 (1972)

162. D-Amino Acid Oxidase, Semi-quinoid Form

SOURCE: Pig kidney
SOLVENT: 17 mM Sodium pyrophosphate, pH 8.3
COMMENTS: Spectral change due to oxidation with phenazine methosulfate
REFERENCE: N. A. Rao, M. Nishikimi, and K. Yagi, Biochim. Biophys. Acta, 276, 350 (1972)

163. D-Amino Acid Oxidase

SOURCE: Pig kidney
COMMENTS: Spectra before and after addition of β-chloro-D-alanine under aerobic conditions and anaerobic conditions

REFERENCE: Y. Miyake, T. Abe, and T. Yamano, in International Symposium on Oxidases and Related Redox Systems, 1971 (T. E. King, H. S. Mason, and M. Morrison, eds.), Vol. 1, 209 (1973), University Park Press, Baltimore

164. D-Amino Acid Oxidase

SOURCE: Pig kidney
SOLVENT: 0.01 M Potassium pyrophosphate, pH 8.3, 8.0 M Urea, and 5 M KCl
COMMENTS: Spectra of apoenzyme and chlorinated apoenzyme
REFERENCE: N. G. Rudie, D. J. T. Porter, and H. J. Bright, J. Biol. Chem., 255, 498 (1980)

165. D-Amino Acid Oxidase, deazaFAD-

SOURCE: Pig kidney
SOLVENT: 0.1 M Sodium phosphate buffer, pH 7.0 and 7.6, containing 0.3 mM EDTA
COMMENTS:
1. Spectral changes upon addition of sulfite
2. Spectral changes upon addition of benzoate to enzyme-sulfite complex
3. Spectral changes upon addition of cyanide and additional changes upon addition of benzoate

REFERENCE: M. S. Jorns and L. B. Hersh, J. Biol. Chem., 251, 4872 (1976)

166. D-Amino Acid Oxidase

SOURCE: Chlorella vulgaris
SOLVENT: 0.13 M Phosphate buffer, pH 7.0
REFERENCE: E. K. Pistorious and H. Voss, Biochim. Biophys. Acta, 481, 395 (1977)

167. L-Amino Acid Oxidase

SOURCE: Sand viper (Viper ammodyte)
SOLVENT: 0.1 M Phosphate buffer, pH 7.5
COMMENTS: Spectral changes after addition of L-phenylalanine and after 1 N HCl to pH 3
REFERENCE: J. Kurth and H. Aurich, Acta Biol. Med. Germ., 31, 641 (1973)

168. L-Amino Acid Oxidase

SOURCE: Crotalus adamenteus
SOLVENT: 100 mM Tris-HCl, pH 8.0

COMMENTS: Spectral changes due to reaction with D,L-vinyl glycine
REFERENCE: P. Marcotte and C. Walsh, Biochemistry, 15, 3070 (1976)

169. L-Amino Acid Oxidase

SOURCE: Crotalus adamanteus
SOLVENT: 55 mM Sodium phosphate, pH 7.5
COMMENTS: Spectral changes due to heat and cold
REFERENCE: C. J. Coles, D. E. Edmondson, and T. P. Singer, J. Biol. Chem., 252, 8035 (1977)

170. δ-Aminolevulinate Dehydrase

SOURCE: Human erythrocytes
SOLVENT: 0.05 M Piperazine-N,N'-bis(2-ethane sulfonic acid) buffer, pH 6.8
REFERENCE: P. M. Anderson and R. J. Desnick, J. Biol. Chem., 254, 6924 (1979)

171. 5-Aminolevulinic Acid Synthetase

SOURCE: Rhodopseudomonas spheroides Y
SOLVENT: pH 5.4-5.6
COMMENTS:
a. Spectra of fractions I(I) and II(II)
b. Effect of sodium borohydride
c. Effect of pH
d. Effect of addition of succinyl-CoA, succinyl-CoA + glycine, and glycine

REFERENCE: M. Fanica-Gaigner and J. Clement-Metral, Eur. J. Biochem., 40, 13 (1973)

172. δ-Aminolevulinic Acid Synthase

SOURCE: Chicken embryo liver
SOLVENT: 20 mM Sodium phosphate buffer, pH 7.6, 0.5 mM dithiothreitol, 0.01 mM pyridoxal 5'-phosphate
COMMENTS:
a. Change in spectrum on addition of glycine
b. Change in spectrum on addition of sodium borohydride

REFERENCE: M. J. Whiting and S. Granick, J. Biol. Chem., 251, 1340 (1976)

173. δ-Aminolevulinic Acid Synthase

SOURCE: Rhodopseudomonas spheroides
COMMENTS:
a. Effect of pyridoxal 5'-phosphate
b. Effect of pyridoxal 5'-phosphate and glycine

c. Effect of pyridoxal-5'-phosphate, glycine, and succinyl-CoA on spectrum
REFERENCE: D. L. Nandi, J. Biol. Chem., 253, 8872 (1978)

174. δ-Aminolevulinic Acid Synthase, apo-

SOURCE: Rhodopseudomonas spheroides
SOLVENT: 0.05 M Potassium phosphate, pH 7.2
COMMENTS: a. Spectrum of apoenzyme in 0.001 M 2-mercapto-ethanol, 10% glycerol buffer
b. Spectrum of enzyme-pyridoxal 5'-phosphate complex
c. Spectrum of "b" reduced with sodium borohydride
REFERENCE: D. L. Nandi, Arch. Biochem. Biophys., 188, 266 (1978)

175. Aminotransferase, Alanine

SOURCE: Glycine hispida seeds
COMMENTS: Spectra of apo- and holenzymes
REFERENCE: I. F. Dumitru, D. Iordachescu, and S. Niculescu, Enzymologia, 47, 79 (1972)

176. Aminotransferase, Alanine

SOURCE: Pig heart
SOLVENT: 50 mM Sodium cacodylate, 5 mM sodium α-ketoglutarate, pH 6.5
COMMENTS: Spectra taken at zero, 0.5 hour and 12 hours after addition of β-chloroalanine
REFERENCE: A. Golichowski and W. T. Jenkins, Arch. Biochem. Biophys., 189, 109 (1978)

177. Aminotransferase, Alanine

SOURCE: Pig heart
SOLVENT: 20 mM Potassium phosphate buffer, pH 7.0 containing 0.1 mM EDTA and 0.1 mM dithiothreitol
COMMENTS: Spectra of reaction with 3-chloro-L-alanine at different times
REFERENCE: Y. Morino, H. Kojima, and S. Tanase, J. Biol. Chem., 254, 279 (1979)

178. Aminotransferase, D-Amino Acid

SOURCE: Bacillus sphaericus
SOLVENT: a. 0.02 M Potassium phosphate, pH 7.4
b. 0.05 M Potassium phosphate, pH 7.4

COMMENTS: Spectra of native enzyme, after reduction with sodium borohydride, after reduction with sodium borohydride and denaturation with 6 M guanidine-HCl, and of the reduced (sodium borohydride) and denatured (6 M guanidine-HCl) semiapoenzyme
REFERENCE: K. Yonaha, H. Misono, Y. Yamamoto, and K. Soda, J. Biol. Chem., 250, 6983 (1975)

179. Aminotransferase, D-Amino Acid

SOURCE: Rhizobium japonicum
SOLVENT: a. 50 mM Potassium phosphate buffer containing 5 mM 2-mercaptoethanol, pH 6.90
b. Solvent (a) raised to pH 8.90 by addition of sodium carbonate
COMMENTS: Spectra of enzyme with and without addition of D-alanine
REFERENCE: J. P. Gosling and P. F. Fottrell, Biochim. Biophys. Acta, 522, 84 (1978)

180. Aminotransferase, ω-Amino Acid

SOURCE: Pseudomonas Sp.
SOLVENT: 0.05 M Potassium phosphate buffer, pH 7.4
REFERENCE: K. Yonaha, S. Toyama, M. Yasuda, and K. Soda, FEBS Lett., 71, 21 (1976)

181. Aminotransferase, ω-Amino Acid: Pyruvate

SOURCE: Pseudomonas species F-126
COMMENTS: Spectra of native and gabaculine-inactivated enzymes
REFERENCE: G. Burnett, K. Yonaha, S. Toyama, K. Soda, and C. Walsh, J. Biol. Chem., 255, 428 (1980)

182. Aminotransferase, 4-Aminobutyrate

SOURCE: Rat brain tissue
SOLVENT: a. 0.1 M Acetate buffer, pH 5.0
b. 0.1 M Tris buffer, pH 8.5
COMMENTS: Spectra taken after addition of hydroxylamine, glutamate, and α-ketoglutarate
REFERENCE: Y. Yu. Vasilev and V. P. Eremin, Biokhimiya (Biochemistry), 33, 936 (1968)

183. Aminotransferase, 4-Aminobutyrate

SOURCE: Rabbit brain
SOLVENT: 0.1 M Tris-HCl buffer, pH 7.5

COMMENTS: Spectra at different concentrations of 4-aminobutyrate and glutamate
REFERENCE: R. A. John and L. J. Fowler, Biochem. J., 155, 645 (1976)

184. Aminotransferase, 4-Aminobutyrate

SOURCE: Pig heart
SOLVENT: 0.1 M Sodium acetate, pH 5.5
COMMENTS: Effect on spectrum of enzyme caused by addition of amino-oxyacetate
REFERENCE: R. A. John, A. Charteris, and L. J. Fowler, Biochem. J., 171, 771 (1978)

185. Aminotransferase, 4-Aminobutyrate

SOURCE: Pig heart
SOLVENT: Tris-acetate buffer, pH 8.3
COMMENTS: Spectral changes upon addition of aminohexynoate with respect to time
REFERENCE: R. A. John, E. D. Jones, and L. J. Fowler, Biochem. J., 177, 721 (1979)

186. Aminotransferase, Aspartate

SOURCE: Pig brain, mitochondria and cytoplasm
SOLVENT: Tris-acetate buffer, pH 8.2
COMMENTS: Effect of phenylalanine and phenylalanine and phenylpyruvate on spectrum
REFERENCE: E. Shrawder and M. Martinez-Carrion, J. Biol. Chem., 247, 2486 (1972)

187. Aminotransferase, Aspartate

SOURCE: Pig kidney
SOLVENT:
a. 0.05 M Phosphate buffer, pH 5.5
b. 0.05 M Carbonate buffer, pH 8.5

COMMENTS: Spectra taken after addition of:
a. aspartate
b. aspartate and formaldehyde
c. erythro-DL-β-hydroxyaspartate
d. erythro-DL-β-hydroxyaspartate and formaldehyde
e. threo-DL-β-hydroxyaspartate
f. threo-DL-β-hydroxyaspartate and formaldehyde

REFERENCE: A. E. Evangelopoulos and I. W. Sizer, in Pyridoxal Catalysis: Enzymes and Model Systems (E. E. Snell, A. E. Braunstein, E. E. Severin, and Yu. M. Turchinsky, eds.), 231 (1968), Interscience, New York

188. Aminotransferase, Aspartate

SOURCE: Pig heart
SOLVENT: 0.05 M Acetate buffer, pH 5.4
COMMENTS: Spectral changes associated with incubation with ammonium sulfate for different time periods
REFERENCE: G. Marino, M. Paterno, and M. DeRosa, FEBS Lett., 21, 53 (1972)

189. Aminotransferase, Aspartate

SOURCE: Pig heart
SOLVENT: a. 0.05 M Acetate buffer, pH 5.4
b. 0.05 M Tris-HCl buffer, pH 8.2
COMMENTS: Changes in spectrum due to addition of erythro-5-carboxyaminomethyliso-oxazolidone-3 (EI_{380}), and reduction with potassium borohydride
REFERENCE: G. K. Kovaleva and E. S. Severin, Biokhimiya (Biochemistry), 37, 1072 (1972)

190. Aminotransferase, Aspartate

SOURCE: Pig heart muscle mitochondria
SOLVENT: 50 mM Cacodylate buffer, pH 6.8
COMMENTS: a. Spectral change upon addition of β-bromopropionate
b. Spectral change of native and alkylated protein after sodium borohydride reduction
c. Spectral change of native and alkylated protein after addition of cysteinesulfinate
REFERENCE: M. Okamoto and Y. Morino, Biochemistry, 11, 3188 (1972)

191. Aminotransferase, Aspartate

SOURCE: Pig heart muscle mitochondria
SOLVENT: pH 7.2
COMMENTS: Spectral change with time upon incubation with [^{14}C] β-bromopropionate
REFERENCE: Y. Morino and M. Okamoto, Biochemistry, 11, 3196 (1972)

192. Aminotransferase, Aspartate

SOURCE: Pig heart muscle
SOLVENT: pH 5.2, pH 7.1, pH 8.2
COMMENTS: Spectra of enzyme and aspartate at equilibrium, of apoenzyme, of aldimine form, and of amino form
REFERENCE: V. I. Ivanov, A. L. Bocharov, M. V. Volkenstein, M. Y. Karpeisky, S. Mora, E. I. Okina, and L. V. Yudina, Eur. J. Biochem., 40, 519 (1973)

193. Aminotransferase, Aspartate

SOURCE: Pig heart, cytoplasmic
SOLVENT: 0.05 M Sodium phosphate, pH 7.5
COMMENTS: Spectra of pyridoxal and pyridoxamine forms modified with either N-ethylmaleimide, 5,5'-dithiobis (2-nitrobenzoate), or cyanide
REFERENCE: W. Birchmeier, K. J. Wilson, and P. Christen, J. Biol. Chem., 248, 1751 (1973)

194. Aminotransferase, Aspartate

SOURCE: Pig heart
SOLVENT: pH 5.3
COMMENTS: Spectra of α, β, γ, and ω subforms
REFERENCE: G. F. Denisova and O. L. Polyanovsky, FEBS Lett., 35, 129 (1973)

195. Aminotransferase, Aspartate

SOURCE: Pig heart muscle supernatant and mitochondria
SOLVENT:
a. 0.1 M Sodium cacodylate buffer, pH 7.0
b. 0.1 M Tris-HCl buffer, pH 8.0
c. 50 mM Tris-HCl buffer, pH 8.0

COMMENTS:
a. Spectral change upon addition of β-bromopyruvate to pyridoxamine form
b. Spectra of native pyridoxal form, inactivated form, and pyridoxamine form
c. Spectra of native and inactivated protein in 6 M guanidine HCl, pH 8.0, pyridoxal form and inactivated enzyme
d. Spectra after reaction of native and inactive pyridoxal forms and pyridoxamine form with potassium borohydride and dialysis

REFERENCE: M. Okamoto and Y. Morino, J. Biol. Chem., 248, 82 (1973)

196. Aminotransferase, Aspartate

SOURCE: Pig heart
SOLVENT: 0.1 M Phosphate buffer, pH 7.4
COMMENTS: Spectra of apoenzyme with pyridoxal 5'-phosphate or pyridoxal 5'-phosphate benzylhydrazone; same with glutamate added; same with erythro-β-hydroxyaspartate
REFERENCE: M. Camier, P. Gonnard, and M. E. Goldberg, Biochemie, 55, 1011 (1973)

197. Aminotransferase, Aspartate

SOURCE: Pig heart cytoplasma, α-subform
SOLVENT: pH 8.2 and pH 5.0
COMMENTS: Spectral change upon addition of pyridoxal 5'-sulfate
REFERENCE: I. Y. Yang, R. M. Khomutov, and D. E. Metzler, Biochemistry, 13, 3877 (1974)

198. Aminotransferase, Aspartate

SOURCE: Pig heart cytoplasm, α-form
COMMENTS: Spectra of active, inactive (reacted with DL-2-amino-3-butenoic acid), and resolved enzyme
REFERENCE: R. R. Rando, Biochemistry, 13, 3859 (1974)

199. Aminotransferase, Aspartate

SOURCE: Pig heart supernatant
SOLVENT: 0.1 M and 0.05 M Sodium pyrophosphate buffer, pH 7.8
COMMENTS:
a. Spectral changes upon addition of β-chloro-L-alanine
b. Spectral changes upon addition of formate
c. Spectral changes upon addition of formate and β-chloro-L-alanine
d. Spectral changes upon addition of potassium borohydride to inactivated enzyme

REFERENCE: Y. Morino, A. M. Osman, and M. Okamoto, J. Biol. Chem., 249, 6684 (1974)

200. Aminotransferase, Aspartate, Phosphopyridoxal Form

SOURCE: Pig heart
SOLVENT:
a. pH 7.4
b. 0.02 M Acetate buffer, pH 5.3
c. 0.05 M Phosphate buffer, pH 6.9
d. 0.05 M Bicarbonate buffer, pH 8.9

COMMENTS: Spectral changes seen upon addition of potassium iodide and potassium iodide and glutarate
REFERENCE: N. Burridge and J. E. Churchich, Eur. J. Biochem., 41, 533 (1974)

201. Aminotransferase, Aspartate

SOURCE: Pig heart muscle
SOLVENT: 0.1 M Potassium phosphate, pH 6.0
COMMENTS: Spectral changes caused by addition of L-2-amino-4-methoxy-trans-3-butenoic acid
REFERENCE: R. Rando, Nature, 250, 586 (1974)

202. Aminotransferase, Aspartate

SOURCE:	Pig heart muscle
SOLVENT:	pH 5.2
COMMENTS:	Spectrum with 5'-methylpyridoxal-phosphate
REFERENCE:	V. G. Tumanyan, O. K. Mamaeva, A. L. Bocharov, V. I. Ivanov, M. Y. Karpeisky and G. L. Yakovlev, Eur. J. Biochem., 50, 119 (1974)

203. Aminotransferase, Aspartate

SOURCE:	Pig heart muscle cytoplasm
SOLVENT:	0.05 M Sodium phosphate, pH 7.5
COMMENTS:	Spectral changes of pyridoxal and pyridoxamine forms modified with N-ethylmaleimide on reacting with β-bromopyruvate
REFERENCE:	W. Birchmeier and P. Christen, J. Biol. Chem., 249, 6311 (1974)

204. Aminotransferase, Aspartate

SOURCE:	Pig heart cytoplasm
SOLVENT:	pH 5
COMMENTS:	α, β, and γ subforms
REFERENCE:	R. W. Evans and J. J. Holbrook, Biochem. J., 143, 643 (1974)

205. Aminotransferase, Aspartate

SOURCE:	Pig heart cytoplasm
COMMENTS:	Spectral changes seen upon incubation of α form with L-ascorbic acid for 0-24 hours
REFERENCE:	F. Bossa, A. Giartosio, R. Petruzzelli, and P. Fasella, FEBS Lett., 44, 31 (1974)

206. Aminotransferase, Aspartate

SOURCE:	Pig heart muscle supernatant
SOLVENT:	pH 8.00
COMMENTS:	Spectral change of α-form upon addition of D- or DL-α-methylaspartate
REFERENCE:	W. R. Melander, FEBS Lett., 51, 5 (1975)

207. Aminotransferase, Aspartate

SOURCE:	Pig heart muscle cytoplasm
SOLVENT:	0.025 M Sodium arsenate, 0.1 mM EDTA, pH 7.8
COMMENTS:	Spectra of pyridoxal-phosphate and pyridoxamine-phosphate forms in presence of either glutamate, or glutamate plus quinolinate, or keto-glutarate, or keto-glutarate plus quinolinate.

REFERENCE: S. L. Hsu and L. A. Fahien, Arch. Biochem. Biophys., 177, 217 (1976)

208. Aminotransferase, Aspartate

SOURCE: Pig heart muscle
SOLVENT: 0.10 M Tris-HCl buffer, pH 8.1
COMMENTS: Complex of apoenzyme plus phosphopyridoxyl trifluoroethylamine
REFERENCE: M. Martinez-Carrion, J. C. Slebe, B. Boettcher, and A. M. Relimpo, J. Biol. Chem., 251, 1853 (1976)

209. Aminotransferase, Aspartate

SOURCE: Pig heart muscle mitochondria and cytosol
SOLVENT: 0.1 M Potassium pyrophosphate, pH 7.7 and pH 8.0
COMMENTS: Spectral change upon addition of L-propargylglycine to pyridoxal form
REFERENCE: S. Tanase and Y. Morino, Biochem. Biophys. Res. Comm., 68, 1301 (1976)

210. Aminotransferase, Aspartate

SOURCE: Pig heart muscle
SOLVENT: 0.1 M Potassium phosphate, pH 7.0
COMMENTS: Spectral change upon addition of sodium borohydride to L-2-amino-4-methoxy-trans-3-butenoic acid treated enzyme
REFERENCE: R. R. Rando, N. Relyea, and L. Cheng, J. Biol. Chem., 251, 3306 (1976)

211. Aminotransferase, Aspartate

SOURCE: Pig heart muscle
SOLVENT: 0.1 M Tris-HCl buffer, pH 7.5
COMMENTS: Spectra of mitochondrial and cytosolic enzyme reacted with chloropyruvate in presence and absence of L-alanine
REFERENCE: A. M. Osman, T. Yamano, and Y. Morino, Biochem. Biophys. Res. Comm., 70, 153 (1976)

212. Aminotransferase, Aspartate

SOURCE: Rat liver
SOLVENT: a. 0.05 M Potassium phosphate buffer, pH 7.0 containing 0.5 mM DTT
b. pH 3.25, 3.35, 4.30, 5.0, 8.3

COMMENTS: a. Spectra of enzyme from 5-, 52-, and 150-week-old rats
b. Spectra of pyridoxal and pyridoxamine forms
REFERENCE: S. K. Patnaik and M. S. Kanungo, Indian J. Biochem. Biophys., 13, 117 (1976)

213. Aminotransferase, Aspartate

SOURCE: Pig heart
SOLVENT: 0.05 M Potassium phosphate, pH 4.4
COMMENTS: Spectra obtained upon addition of L-canaline or hydroxylamine to pyridoxal-5'-phosphate form
REFERENCE: K. Kito, Y. Sanada, and N. Katunma, J. Biochem. Tokyo, 83, 201 (1978)

214. Aminotransferase, Aspartate

SOURCE: Pig heart
SOLVENT: 0.2 M Succinate, pH 7
COMMENTS: Spectra of α form taken after 4, 8, 18, and 27 days at 25°
REFERENCE: J. A. Williams and R. A. John, Biochem. J., 177, 121 (1979)

215. Aminotransferase, Aspartate

SOURCE: Pig heart muscle
COMMENTS: a. Spectrum of enzyme-threo-α-cycloglutamic acid complex
b. Spectrum of pyridoxamine form
REFERENCE: G. K. Kovaleva and E. S. Severin, Biokhimiya (Biochemistry), 37, 393 (1972)

216. Aminotransferase, Aspartate

SOURCE: Pig heart
SOLVENT: 0.04 M N-2-Hydroxyethylpiperazine-N'-2-ethanesulfonic acid
COMMENTS: Spectral changes upon addition of either amino-oxyacetate or cysteinsulfinate to enzyme-amino-oxyacetate complex
REFERENCE: R. A. John, A. Charteris, and L. J. Fowler, Biochem., J., 171, 771 (1978)

217. Aminotransferase, Aspartate

SOURCE: Pig heart
SOLVENT: Solid crystal used

COMMENTS: Polarized absorption spectra of complexes of enzyme with either L-glutamate, DL-α-methylaspartate, or erythro-β-hydroxyaspartate
REFERENCE: C. M. Metzler, D. E. Metzler, D. S. Martin, R. Newman, A. Arnone, and P. Rogers, J. Biol. Chem., 253, 5251 (1978)

218. Aminotransferase, Aspartate

SOURCE: Chicken heart mitochondria and supernatant
SOLVENT: a. 0.1 M Sodium acetate, pH 5.3
b. 0.1 M Tris-acetate, pH 8.0
COMMENTS: Effect of cysteine sulfinate on spectra
REFERENCE: E. J. Shrawder and M. Martinez-Carrion, J. Biol. Chem., 248, 2140 (1973)

219. Aminotransferase, Aspartate

SOURCE: Chicken heart mitochondria
SOLVENT: a. Single crystals wet with polyethyleneglycol solutions of either oxalacetate (pyridoxal form), cysteine sulfinate (pyridoxamine form), 3-hydroxyaspartate, or 2-methylaspartate
b. 50 mM Sodium phosphate, pH 7.1 and pH 7.5
COMMENTS: Single crystal titration with polyethyleneglycol, 50 mM sodium acetate/50 mM sodium phosphate: pH 5.1-pH 7.7
REFERENCE: G. Eichele, D. Karabelnick, R. Halonbrenner, J. N. Jansonius, and P. Christen, J. Biol. Chem., 253, 5239 (1978)

220. Aminotransferase, Aspartate

SOURCE: Chicken heart cytosol
SOLVENT: 0.2 M Acetate buffer, pH 5.0
COMMENTS: Spectra in the presence and absence of sodium α-ketoglutarate
REFERENCE: V. M. Kochkina, A. V. Azaryan, M. L. Mekhanik, L. N. Zakomyrdina, N. I. Sinitsyna, I. A. Bolotina, T. A. Egorov, and Y. M. Torchinski, Biokhimiya (Biochemistry), 43, 1162 (1978)

221. Aminotransferase, Aspartate

SOURCE: Beef brain and beef heart cytoplasm
SOLVENT: 50 mM Acetate, pH 5.4
REFERENCE: M. L. Krista and M. L. Fonda, Biochim. Biophys. Acta, 309, 83 (1973)

222. Aminotransferase, Aspartate

SOURCE:	Beef kidney mitochondria
SOLVENT:	a. 0.1 M Acetate buffer, pH 5.0 b. 0.05 M Tris-HCl buffer, pH 8
COMMENTS:	Changes in spectrum with addition of glutamate or 2-oxoglutarate or both
REFERENCE:	R. Scandurra and C. Cannella, Eur. J. Biochem., 26, 196 (1972)

223. Aminotransferase, Aspartate

SOURCE:	Beef heart
COMMENTS:	a. Spectrum of dialysed preparation b. Spectrum of ion-exchanged preparation
REFERENCE:	V. Scardi, 1968 in Pyridoxal Catalysis: Enzymes and Model Systems (E. E. Snell, A. E. Braunstein, E. E. Severin, and Yu. M. Torchinsky, eds.), 179 (1968), Interscience, New York

224. Aminotransferase, Aspartate

SOURCE:	Male Wistar strain rat liver
SOLVENT:	0.05 M Potassium phosphate buffer, pH 7.4
COMMENTS:	Spectra of pyridoxal phosphate form, pyridoxamine phosphate form, and apoenzyme from control animals and pyridoxine deficient animals
REFERENCE:	M. Okada and M. Hirose, Arch. Biochem. Biophys., 193, 294 (1979)

225. Aminotransferase, Aspartate, nitro derivative

SOLVENT:	a. pH 8.4 b. pH 9.6
REFERENCE:	S. V. Shlyapnikov, A. L. Bocharov, J. A. Jejelava, and M. Y. Karpeisky, Mol. Biol., 3, 709 (1969)

226. Aminotransferase, Aspartate

SOLVENT:	a. pH 8.4 b. pH 9.4
REFERENCE:	A. V. Borodavkin, M. Y. Karpeisky, and Y. V. Morozov, Mol. Biol., 5, 577 (1971)

227. Aminotransferase, Cysteine

SOURCE:	Rat liver
SOLVENT:	a. 70 mM Tris-HCl buffers, pH 7.5-8.3 b. 70 mM Borax-NaOH buffer, pH 10.0

COMMENTS: Changes in spectrum in system containing enzyme, L-cysteine, and pyridoxal 5'-phosphate
REFERENCE: M. P. C. Ip, R. J. Thibert, and D. E. Schmidt, Can. J. Biochem., 55, 958 (1972)

228. Aminotransferase, Glutamic Acid

SOURCE: Escherichia coli
SOLVENT: 0.023 M Acetate buffer, pH 4.25-8.50 (due to addition of KOH)
COMMENTS: 1% ampholines present, pH 5.90-pH 8.50
REFERENCE: S. Chesne and J. Pelmont, Biochimie, 56, 631 (1974)

229. Aminotransferase, Glutamine

SOURCE: Rat liver
SOLVENT: 5 mM Potassium phosphate buffer, pH 7.2
REFERENCE: A. J. L. Cooper and A. Meister, Biochemistry, 11, 661 (1972)

230. Aminotransferase, Glutamine

SOURCE: Rat brain
SOLVENT: 50 mM Phosphate buffer containing 2 mM 2-mercapto-ethanol, pH 7.6
REFERENCE: F. van Leuven, Eur. J. Biochem., 58, 153 (1975)

231. Aminotransferase, Lysine

SOURCE: Flavobacterium lutescens (Achromobacter liquidum)
SOLVENT: 10 mM Potassium phosphate buffer, pH 7.4
COMMENTS: Spectra of fresh enzyme, aged enzyme, and after reduction with sodium borohydride
REFERENCE: H. Misono and K. Soda, J. Biochem. Tokyo, 82, 535 (1977)

232. Aminotransferase, Ornithine

SOURCE: Rat liver
SOLVENT: pH 8.0
COMMENTS: Spectra of phosphopyridoxal and phosphopyridoxamine enzyme in presence and absence of L-valine
REFERENCE: T. Matsuzawa, J. Biochem. Tokyo, 75, 601 (1974)

233. Aminotransferase, Ornithine

SOURCE: Rat liver
SOLVENT: 0.05 M Potassium phosphate buffer, pH 7.5

COMMENTS: a. Spectrum of apo-form I before and after addition of 2-oxoglutarate
b. Spectrum of holo-enzyme before and after addition of L-ornithine
c. Spectrum of apo-form II
d. Spectrum of holo-form II before and after addition of L-ornithine

REFERENCE: Y. Sanada, T. Shiotani, E. Okuno, and N. Katunuma, Eur. J. Biochem., 69, 507 (1976)

234. Aminotransferase, Ornithine

SOURCE: Rat liver
SOLVENT: 0.05 M Potassium phosphate buffer, pH 8.0
COMMENTS: Absorption spectra after addition of either L-ornithine, α-N-acetyl-L-ornithine, L-aminovaleric acid, L-norvaline, hydroxylamine, or L-canaline
REFERENCE: K. Kito, Y. Sanada, and N. Katunuma, J. Biochem. Tokyo, 83, 201 (1976)

235. Aminotransferase, Ornithine

SOURCE: Rat liver
SOLVENT: 10 mM Tris-HCl, pH 8.0
COMMENTS: Spectral changes upon addition of aminohexanoate (0-22 minutes) and the effect of 2-oxoglutarate on spectra
REFERENCE: R. A. John, E. D. Jones, and L. J. Fowler, Biochem. J., 177, 721 (1979)

236. Aminotransferase, Ornithine

SOURCE: Vitamin B_6 deficient rats, small intestine
SOLVENT: 0.05 M Potassium phosphate buffer, pH 7.5
COMMENTS: Spectra of apo-form II and product formed by proteolysis after incubation with pyridoxal 5'-phosphate
REFERENCE: N. Katunuma, Rev. Physiol. Biochem. Pharm., 72, 83 (1975)

237. Aminotransferase, Phenylalanine

SOURCE: Pseudomonas fluorescens
SOLVENT: 0.1 M Tris-HCl, pH 8.5
COMMENTS: Spectra of native enzyme and after treatment with phenylhydrazine HCl
REFERENCE: G. W. Jack and P. C. McMahon, Biochim. Biophys. Acta, 523, 344 (1978)

238. Aminotransferase, Taurine

SOURCE: Achromobacter superficialis
SOLVENT: 0.01 M Potassium phosphate, pH 7.0
COMMENTS: Spectra of inactivated, reactivated, and aged enzyme
REFERENCE: S. Toyama, H. Misono, and K. Soda, Biochim. Biophys. Acta, 523, 75 (1978)

239. Aminotransferase, Tyrosine

SOURCE: Rat brain
SOLVENT: 0.1 M Phosphate buffer, pH 7.6
COMMENTS: Spectra of native enzyme and native enzyme plus tyrosine
REFERENCE: P. Mandel and D. Aunis, CIBA Found. Symp., 22, 67 (1974)

240. Aminotransferase B

SOURCE: Escherichia coli
SOLVENT: a. 20 mM Phosphate buffer, pH 6.8
b. 0.2 M Acetate buffer, pH 4.6
c. 0.1 M Tris-HCl buffer, pH 9.0
COMMENTS: Spectra of the native enzyme, after the addition of L-phenylalanine, after the addition of L-valine, after the addition of L-glutamic acid, and after the addition of L-glutamic acid followed by α-keto-glutarate
REFERENCE: N. Monnier, A. Montmitonnet, S. Chesne, and J. Pelmont, Biochimie, 58, 663 (1976)

241. Amylase

SOURCE: Commercial, Taka-amylase
SOLVENT: 0.1 M Potassium/sodium phosphate buffer, pH 7.0
COMMENTS: Spectra of 5-nitroponyl-derivative and of denitroponylation with hydrazine at pH 9.0
REFERENCE: H. Tamaoki, Y. Murase, S. Minato, and K. Nakanishi, J. Biochem. Tokyo, 62, 7 (1967)

242. β-Amylase

SOURCE: Pirkka barley
SOLVENT: a. 0.1 M NaOH
b. 0.05 M Sodium phosphate, pH 7
c. 0.1 M HCl
d. 6 M Guanidine HCl, pH 6.5
e. 6 M Guanidine HCl, pH 12.5
REFERENCE: K. Visuri and M. Nummi, Eur. J. Biochem., 28, 555 (1972)

243. Amylase, Alkaline

SOURCE: Bacillus No. A-40-2
SOLVENT: 0.1 M Tris-HCl buffer, pH 7.0
REFERENCE: K. Horikoshi, J. Biochem. Tokyo, 35, 1783 (1971)

244. α-Amylase

SOURCE: Human urine
REFERENCE: N. Minamiura, Y. Kimura, K. Tsujino, and T. Yamamoto, J. Biochem. Tokyo, 77, 163 (1975)

245. Amylase

SOURCE: Pig pancreas
SOLVENT: 40 mM Tris-HCl buffer, pH 8.0, containing 1% of sodium dodecylsulfate
COMMENTS: Spectra of disulfide-bridged amylase and derivative from reaction with 5,5'-dithiobis (2-nitrobenzoic acid)
REFERENCE: G. Pommer, P. Cozzone, and G. Marchis-Mouren, Biochim. Biophys. Acta, 350, 71 (1974)

246. Amylase, Tri(dinitrophenyl)-

SOURCE: Pig pancreas
REFERENCE: R. Lifshitz and A. Levitzki, Biochemistry, 15, 1987 (1976)

247. Amyloid

SOURCE: Human spleen
SOLVENT: Water, pH 6.4
REFERENCE: M. Pras, M. Schubert, D. Zucker-Franklin, A. Rimon, and E. C. Franklin, J. Clin. Inves., 47, 924 (1968)

248. Amyloid

SOURCE: Human livers and spleens
SOLVENT: 0.02 M HCl
REFERENCE: M. Pras and T. Reshef, Biochim. Biophys. Acta, 271, 193 (1972)

249. Amylose Isomerase

SOURCE: Rabbit muscles
REFERENCE: R. D. Filippova and A. N. Petrova, Biokhimiya (Biochemistry), 36, 1038 (1971)

250. Anthranilate Synthase - Phosphoribosyl Transferase Complex

SOURCE: Salmonella typhimurium
SOLVENT: 50 mM Potassium phosphate, pH 7.4
REFERENCE: T. H. Grove and H. R. Levy, Biochim. Biophys. Acta, 445, 464 (1976)

251. Antigen, Hepatitis B, Surface

SOURCE: Human plasma
SOLVENT: 8.0 M Urea, pH 4.0
COMMENT: Spectra of protein before and after treatment with N-bromosuccinimide
REFERENCE: K. R. Rao and G. N. Vyas, J. Gen. Virol., 24, 571 (1974)

252. Antigen, Hepatitis B

SOURCE: Human plasma
REFERENCE: B. L. Wong and S. E. Charm in Biomedical Applications of Immobilized Enzymes and Proteins (T. M. S. Chang, ed.), Vol. 2, 131 (1977), Plenum Press, New York

253. Antigen (Allergen)

SOURCE: Timothy N pollen
SOLVENT: a. Water
b. 0.1 M HCl
c. 0.1 M NaOH
REFERENCE: O. Grimmer, I. Andersen, S. Dale, S. Oseid, N. Thorsdalen, B. Skalhegg, and K. Aas, Int. Arch. Allergy Appl. Immunol., 55, 29 (1977)

254. Antithrombin

SOURCE: Human plasma
SOLVENT: 0.025 M Tris-HCl containing 0.2 M glycine and 0.03 M NaCl, pH 7.4
REFERENCE: R. Einarsson, Biochim. Biophys. Acta, 446, 124 (1976)

255. α_1-Antitrypsin

SOURCE: Human plasma
SOLVENT: 0.05 M Tris, 0.075 M NaCl, pH 7.5
REFERENCE: J. W. Bloom and M. J. Hunter, J. Biol. Chem., 253, 547 (1978)

256. α_1-Antitrypsin, 3-carboxy-4-nitrobenzene-thio-

SOURCE: Human plasma
SOLVENT: 0.1 M Phosphate buffer, pH 6.0
REFERENCE: C. B. Laurell, J. Pierce, U. Persson, and E. Thulin, Eur. J. Biochem., 57, 107 (1975)

257. Aphanocapsa 6701

SOURCE: Aphanocapsa 6701
COMMENTS: In vivo spectrum at -196° C
REFERENCE: C. Lemasson, N. Tandeau de Marsac, and G. Cohen-Bazaire, Proc. Nat. Acad. Sci., 70, 3130 (1973)

258. Apyrase

SOURCE: Potato
SOLVENT: pH 4.0, pH 7.0, pH 7.5
COMMENTS: Spectra of acetylated, non-acetylated enzyme, and 2-hydroxy-5-nitrobenzylated enzyme
REFERENCE: M. A. Valenzuela, G. Delcampo, E. Marin, and A. Traverso-Cori, Biochem. J., 133, 755 (1973)

259. Aqueous Humor

SOURCE: Calf eye
COMMENTS: Spectra of untreated, diluted aqueous humor (1:10) and U.V.-irradiated material (1:10)
REFERENCE: S. Zigman, J. Schultz, T. Yulo, and D. Grover, Israel J. Med., 1591 (1972)

260. Arabinogalactanase

SOURCE: Bacillus subtilis K-50
SOLVENT: a. 0.01 M Phosphate buffer, pH 7.0
b. 0.1 M NaOH
REFERENCE: S. Emi, J. Fukumoto, and T. Yamamoto, J. Biochem. Tokyo, 35, 1891 (1971)

261. Arginine Decarboxylase (Biosynthetic)

SOURCE: Escherichia coli
SOLVENT: a. 0.1 M N-2-Hydroxyethylpiperazine-N'-2-ethanesulfonic acid buffer, pH 8.0
b. 0.1 M Piperazine-N,N'-bis (2-ethanesulfonic acid) buffer, pH 6.0
COMMENTS: Spectra of apoenzyme, after addition of pyridoxal 5'-phosphate, and after subsequent addition of magnesium sulfate

REFERENCE: W. H. Wu and D. R. Morris, J. Biol. Chem., 248, 1687 (1973)

262. Arginine Esterase (Coagulation Principle)

SOURCE: Snake (Agkistrodon acutus) venom
REFERENCE: H. Sugihara, T. Nikai, H. Oda, and T. Tanaka, Yukugaku Zasshi, 98, 832 (1978)

263. Aromatic Amino Acid Decarboxylase

SOURCE: Micrococcus percitreus
SOLVENT: 0.05 M Potassium phosphate buffer containing 0.0001 M pyridoxal 5'-phosphate, 0.001 M 2-mercaptoethanol, 0.0001 M EDTA, pH 7.0
REFERENCE: H. Nakazawa, H. Kumagai, and H. Yamada, Biochem. Biophys. Res. Comm., 61, 75 (1974)

264. Arylsulfatases A and B

SOURCE: Human liver
SOLVENT: 0.05 M Tris-HCl buffer, pH 8.0
REFERENCE: E. Shapira and H. L. Nadler, Arch. Biochem. Biophys., 170, 179 (1975)

265. Ascorbate Oxidase

SOURCE: Green zucchini
SOLVENT: 0.1 M Acetate buffer, pH 5.2
COMMENTS: a. Spectra of enzyme before and after treatment with EDTA-dimethylglyoxime
b. Spectrum after recombination of copper to copper-depleted protein
REFERENCE: L. Avigliano, A. Desideri, S. Urbanelli, B. Mondovi and A. Marchesini, FEBS Lett., 100, 318 (1979)

266. Ascorbate Oxidase

SOURCE: Yellow crookneck summer squash (Cucurbita pepo condensa)
COMMENTS: Spectra of oxidized enzyme, ascorbate-reduced enzyme, and reductate-reduced enzyme
REFERENCE: M. Katz and J. Westley, J. Biol. Chem., 254, 9142 (1979)

267. Ascorbic Acid Oxidase

SOURCE: Green zucchini squash (Cucurbita pepo medullosa)

REFERENCE: L. Avigliano, P. Gerosa, G. Rotillo, A. Finazzi Agro, L. Calabrese, and B. Mondovi, Ital. J. Biochem., 21, 248 (1972)

268. Ascorbic Acid Oxidase

SOURCE: Green zucchini squash (Cucurbita pepo medullosa)
SOLVENT: a. 0.01 M Phosphate buffer, pH 7.6
b. McIlvaine's buffer, pH 5.6
COMMENTS: Spectra of native, reduced, oxidized and apoenzyme
REFERENCE: M. H. Lee and C. R. Dawson, J. Biol. Chem., 248, 6603 (1973)

269. Ascorbic Acid Oxidase

SOURCE: Green zucchini squash (Cucurbita pepo medullosa)
SOLVENT: 0.1 M Phosphate buffer, pH 7.0
COMMENTS: Spectra of reduced and oxidized enzyme
REFERENCE: J. Deinum, B. Reinhammer, and A. Marchesini, FEBS Lett., 42, 241 (1974)

270. Ascorbic Acid Oxidase

SOURCE: Green zucchini squash (Cucurbita pepo medullosa)
SOLVENT: 100 mM Potassium phosphate buffer, pH 7.0
COMMENTS: Effect of increasing amounts of nitric oxide on spectrum
REFERENCE: F. X. R. van Leeuwen, R. Wever, B. F. van Gelder, L. Avigliano, and B. Mondovi, Biochim. Biophys. Acta, 403, 285 (1975)

271. Ascorbic Acid Oxidase

SOURCE: Green zucchini squash (Cucurbita pepo medullosa)
SOLVENT: 0.08 M Phosphate buffer, pH 6.0
COMMENTS: Spectra in presence and absence of sodium azide
REFERENCE: B. Mondovi, L. Avigliano, G. Rotillo, A. Finazzi Agro, P. Gerosa, and C. Giovagnoli, Mol. Cell Biochem., 7, 131 (1975)

272. Asparaginase, Extracellular

SOURCE: Candida utilis
REFERENCE: T. Sakamoto, C. Araki, T. Beppu, and K. Arima, Agr. Biol. Chem., 41, 1359 (1971)

273. Asparaginase

SOURCE: a. Escherichia coli
b. Erwinia carotovora
SOLVENT: 0.1 Ionic strength carbonate buffer, pH 8.5
COMMENTS: Spectra of Fremy's salt-modified proteins
REFERENCE: R. B. Homer, S. R. Allsop, and J. E. Arrieta, FEBS Lett., 59, 173 (1975)

274. Asparaginase

SOURCE: Escherichia coli
SOLVENT: 0.05 M Sodium borate buffer, pH 8.5
COMMENTS: Spectra of native enzyme and after addition of N-bromosuccinimide in presence and absence of glutamine
REFERENCE: U. Menge and L. Jaenicke, Hoppe-Seyler's Z. Physiol. Chem., 355, 603 (1974)

275. Asparaginase: 3-Nitrotyrosine- and 3-Aminotyrosine-

SOURCE: Escherichia coli
REFERENCE: R. K. Blumark, E. A. Vina, I. P. Kirstukas, I. A. Milman, I. I. Geiman, and R. A. Zhagat, Bioorg. Chem., 4, 1264 (1978)

276. Asparaginase A (and B)

SOURCE: Escherichia coli
SOLVENT: a. Phosphate buffer, pH 7.0
b. 0.1 N NaOH
COMMENTS: Spectrum of asparaginase B not significantly different from that of asparaginase A
REFERENCE: A. Arens, E. Rauenbusch, E. Irion, O. Wagner, K. Bauer, and W. Kaufmann, Hoppe-Seyler's Z. Physiol. Chem., 351, 197 (1970)

277. D-Aspartate Oxidase

SOURCE: Octopus vulgaris Lam hepatopancreas
SOLVENT: Potassium phosphate/sodium succinate buffer, pH 6.7
REFERENCE: A. D'Aniello and E. Rocca, Comp. Biochem. Physiol., 41B, 625 (1972)

278. Aspartate Transcarbamylase

SOURCE: Escherichia coli
SOLVENT: 0.05 M Tris-HCl, pH 7.05
COMMENTS: Spectra of apo-regulatory subunit and the zinc- and mercury-containing regulatory subunits

REFERENCE: M. E. Nelbach, V. P. Pigiet, Jr., J. C. Gerhart, and H. K. Schachman, Biochemistry, 11, 315 (1972)

279. Aspartate Transcarbamylase

SOURCE: Escherichia coli
SOLVENT: 0.04 M Potassium phosphate buffer containing 2 mM 2-mercaptoethanol, and 0.2 mM sodium EDTA or zinc acetate, pH 7.0
COMMENTS: Spectra of the enzyme, the regulatory subunits, the catalytic subunits, and the 2-thiouracil-modified derivatives
REFERENCE: D. Kerbiriou, G. Herve, and J. H. Griffin, J. Biol. Chem., 252, 2881 (1977)

280. Aspartylkinase

SOURCE: Escherichia coli
SOLVENT: 10 mM Potassium phosphate-1 mM EDTA-2 mM magnesium chloride·6 H_2O-1 mM lysine-0.1 mM dithiothreitol, pH 6.8
COMMENTS: Protein from mid-log phase cells and late-stationary phase cells
REFERENCE: E. G. Niles and E. W. Westhead, Biochemistry, 12, 1715 (1973)

281. Azoreductase

SOURCE: Rat liver (3-methyl cholanthrene-treated)
SOLVENT: 50 mM Potassium phosphate buffer, pH 7.4-pH 7.7
COMMENTS:
a. Spectra in presence and absence of NADPH under anaerobic conditions
b. Spectra in presence and absence of dicumarol with or without NADPH
REFERENCE: M. T. Huang, G. T. Miwa, and A. Y. H. Lu, J. Biol. Chem., 254, 3920 (1979); M. T. Huang, G. T. Miwa, N. Cronheim, and A. Y. H. Lu, J. Biol. Chem., 254, 11223 (1979)

282. Azurin

SOURCE: Paracoccus denitrificans (ATCC 13543)
SOLVENT: 0.05 M Tris-cacodylate, pH 7.2
REFERENCE: K. Matinkus, P. J. Kennelly, T. Rea, and R. Timkovich, Arch. Biochem. Biophys., 199, 465 (1980)

283. Azurin

SOURCE: Pseudomonas fluorescens
SOLVENT: a. pH 6.8
b. pH 12.8
COMMENTS: Spectra taken immediately, after 40 minutes, and after 1 hour after adding KOH
REFERENCE: L. Avigliano, P. Guerrieri, L. Calabrese, M. P. Vallogini, G. Rotilio, B. Mondovi, and A. Finazzi Agro, Ital. J. Biochem., 19, 125 (1978)

284. Azurin

SOURCE: Achromobacter cycloelastes
SOLVENT: pH 7
REFERENCE: H. Iwasaki and T. Matsubara, J. Biochem. Tokyo, 73, 659 (1973)

285. Azurin

SOURCE: Pseudomonas aeruginosa
COMMENTS: Spectra of oxidized and dithionite reduced protein
REFERENCE: M. G. Kamalyan and R. M. Nalbandyan, Biokhimiya (Biochemistry), 42, 171 (1977)

286. Bacteriochlorophyll a-Protein

SOURCE: Chlorobium limicola and Chlorobium thiosufatophilum
COMMENTS: Room temperature and liquid nitrogen temperature
REFERENCE: C. F. Fowler, B. H. Gray, N. A. Nugent, and R. C. Fuller, Biochim. Biophys. Acta, 292, 692 (1973)

287. Bacteriochlorophyll a-Protein

SOURCE: Prostecochloris aestaurii
COMMENTS: Spectra taken at room temperature and liquid nitrogen temperature
REFERENCE: W. B. Whitten, J. A. Nairn, and R. M. Pearlstein, Biochim. Biophys. Acta, 503, 251 (1978)

288. Bacteriochlorophyll a-Reaction Center Complex

SOURCE: Chlorobium limicola and Chlorobium thiosulfatophilum
COMMENTS: Spectra taken at room temperature and liquid nitrogen temperature
REFERENCE: C. F. Fowler, B. H. Gray, N. A. Nugent, and R. C. Fuller, Biochim. Biophys. Acta, 292, 692 (1973)

289. Bacteriocin, Pyocin Fl

SOURCE: Pseudomonas aeruginosa
REFERENCE: K. Kuroda and M. Kageyama, J. Biochem. Tokyo, 85, 7 (1979)

290. Bacteriocuprein

SOURCE: Photobacterium leiognathi
SOLVENT: 0.005 M Phosphate, pH 7.4
REFERENCE: K. Paget and A. M. Michelson, Biochem. Biophys. Res. Comm., 58, 830 (1974)

291. Bacteriophage fd

SOLVENT: a. 0.1 M Tris, 0.14 M NaCl, pH 7.6
b. 7.6 M Guanidine-HCl, 0.14 M NaCl, 0.1 M Tris, pH 7.6
COMMENTS: Spectra of intact and disassembled phage
REFERENCE: K. Ikehara, H. Utiyama, and M. Kurata, Virology, 66, 306 (1975)

292. Bacteriophage fd

SOLVENT: 0.02 M Tris, 0.14 M NaCl, pH 7.6
COMMENTS: Effect of different amounts of 2-chloroethanol on spectra
REFERENCE: K. Ikehara and H. Utiyama, Virology, 66, 316 (1975)

293. Bacteriophage ϕN S11

SOLVENT: pH 4.0
COMMENTS: Spectra of intact phage and phage disrupted by 1% sodium dodecylsulfate
REFERENCE: Y. Sakaki, K. Yamada, M. Oshima, and T. Oshima, J. Biochem. Tokyo, 82, 1451 (1977)

294. Bacteriophage G, Bacillus Megatherium KM

REFERENCE: J. S. Murphy and L. Philipson, J. Gen. Physiol., 45, 155 (1962)

295. Bacteriophage T2L and Heads

SOLVENT: 0.01 M Phosphate buffer
REFERENCE: V. V. Mesyanzhinov, T. A. Bogomolova, and B. F. Poglazov, Biokhimiya (Biochemistry), 36, 1001 (1971)

296. Bacteriophage PM2

SOLVENT: 0.05 M Tris, 1 M NaCl, pH 7.4
COMMENTS: Effect of Sarkosyl on spectrum
REFERENCE: R. D. Camerini-Otero and R. M. Franklin, Eur. J. Biochem., 53, 343 (1975)

297. Bacteriorhodopsin

SOURCE: Halobacterium halobium
SOLVENT: Water
REFERENCE: E. Hildebrand and N. Dencher, Nature, 257, 47 (1975)

298. Benzene Dioxygenase Components

SOURCE: Pseudomonas sp.
SOLVENT: 25 mM Potassium phosphate buffer, pH 7.4
COMMENTS: Spectra of protein A1, protein A2, protein B, and protein A1 after reduction with dithionite
REFERENCE: B. C. Axcell and P. J. Geary, Biochem. J., 146, 173 (1975)

299. Benzene Dioxygenase Components

SOURCE: Pseudomonas putida
SOLVENT: 25 mM Potassium phosphate, pH 7.3
COMMENTS: Spectra of protein A1, oxidized and NADH reduced, and protein B, oxidized and NADH reduced
REFERENCE: S. E. Crutcher and P. J. Geary, Biochem. J., 177, 393 (1979)

300. Benzylamine Oxidase

SOURCE: Pig plasma
SOLVENT: 0.06 M Phosphate buffer, pH 7.4
COMMENTS: Spectra of untreated, borohydride reduced, cyanide treated, hydroxylamine treated, and borohydride reduced, histamine-binding protein
REFERENCE: F. Buffoni, in Pyridoxal Catalysis in Enzyme and Model Systems (E. E. Snell, A. E. Braunstein, E. E. Severin, and Yu. M. Torchinsky, eds.), 336 (1968), Interscience, New York

301. Benzylamine Oxidase

SOURCE: Pig plasma
COMMENTS: Spectra taken of enzyme, enzyme and phenylhydrazine, and enzyme and Catran

REFERENCE: A. Lindstrom and G. Pettersson, Eur. J. Biochem., 34, 564 (1973)

302. Benzylamine Oxidase

SOURCE: Pig plasma
SOLVENT: 0.1 M Phosphate buffer, pH 7
COMMENTS: Effect of azide on absorption spectrum of benzylamine oxidase
REFERENCE: A. Lindstrom, B. Olsson, and G. Pettersson, Eur. J. Biochem., 48, 237 (1974)

303. Biotin Carboxyl Carrier Protein$_{SC}$

SOURCE: Escherichia coli
SOLVENT: pH 7 and pH 12
REFERENCE: R. R. Fall, M. Glaser, and P. R. Vagelos, J. Biol. Chem., 251, 2063 (1976)

304. Bisulfite Reductase

SOURCE: Desulfotomaculum nigrificans
SOLVENT: 0.1 M Phosphate buffer, pH 7.0
COMMENTS: Spectra of reduced and oxidized forms
REFERENCE: J. M. Akagi and V. Adams, J. Bacteriol., 116, 392 (1973)

305. Blastokinin

SOURCE: Rabbit
SOLVENT: pH 5.6 and pH 11.5
REFERENCE: R. W. McGaughey and F. A. Murray, Fertil. and Steril., 23, 399 (1972)

306. Blood

SOURCE: Human erythrocytes
COMMENTS: Spectra of normal and acatalasemic hemolysates (diluted 1:1000) after treatment with hydrogen peroxide
REFERENCE: S. Takohara, in Hereditary Disorders of Erythrocyte Metabolism (E. Beutler, ed.), 21 (1968), Grune and Stratton, New York

307. Bromelain, Stem

SOURCE: Pineapple
SOLVENT: pH 8.0
COMMENTS: Spectra made before and after photooxidation with methylene blue

REFERENCE: T. Murachi, T. Tsudzuki, and K. Okumura, Biochemistry, 14, 249 (1975)

308. α-Bungarotoxin, 4-azido-2-nitrophenyl-

SOURCE: Commercial
COMMENTS: Spectral changes upon 0-10 minutes irradiation
REFERENCE: F. Hucho, FEBS Lett., 103, 27 (1979)

309. β_1-Bungarotoxin

SOURCE: Bungarus multicinctus venom
SOLVENT: a. H_2O
b. 0.02 M Ammonium hydroxide
c. 0.1 M Formic acid
COMMENTS: Spectra of intact and p-bromophenacyl-modified protein and its modified carboxymethylated A and B chains
REFERENCE: K. Kondo, H. Toda, and K. Narita, J. Biochem. Tokyo, 84, 1301 (1978)

310. Butyryl-CoA Dehydrogenase

SOURCE: Peptostreptococcus elsdenii
SOLVENT: 0.1 M Potassium phosphate buffer, pH 7
COMMENTS: Spectra of intact enzyme, after dithionite reduction, after borohydride reduction, after reoxidation after reduction, after addition of crotonyl-CoA, after addition of acetoacetyl-CoA, after addition of acryloyl-CoA, after addition of butyryl-CoA, after addition of pent-2-enoyl-CoA, and after addition of penta-2,4-dienoyl-CoA
REFERENCE: P. C. Engel and V. Massey, Biochem. J., 125, 889 (1971)

311. Butyryl-CoA Dehydrogenase

SOURCE: Peptostreptococcus elsdenii
SOLVENT: 0.046 M Potassium phosphate buffer, pH 7.0
COMMENTS: Spectra after addition of D-lactate, after addition of NADH and electron-transferring protein, and after addition of D-lactate dehydrogenase, D-lactate, and electron-transferring protein
REFERENCE: H. L. Brockman and W. A. Wood, J. Bacteriol., 124, 1447 (1975)

312. Byssus Threads

SOURCE: Mollusc (Mytilus edulis)
COMMENTS: Spectra of native threads, pepsin-solubized protein, acid-soluble protein, and residue after pepsin extraction
REFERENCE: G. Chandrakasan, B. Geetha, G. Kashnan, and K. T. Joseph, Ind. J. Biochem. Biophys., 14, 132 (1977)

313. Caeruloplasmin

SOURCE: Human plasma
SOLVENT: 0.25 M Sodium acetate buffer, pH 5.0
COMMENTS: Changes in the spectrum upon the addition of sodium azide, potassium cyanate, or potassium thiocyanate
REFERENCE: W. Byers, G. Curzon, K. Garbett, B. E. Speyer, S. N. Young, and R. J. P. Williams, Biochim. Biophys. Acta, 310, 38 (1950)

314. Calmodulin, Activator of Cyclic Nucleotide Phosphodiesterase

SOURCE: Pig brain
SOLVENT: a. 0.025 M Tris-HCl buffer containing either 0.05 M NaCl or 1.0 M NaCl
b. 0.025 M Potassium phosphate, pH 7.0
c. 0.02 M Tris-HCl buffer, pH 7.7 containing either 0.05 M NaCl or 1.0 M NaCl
COMMENTS: Spectra of protein and nitrated protein and changes in spectra due to addition of Ca^{2+} and Mg^{2+} to nitrated and unmodified protein
REFERENCE: P. G. Richman and C. B. Klee, J. Biol. Chem., 254, 5372 (1979)

315. Calmodulin

SOURCE: Chicken fibroblast
SOLVENT: 0.10 M Imidazole-HCl, 0.001 M EDTA, pH 7.0
REFERENCE: L. J. Van Eldik and D. M. Watterson, J. Biol. Chem., 254, 10250 (1979)

316. Capsid

SOURCE: Bacteriophage F2
SOLVENT: pH 6.2, 0.1 M NaCl, and 1.0 M NaCl
REFERENCE: R. W. Henkens and V. L. Middlerook, Biochemistry, 12, 2910 (1973)

317. Carbonic Anhydrase, Cu^{II}

SOURCE: Beef erythrocytes
SOLVENT: pH 6 and 20 mM Carbonate buffer, pH 9.9
COMMENTS: Enzyme spectra before and after addition of either cyanide, azide, or iodide
REFERENCE: L. Morpurgo, G. Rotilio, A. F. Agro, and B. Mondovi, Arch. Biochem. Biophys., 170, 360 (1975)

318. Carbonic Anhydrase, Cu^{II}

SOURCE: Beef erythrocytes
SOLVENT: pH 6.0
COMMENTS: Enzyme spectra before and after addition of either cyanide, 2-mercaptoethanol, azide, or iodide
REFERENCE: B. Mondovi, L. Morpurgo, G. Rotilio, and A. Finazzi-Agro, Adv. Exp. Med. Biol., 74, 424 (1976)

319. Carbonic Anhydrase, Cu^{II}

SOURCE: Beef erythrocytes
SOLVENT: pH 6.0 and 0.05 M Tris-HCl buffer, pH 7.6
COMMENTS: Enzyme spectra before and after addition of 2-mercaptoethanol or sodium disulfide
REFERENCE: L. Morpurgo, A. F. Agro, G. Rotilio, and B. Mondovi, Eur. J. Biochem., 64, 453 (1976)

320. Carbonic Anhydrase B, Co (II)-, Complexes

SOURCE: Beef erythrocytes
COMMENTS:
1. Spectra of Co(II)-carbonic anhydrase-complexes with dithionite, borohydride, p-toluenesulfonamide, and cyanide
2. Spectral changes during hydrogen peroxide oxidation of Co(II)-carbonic anhydrase-cyanide complex

REFERENCE: H. Shinar and G. Navon, Eur. J. Biochem., 93, 313 (1979)

321. Carbonic Anhydrase

SOURCE: Chicken erythrocytes
SOLVENT:
a. 0.01 M Potassium phosphate, 0.1 M KCl, 1 mM dithiothreitol, pH 6
b. 0.1 M NaOH, pH 13

REFERENCE: R. S. Bernstein and R. Schraer, J. Biol. Chem., 247, 1306 (1972)

322. Carbonic Anhydrase Co^{II}

SOURCE: Human erythrocytes
SOLVENT: 0.025 M Tris, pH 9.0
COMMENTS: Spectra of enzyme and of 1:1 cyanide complex
REFERENCE: J. E. Coleman and R. V. Coleman, J. Biol. Chem., 247, 4718 (1972)

323. Carbonic Anhydrase B, Co^{II}, carboxamidomethyl

SOURCE: Human erythrocytes
SOLVENT: 25 mM Sodium sulfate
COMMENTS: a. Changes in spectrum due to pH change from pH 7.1 to pH 11.4
b. Spectra of unmodified and carboxamidomethyl-modified enzyme
REFERENCE: P. L. Whitney, J. Biol. Chem., 248, 2785 (1973)

324. Carbonic Anhydrase, Co^{II}

SOURCE: Human erythrocytes
SOLVENT: 0.025 Tris-HCl, pH 9.0
COMMENTS: Spectra of enzyme with different amounts of cyanide at 25°C and at -160°C
REFERENCE: P. H. Haffner and J. E. Coleman, J. Biol. Chem., 250, 996 (1975)

325. Carbonic Anhydrase B, Co^{II}

SOURCE: Human erythrocytes
SOLVENT: 20 mM Bis-Tris-sulfate at different ionic strengths and different pH values
COMMENTS: Spectra of enzyme, carboxymethyl-enzyme, and carboamidomethyl enzyme and the changes due to the addition of halide ions
REFERENCE: P. L. Whitney and H. Brandt, J. Biol. Chem., 251, 3862 (1976)

326. Carbonic Anhydrase Co^{II}

SOURCE: Human erythrocytes
SOLVENT: pH 6.5, 46% Sucrose and pH 11.0, 48% sucrose
REFERENCE: R. Bauer, P. Limkilde, and J. T. Johansen, Biochemistry, 15, 334 (1976)

327. Carbonic Anhydrase

SOURCE: Turkey erythrocytes (Mele agris gallopavo)
SOLVENT: 0.1 M Tris-sulfate buffer, 0.01 M 2-mercaptoethanol, pH 8.0

REFERENCE: P. R. Lemke and G. Graf, Mol. Cell Biochem., 4, 141 (1974)

328. Carboxypeptidase and Other Proteins

SOURCE: Commercial
COMMENTS: Spectra of carboxypeptidase, insulin and serum albumin, all of bovine origin, and ovomucoid from egg white, and of their acetylated derivatives
REFERENCE: J. F. Riordan, W. E. C. Wacker, and B. L. Vallee, Biochemistry, 4, 1758 (1965)

329. Carboxypeptidase A_α, arsanilazotyrosine 248-, Zinc

SOURCE: Beef pancreatic juice
SOLVENT: 0.02 M Tris-HCl pH 8.2, pH 7.4 and water
COMMENTS: Changes in spectrum upon addition of NaCl
REFERENCE: F. A. Quiocho, C. H. McMurray, and W. N. Lipscomb, Proc. Nat. Acad. Sci., 69, 2850 (1972)

330. Carboxypeptidase A, hydroxynitrobenzyl-

SOURCE: Beef
SOLVENT: 0.2 M Tris-Acetate, pH 5.0-pH 12
REFERENCE: V. R. Naik and H. R. Horton, J. Biol. Chem., 248, 6709 (1973)

331. $\text{Carboxypeptidase}_{\alpha,\beta}$, arsanilazotyrosine-248

SOURCE: Beef (commercial)
SOLVENT: 2 mM Tris-HCl-0.5 M NaCl, pH 6.2-10.8
COMMENTS: Spectra in presence and absence of zinc
REFERENCE: J. T. Johansen and B. L. Vallee, Proc. Nat. Acad. Sci., 70, 2006 (1973)

332. Carboxypeptidase A_α, azo-

SOURCE: Commercial
SOLVENT: 0.05 M N-2-Hydroxyethyl piperazine-N'-2-ethanesulfonic acid-1 M NaCl, pH 8.5
COMMENTS: Effect of temperature (14° to -30°C) on spectrum of monoarsanilazotyrosine-248-zinc carboxypeptidase
REFERENCE: J. S. Thompson, H. Gehring, and B. L. Vallee, Proc. Nat. Acad. Sci., 77, 132 (1980)

333. $\text{Carboxypeptidase}_\alpha$, arsanilazotyrosine-248

SOURCE: Beef (commercial)
SOLVENT: 2 mM Tris-HCl-0.5 M NaCl, pH 6.2-10.8

COMMENTS: Spectra in presence and absence of zinc
REFERENCE: J. T. Johansen and B. L. Vallee, Biochemistry, 14, 649 (1975)

334. Carboxypeptidase, monoarsanilazotyrosine-248-, Zinc

SOURCE: Beef (commercial)
SOLVENT: a. 1.0 M NaCl, pH 6.0-10.5
b. 50 mM N-2-hydroxyethylpiperazine-N'-2-ethanesulfonic acid, 1 M NaCl, pH 6.5-8.2
REFERENCE: L. W. Harrison, D. S. Auld, and B. L. Vallee, Proc. Nat. Acad. Sci., 72, 4356 (1975)

335. Carboxypeptidase-, arsanilazotyrosine-248, Zinc

SOURCE: Beef (commercial)
SOLVENT: 0.5 Tris-0.5 M NaCl, pH 8.5
COMMENTS: Changes in spectrum upon addition of L-benzylsuccinate
REFERENCE: J. T. Johansen, A. A. Klyosov, and B. L. Vallee, Biochemistry, 15, 296 (1976)

336. Carboxypeptidase, azo-, and nitroazo-

SOURCE: Beef (commercial)
SOLVENT: 0.1 N NaOH
REFERENCE: L. Cueni and J. R. Riordan, Biochemistry, 17, 1834 (1978)

337. Carboxypeptidase B

SOURCE: Commercial (pig)
SOLVENT: 0.05 M Borate, pH 7.9
COMMENTS: Spectra before and after reaction with phenylglyoxal
REFERENCE: M. M. Weber, M. Moldovan, and M. Sokolovsky, Eur. J. Biochem., 53, 207 (1975)

338. DD-Carboxypeptidase-Endopeptidase

SOURCE: Streptomyces albus G, exocellular
SOLVENT: 10 mM Tris-HCl buffer, pH 7.7 containing 4 mM $MgCl_2$
REFERENCE: C. Duez, J. M. Frere, F. Geurts, J. M. Ghysen, L. Dierickx, and L. Delcambe, Biochem. J., 175, 793 (1978)

339. DD-Carboxypeptidase-Transpeptidase

SOURCE: Streptomyces R39, exocellular
SOLVENT: 0.1 M Tris-HCl buffer, pH 7.7, containing 0.05 M $MgCl_2$ and 0.2 M NaCl
REFERENCE: J. M. Frere, R. Moreno, J. M. Ghysen, H. R. Perkins, L. Dierickx, and L. Delcambe, Biochem. J., 143, 233 (1974)

340. Carcinoembryonic Antigen

SOURCE: Human embryonic tissue, cancer (colon, liver) tissue, and normal intestinal tissue
SOLVENT: Phosphate buffer, pH 4.5 containing 0.85% NaCl
REFERENCE: H. Grossman, G. Heidl, and M. Muller, Acta Biol. Germ., 31, 321 (1973)

341. Carotenoprotein

SOURCE: Labidocera acutifrons
SOLVENT: 0.02 M Phosphate buffer, pH 7
REFERENCE: P. F. Zagalsky and P. J. Herring, Comp. Biochem. Physiol., 41B, 397 (1972)

342. Carotenoprotein

SOURCE: Scenedesmus obliquus
REFERENCE: R. Powls and G. Britton, Biochim. Biophys. Acta, 453, 270 (1976)

343. Carotovoricin Er (A Bacteriocin)

SOURCE: Erwinia caratovora
SOLVENT: 0.05 M Phosphate buffer, pH 7.2
REFERENCE: Y. Itoh, K. Izaki, and H. Takahashi, J. Gen. Appl. Microbiol., 24, 27 (1978)

344. Casein and Paracasein

SOURCE: Milk
REFERENCE: C. Porcher, H. Volkringer, and J. Brigando, Compte Rend. Acad. Sci., 199, 19 (1934)

345. Casein and Paracasein, azo derivatives

SOURCE: Milk
REFERENCE: H. G. Higgins and D. Fraser, Aus. J. Biol. Sci., 7, 89 (1954)

346. Casein, Temperature Sensitive

SOURCE: Milk
SOLVENT: 0.1 N NaOH
REFERENCE: Y. Igarashi and Z. Saito, Jap. J. Zootech. Soc., 41, 262 (1970)

347. Casein and Azocasein

SOURCE: Milk
SOLVENT: 5% Trichloroacetic acid - 0.4% SDS at pH 1-6; pH 7.10 and 11.6 with and without SDS
REFERENCE: J. Langner, A. Wakil, M. Zimmermann, S. Ansorge, P. Bohley, H. Kirschke, and B. Weideranders, Acta. Biol. Med. Germ., 31, 1 (1973)

348. κ-Casein

SOURCE: Cow's milk
SOLVENT: 0.1 M Phosphate buffer, pH 6.1
COMMENTS: Spectra of untreated and diethyl pyrocarbonate-treated protein
REFERENCE: N. M. C. Kaye and P. Jolles, Biochim. Biophys. Acta, 536, 329 (1978)

349. Casein

SOURCE: Milk
COMMENTS: Spectra of unmodified casein and aspartylcasein
REFERENCE: A. J. Puigserver, L. C. Sen, E. Gonzales-Flores, R. E. Feeney, and J. R. Whitaker, Agric. Food Chem., 27, 2098 (1979)

350. Catalase

SOURCE: Human erythrocytes
SOLVENT:
a. pH 2.9
b. pH 5.0
c. pH 7.0
d. pH 12.0

COMMENTS: Spectra of enzyme, cyanide complex, and azide complex (all at pH 7.0) and enzyme and fluoride complex (at pH 5.0)
REFERENCE: M. Palcic, and H. B. Dunford, Can. J. Biochem., 57, 321 (1979)

351. Catalase

SOURCE: Aspergillus niger
REFERENCE: H. Gruft, R. Ruck, and J. Traynor, Can. J. Biochem., 56, 916 (1978)

352. Catalase Complexes

SOURCE: Beef liver
COMMENTS: Spectra of complexes of catalase with either formate, acetate, rhodanate, or cyanate
REFERENCE: H. Rein, O. Ristau, F. Hackenberger, and F. Jung, Biochim. Biophys. Acta, 167, 538 (1968)

353. Catalase, Compound I and Compound II

SOURCE: Beef liver
SOLVENT: 10 mM Sodium cacodylate, pH 5.0
COMMENTS: Spectra of catalase before and after the addition of hydroxymethyl hydroperoxide
REFERENCE: S. Marklund, Biochim. Biophys. Acta, 289, 269 (1972)

354. Catalase

SOURCE: Beef liver and Penicillium notatum
REFERENCE: M. Y. Kamel and M. M. Tahany, Acta Biol. Med. Germ., 30, 13 (1973)

355. Catalase, Glyoxal and Acrylonitrile Derivatives

SOURCE: Beef liver
REFERENCE: K. Abe, M. Hiraga, and F. K. Anan, J. Biochem. Tokyo, 74, 889 (1973)

356. Catalase, acetylated

SOURCE: Beef liver
COMMENTS: Acetylated by N-acetylimidazole or acetic anhydride
REFERENCE: H. Furuta, A. Hachimori, Y. Ohta, and T. Samejima, J. Biochem. Tokyo, 76, 481 (1974)

357. Catalase, acetylated

SOURCE: Beef liver
SOLVENT: 20 mM Tris-HCl buffer, pH 7.5, with or without 1% SDS
REFERENCE: A. Takeda, A. Hachimori, M. Murai, K. Sato, and T. Samejima, J. Biochem. Tokyo, 78, 911 (1975)

358. Catalase

SOURCE: Beef liver
SOLVENT: 6 M Guanidine HCl, 0.05 M phosphate, pH 6.5
REFERENCE: C. Balestrieri, G. Colonna, A. Giovane, G. Irace, and L. Servillo, Eur. J. Biochem., 90, 433 (1972)

359. Catalase I and Catalase V

SOURCE: Beef liver
SOLVENT: 0.1 M Sodium phosphate buffer, 0.2 NaCl, pH 8.0
REFERENCE: G. Kuntz, P. Stockel, and H. G. Heidrich, Hoppe-Seyler's Z. Physiol. Chem., 359, 959 (1978)

360. Catalase

SOURCE: Beef liver
SOLVENT: Phosphate buffer, pH 6.9 and 0.02 M lactate buffer, pH 5.0
COMMENTS: Changes in the spectrum of catalase upon the addition of fluoride, cyanate, thiocyanate, azide, and cyanide
REFERENCE: W. R. Browett and M. J. Stillman, Biochim. Biophys. Acta, 577, 291 (1979)

361. Catalase

SOURCE: Goat liver and beef liver
SOLVENT: 50 mM Phosphate buffer, pH 7.0
COMMENTS: Spectra of native, 2-mercaptoethanol treated, and dithiothreitol treated catalase
REFERENCE: T. Miyahara, A. Takeda, A. Hachimori, and T. Samejima, J. Biochem. Tokyo, 84, 1267 (1978)

362. Catalase

SOURCE: Mouse liver
SOLVENT: 0.1 M Acetate buffer, pH 4.7
REFERENCE: G. L. Jones and C. F. Masters, Arch. Biochem. Biophys., 169, 7 (1975)

363. Catalase

SOURCE: Mouse liver
SOLVENT: 0.02 M Sodium phosphate buffer, pH 7.0
REFERENCE: M. B. Baird, H. R. Massie, and L. S. Birnbaum, Biochem. J., 163, 449 (1977)

364. Catalase, monomer and dimer

SOURCE: Pig erythrocytes
SOLVENT: 50 mM Phosphate buffer, pH 7.0, 1.2 mM sodium dithionite
REFERENCE: A. Takeda and T. Samejima, J. Biochem. Tokyo, 82, 1025 (1977)

365. Catalase

SOURCE: Rat liver
COMMENTS: Changes in spectrum upon addition of cyanide and of hydrogen peroxide
REFERENCE: H. Sies and T. Bucher, Arch. Biochem. Biophys., 154, 101 (1973)

366. Catalase

SOURCE: Rat liver
SOLVENT: 50 mM Potassium phosphate buffer, pH 7.0
COMMENTS: Spectra of compound I and compound II generated by H_2O_2 and by H_2O_2 and p-cresol
REFERENCE: N. Oshino, B. Chance, and H. Sies, Arch. Biochem. Biophys., 159, 704 (1973)

367. Catalase

SOURCE: Rat liver
SOLVENT: 5 mM Sodium-potassium phosphate buffer, pH 7.0 containing 0.1 mM EDTA and 0.1 mM dithiothreitol
COMMENTS: Spectra of catalase before and after the addition of 3-amino-1,2,4-triazole
REFERENCE: Y. Kanai, T. Sugimura, T. Matsushima, and A. Kawamura, J. Biol. Chem., 249, 6505 (1974)

368. Catalase

SOURCE: Neurospora crassa 5297a
SOLVENT: 0.1 M Potassium phosphate, pH 7.3, 0.001 M phenylmethanesulfonyl fluoride, 0.001 M DTT, and 0.0005 M EDTA
COMMENTS: Spectra of enzyme and cyanide-enzyme complex
REFERENCE: G. S. Jacob and W. H. Orme-Johnson, Biochemistry, 18, 2967 (1979)

369. Catalase

SOURCE: Desulfovibrio vulgaris
SOLVENT: 50 mM Potassium phosphate buffer, pH 7.6
REFERENCE: C. E. Hatchikian and J. Le Gall in Superoxide and Superoxide Dismutases (A. M. Michelson, J. M. McCord, and I. Fridovich, eds.), 159 (1977), Academic Press, New York

370. Catalase

SOURCE: Spinach leaf
REFERENCE: H. Mitsuda and K. Yasumatsu, Bull. Agr. Chem. Soc., 19, 208 (1955)

371. Catalase

SOURCE: Toad liver and beef liver
SOLVENT: Phosphate buffer, pH 7.0
COMMENTS: Spectra of fractions taken during purification
REFERENCE: H. Mitsuda and K. Yasumatsu, Bull. Agr. Chem. Soc., 19, 200 (1955)

372. Catechol 1,2-Dioxygenase

SOURCE: Acinetobacter calcoaceticus
SOLVENT: 50 mM Tris-HCl buffer, pH 8.0
REFERENCE: R. N. Patel, C. T. Hou, A. Felix, and M. O. Lillard, J. Bacteriol., 127, 536 (1976)

373. Cathepsin D

SOURCE: Rabbit muscle
SOLVENT: 10 mM Phosphate buffer, pH 7.0 and 0.1 M NaOH
REFERENCE: A. Suzuki, A. Okitani, and M. Fujimaki, Agr. Biol. Chem., 33, 579 (1969)

374. Cathepsin D

SOURCE: Chicken muscle
SOLVENT: 0.05 M Acetate buffer, pH 4.0
REFERENCE: K. Fukushima, G. H. Gnoh, and S. Shinano, Agr. Biol. Chem., 35, 1495 (1971)

375. Cathepsin D

SOURCE: Horse spleen
SOLVENT: 0.005 M Tris buffer, pH 7.0
REFERENCE: A. Ducastaing, J. L. Azanza, J. M. Robin, J. Raymond, and P. Creach, Biochimie, 58, 771 (1976)

376. Cellobiose Oxidase

SOURCE: Sporotrichum pulverulentum
SOLVENT: 50 mM Sodium acetate buffer, pH 5.0
COMMENTS: Spectra of the oxidized and reduced (sodium dithionite) forms
REFERENCE: A. R. Ayers, S. B. Ayers, and K. E. Eriksson, Eur. J. Biochem., 90, 171 (1978)

377. Cellulase, Acid-

SOURCE: Aspergillus niger
REFERENCE: R. Ikeda, T. Yamamoto, and M. Funatsu, Agr. Biol. Chem., 37, 1153 (1973)

378. Ceruloplasmin

SOURCE: Human plasma
SOLVENT: 100 mM Sodium acetate and 250 mM NaCl, pH 7.0
COMMENTS: Changes in the spectrum of oxidized ceruloplasmin due to nitric oxide
REFERENCE: R. Wever, F. X. R. van Leeuwen, and B. F. Vangelder, Biochim. Biophys. Acta, 302, 236 (1973)

379. Ceruloplasmin

SOURCE: Human plasma
COMMENTS: Changes in the spectrum upon the addition of p-phenylenediamine, sodium azide, and histamine
REFERENCE: B. Bozhkov, Enzyme, 14, 65 (1972-1973)

380. Ceruloplasmin

SOURCE: Human plasma
SOLVENT: 0.15 M NaCl
REFERENCE: O. Nakagawa, Int. J. Pept. Prot. Res., 4, 385 (1972)

381. Chelatin

SOURCE: Rat liver
SOLVENT: 0.01 M Potassium phosphate, pH 7.8
COMMENTS: Spectra of Cu-chelatin, apochelatin, and reconstituted Cu-chelatin
REFERENCE: D. R. Winge, R. Premakumar, R. D. Wiley, and K. V. Rajagopalan, Arch. Biochem. Biophys., 170, 253 (1975)

382. Chelatin, Cu-

SOURCE: Rat liver, chicken liver, rabbit liver
SOLVENT: 0.01 M Potassium phosphate, pH 7.8
REFERENCE: R. Premakumar, D. R. Winge, R. D. Wiley, and K. V. Rajagopalan, Arch. Biochem. Biophys., 170, 278 (1975)

383. Chlorocruorin

SOURCE: Potamilla leptochaeta
SOLVENT: 50 mM Sodium phosphate buffer, pH 7.4-7.85
COMMENTS: Changes in spectrum upon addition of SDS, and SDS plus imidazole; spectra of ferri-, ferro-, and oxygenated protein; effect of sodium bisulfite on SDS-treated oxychlorocruorin; Schiff base formation spectra from ferrichlorucruorin

REFERENCE: Y. Orii and N. Washio, J. Biochem. Tokyo, 81, 495 (1977)

384. Chloroperoxidase, Ferric

SOURCE: Caldariomyces fumago
SOLVENT: 0.1 M Potassium phosphate buffer, pH 3.0
COMMENT: Spectra of ferric enzyme, reduced enzyme, and reduced-CO complex
REFERENCE: P. F. Hollenberg and L. P. Hager, J. Biol. Chem., 248, 2630 (1973)

385. Chloroperoxidase

SOURCE: Caldariomyces fumago
SOLVENT: 0.1 Potassium phosphate buffer, pH 6.3
COMMENTS: Spectral titrationswith dithionite, NADPH, and potassium ferricyanide
REFERENCE: R. Chiang, R. Makino, W. E. Spomer, and L. P. Hager, Biochemistry, 14, 4166 (1975)

386. Chloroperoxidase A

SOURCE: Caldariomyces fumago
SOLVENT: Water
REFERENCE: A. S. W. Sae and B. Y. Cunningham, Phytochemistry, 18, 1785 (1979)

387. Chloroperoxidase, ferric and ferrous

SOURCE: Caldariomyces fumago
SOLVENT: 0.1 M Potassium phosphate buffer, pH 3.7 or 0.1 M sodium phosphate buffer, pH 3.0
COMMENTS: Spectral titrations with methyl viologen semiquinone and with carbon monoxide; spectra of the following halide complexes of ferric chloroperoxidase: fluoride, chloride, bromide or iodide
REFERENCE: R. Makino, R. Chiang, and L. P. Hager, Biochemistry, 15, 4748 (1976)

388. Chlorophylla-Protein Complex (CP)

SOURCE: Chlamydomonas reinhardtii strain 137C and thylakoid membranes
COMMENTS: Spectra of CPIII and CPIV at 25°C and -196°C
REFERENCE: P. Delepelaire and N. H. Chua, Proc. Nat. Acad. Sci., 76, 111 (1979)

389. Cholesterol Oxidase

SOURCE: Schizophyllum commune
SOLVENT: 0.1 M Acetate buffer, pH 4.0
COMMENTS: Spectra of enzyme and dehydro-epiandrosterone-reduced enzyme
REFERENCE: M. Fukuyama and Y. Miyake, J. Biochem. Tokyo, 85, 1183 (1979)

390. Cholesterol Oxidase

SOURCE: Beef adrenal cortex
SOLVENT: Phosphate buffer
COMMENTS: Spectra of oxidized, reduced, and CO-reduced forms
REFERENCE: J. Ramseyer and B. W. Harding, Biochim. Biophys. Acta, 315, 306 (1973)

391. Choline Acetylase

SOURCE: Human placenta
COMMENTS: Spectra of native and photoinactivated enzyme
REFERENCE: V. P. Miroshnichenko, I. A. Popova, and S. E. Severin, Biokhimiya (Biochemistry), 36, 1066 (1971)

392. Choline Acetyltransferase

SOURCE: Beef brain
REFERENCE: L. P. Chao and F. Wolfgram, J. Neurochem., 20, 1075 (1973)

393. Choline Oxidase

SOURCE: Arthrobacter globiformis
SOLVENT: 10 mM Phosphate-2 mM EDTA-0.1% KCl buffer, pH 7
COMMENTS: Changes in spectrum of oxidized enzyme after separate additions of choline chloride and sodium hydrosulfite
REFERENCE: S. Ikuta, S. Imamura, H. Misaki, and Y. Horiuti, J. Biochem. Tokyo, 82, 1741 (1977)

394. Choline Oxidase

SOURCE: Fungus (Cylindrocarpon didymum M-1)
SOLVENT: 0.05 M Tris-HCl, pH 8.5, containing 1 mM EDTA and 0.1 mM DTT
REFERENCE: Y. Tani, N. Mori, K. Ogata, and H. Yamada, Agric. Biol. Chem., 43, 815 (1979)

395. Cholinergic Receptor Protein Fraction

SOURCE: Rat brain
SOLVENT: 0.1 M Salt solution used
COMMENTS: Changes in spectrum after addition of either Na^+, K^+, Li^+, or Ca^{++} in presence and absence of acetylcholine
REFERENCE: R. N. Glebov and A. N. Mezentsev, Biokhimiya (Biochemistry), 33, 427 (1968)

396. Chromatin

SOURCE: Human leukemic (CCRF-CEM) cells
COMMENTS: Almost the same spectrum seen with chromatin from normal (SLT, TOH) infectious mononucleosis (RKB, SB) and leukemic (CEM, HSB) lymphocytic cells
REFERENCE: L. S. Desai, U. C. Wulff, and G. E. Foley, Biochimie, 57, 315 (1975)

397. Chromatin

SOURCE: Chlamydomonas reinhardtii
REFERENCE: G. Robreau and Y. Le Gal, Biochimie, 57, 703 (1975)

398. Chromatin

SOURCE: Salmon liver: sexually immature and sexually mature; before and after spawning
REFERENCE: V. A. Rasskazov, V. F. Nesterenko, and V. V. Galkin, Comp. Biochem. Physiol., 53B, 149 (1976)

399. Chromatin, sheared

SOURCE: Chicken oviduct
COMMENTS: Effect of storage on ice for 1-28 days on spectrum
REFERENCE: R. T. Francschi and K. H. Kim, Arch. Biochem. Biophys., 176, 489 (1976)

400. Chromatin

SOURCE: Dictyostelium discoideum
SOLVENT: 0.1 M NaCl-0.01 M EDTA-0.01 M Tris HCl, pH 8.0
REFERENCE: T. Pederson, Biochemistry, 16, 2771 (1977)

401. Chymoelastase

SOURCE: Streptomyces griseus strain K-1
COMMENTS: Spectra of native and acetylated lysine-free, and guanidine-stable protease

REFERENCE: S. Siegel and W. M. Awad, J. Biol. Chem., 248, 3233 (1973)

402. Chymotrypsin, N-α-Dimethylaminonapthylsulfonyl-L-phenylalanyl Chymotrypsino (H_{15-57}) Methane

SOLVENT: 0.05 M Sodium phosphate buffer, pH 7.0
REFERENCE: G. Schoellman, Int. J. Pept. Prot. Res., 4, 221 (1972)

403. Chymotrypsin, acyl-

SOLVENT: Phosphate buffer, pH 6.95
COMMENTS: Acyl-enzyme prepared by reaction with o-(4-nitrophenylene) carbonate
REFERENCE: T. H. Fife, J. E. C. Hutchins, and D. McMahon, J. Am. Chem. Soc., 94, 1316 (1972)

404. α-Chymotrypsin

SOLVENT: pH 6.8
COMMENTS: Spectra of pulse-radiolysed nitrogen saturated and N_2O-saturated solutions of enzyme, and enzyme and t-butanol
REFERENCE: G. E. Adams, K. F. Baverstock, R. B. Cundall, and J. L. Redpath, Rad. Res., 54, 375 (1973)

405. Chymotrypsin, dibromofluoresceinisothiocarbamyl labeled (CHY-F) and α_2-Macroglobulin-chy-F Complex

SOURCE: Commercial (beef pancreas)
SOLVENT: 0.1 M Sodium acetate, pH 6.0
REFERENCE: F. Pochon, B. Amand, D. Lavalette, and J. Bieth, J. Biol. Chem., 253, 7496 (1978)

406. α-Chymotrypsin

SOLVENT: 0.001 M HCl
COMMENTS: Spectra of enzyme and of enzyme modified with 2-bromoethyl-3,1-benzoaxazin-4-one
REFERENCE: R. Alazard, J. J. Bechet, A. Dupaix, and J. Yon, Biochim. Biophys. Acta, 309, 379 (1973)

407. α-Chymotrypsin

SOLVENT: 0.1 M Phosphate buffer, pH 7.8
COMMENTS: Spectra of enzyme and of enzyme modified with active orange KKh
REFERENCE: V. B. Yanushauskaite, L. V. Kozlov, and V. K. Antonov, Biokhimiya (Biochemistry), 39, 372 (1974)

408. α-Chymotrypsin

SOURCE: Beef pancreas
SOLVENT: a. 0.01 M Citrate and 0.2 M KCl, pH 3
b. 0.01 M Potassium phosphate and 0.2 KCl, pH 7
COMMENTS: Spectra of enzyme, 1-napthacyl-enzyme, and 2-napthacyl-enzyme
REFERENCE: G. L. Glover, P. S. Mariano, T. J. Wilkinson, R. A. Hildreth, and T. W. Lowe, Arch. Biochem. Biophys., 162, 73 (1974)

409. α-Chymotrypsin, Modified at Tyrosine 146

SOLVENT: a. 0.1 M NaCl, 0.1 M sodium acetate, pH 4.8
b. pH 11
COMMENTS: Spectra of enzyme modified with p-diazo-N^{α}-acetyl-D-phenylanine methyl ester, p-diazo-N^{α}-benzoyl-D-phenylalanine methyl ester, and p-nitro-N^{α}-(4-diazobenzoyl)-D-phenylalanine methyl ester
REFERENCE: M. Gorecki, M. Wilchek, and S. Blumberg, Biochim. Biophys. Acta, 535, 90 (1978)

410. Chymotrypsinogen

COMMENTS: Spectra of zymogen, and of zymogen covalently linked to effector molecule by reaction with 2,2'-bis[α(o-bromomethylbenzyl-dimethyl ammonium) methyl] azo benzene dibromide
REFERENCE: B. F. Erlanger, N. H. Wasserman, A. G. Cooper, and R. J. Monk, Eur. J. Biochem., 61, 287 (1976)

411. Clostocin O (A Bacteriocin)

SOURCE: Clostridium saccharoperbutylacetonicum
REFERENCE: S. Ogata, F. Kato, M. Miyanaga, S. Hayashida, and M. Hongo, J. Gen. Appl. Microbiol., 24, 223 (1978)

412. Clupein, dinitrophenyl-

SOURCE: Herring (Clupea pallasaii)
SOLVENT: 1 N HCl
REFERENCE: T. Ando, M. Yamasaki, E. Abukumagawa, S. Ishii, and Y. Nagai, J. Biochem. Tokyo, 45, 429 (1958)

413. Coagulogen

SOURCE: Horse shoe crab (Tachypleus tridentatus)
SOLVENT: 0.1 M Ammonium formate buffer, pH 6.5
REFERENCE: S. Nakamura, S. Iwanaga, T. Harada, and M. Niwa, J. Biochem. Tokyo, 80, 1011 (1976)

414. Cobrotoxin

SOURCE: Taiwan cobra (Naja naja atra) venom
SOLVENT: a. 0.2 M Acetic acid
b. 0.05 M Sodium bicarbonate buffer, pH 10.0
COMMENTS: Spectra of native, 2-hydroxy-5-nitrobenzyl, 2-nitrophenylsulfonyl-, and N-bromosuccinimide-modified enzymes
REFERENCE: C. C. Chang and C. C. Yang, Biochim. Biophys. Acta, 295, 595 (1973)

415. Cobrotoxin, phenylglyoxal-modified

SOURCE: Taiwan cobra (Naja naja atra) venom
SOLVENT: 0.01 M Ammonium bicarbonate
COMMENTS: Spectra of native and phenylglyoxal modified protein
REFERENCE: C. C. Yang, C. C. Chang, and I. F. Liou, Biochim. Biophys. Acta, 365, 1 (1974)

416. Co-enzyme A Transferase

SOURCE: Peptostreptococus elsdenii
SOLVENT: Phosphate buffer, pH 7.0
COMMENTS: Spectra of enzyme and putative enzyme-Co A intermediate
REFERENCE: K. K. Tung and W. A. Wood, J. Bacteriol., 124, 1462 (1975)

417. Co-enzyme A Transferase

SOURCE: Escherichia coli
COMMENTS: Spectra of native and pyridoxal 5'-phosphate-sodium borohydride reduced enzymes
REFERENCE: F. E. Frerman, P. Andreone, and D. Mielke, Arch. Biochem. Biophys., 181, 508 (1977)

418. Colicine V

SOURCE: Escherichia coli K357 L-T
SOLVENT: Tris buffer, pH 8.0
REFERENCE: J. J. Hutton and W. F. Goebel, J. Gen. Physiol., 45, 125 (1962)

419. Coliphage, N4 and Protein Subunits

SOURCE: N4 bacteriophage
SOLVENT: a. 0.1 M Ammonium acetate buffer, pH 7.0
b. 0.1 M Tris buffer, pH 7.2
REFERENCE: G. C. Schito, Giorn. Microbiol., 14, 75 (1966)

420. Collagen

SOURCE: Calf skin
COMMENTS: Spectra of crude collagen extract and of reprecipitated collagen
REFERENCE: H. C. Grillo and J. Gross, J. Surg. Res., 2, 69 (1962)

421. Collagen, α1, α2, α3 Chains

SOURCE: Codfish skin
SOLVENT: 0.1 N NaOH
REFERENCE: K. A. Piez, Biochemistry, 4, 2590 (1965)

422. Collagen

SOURCE: Carp swim bladder
SOLVENT: 0.01 M Phosphate buffer, pH 3.8
COMMENTS: Spectra taken at 23°C and 40°C
REFERENCE: P. H. von Hippel in Treatise on Collagen (G. N. Ramachandran, ed.), 253 (1967), Academic Press, London

423. Complexes, Pigment-protein

SOURCE: Rhodopseudomonas sphaeroides
COMMENTS: Mutants used: Mutant-8, Mutant-8-29, and Mutant-8-47
REFERENCE: W. R. Richards, R. B. Wallace, M. S. Tsao, and E. Ho, Biochemistry, 14, 5554 (1975)

424. Compound B

SOURCE: Human plasma
REFERENCE: S. Takahashi and K. Schmid, Biochim. Biophys. Acta, 82, 627 (1964)

425. Concanavalin, Co^{2+}

SOURCE: Jack bean
SOLVENT: 0.1 M Sodium acetate, 1 M NaCl, pH 5.2
REFERENCE: A. J. Kalb and I. Pecht, Biochim. Biophys. Acta, 303, 264 (1973)

426. Concanavalin A

SOURCE: Jack bean (Canavalia ensiformis)
SOLVENT: 0.1 M D-Mannose, pH 5.0
REFERENCE: M. Beppu, T. Terao, and T. Osawa, J. Biochem. Tokyo, 78, 1013 (1975)

427. Concanavalin A, Activated and Irradiated

SOURCE: Jack bean
COMMENTS: Spectra of Con A imidated with methyl 4-azidobenzo-imidate and then exposed to ultraviolet irradiation
REFERENCE: T. H. Ji, J. Biol. Chem., 252, 1566 (1977)

428. Connectin

SOURCE: Carp (Cyprinus carpio)
SOLVENT: 0.1 N NaOH
REFERENCE: S. Kimura, Y. Akashi, and M. Kubota, J. Biochem. Tokyo, 83, 321 (1978)

429. Creatine Kinase, 2-hydroxy-5-nitrobenzyl-

SOURCE: Rabbit muscle
SOLVENT: 0.05 M Tris-acetate buffer, pH 4.1-11.3
REFERENCE: C. Roustan, A. Brevet, L. A. Pradel, Eur. J. Biochem., 39, 371 (1973)

430. Creatine Kinase, 2-nitro-5-thiocyanobenzoic acid modified

SOURCE: Rabbit muscle
SOLVENT: 0.02 M Tris-acetate buffer, pH 7.0
REFERENCE: Y. Degani and C. Degani, Biochemistry, 18, 5917 (1979)

431. Creatine Kinase

SOURCE: Rabbit muscle
SOLVENT: pH 7.3 and pH 12
COMMENTS: Spectrum of 2-hydroxy-5-nitro-benzylated, iodinated enzyme
REFERENCE: A. Fattoum, R. Kassab, and L. A. Pradel, Biochim. Biophys. Acta, 405, 324 (1975)

432. Creatine Kinase

SOURCE: Rabbit muscle
COMMENTS: Spectra of product of reaction of enzyme with 2-nitro-5-thiocyanatobenzoic acid in presence and absence of KCN
REFERENCE: N. C. Price, Biochem. J., 159, 177 (1976)

433. Creatine Kinase, Modified Enzyme

SOURCE: Rabbit muscle
SOLVENT: 0.05 M Tris buffer, pH 7.1

COMMENTS: Spectra of enzyme modified with mixed disulfide of 2-thiopyridine and 2-thiobenzyldiazoacetate before and after photolysis
REFERENCE: J. Henkin, J. Biol. Chem., 252, 4293 (1977)

434. p-Cresol Methylhydroxylase

SOURCE: Pseudomonas putida
SOLVENT: a. 50 mM Phosphate buffer, pH 6.8
b. 0.1 M Tris base, pH 12
COMMENTS: Spectra of protein after extraction with acidified acetone and after addition of sodium dithionite
REFERENCE: D. J. Hopper and D. G. Taylor, Biochem. J., 167, 155 (1977)

435. p-Cresol Methylhydroxylase A

SOURCE: Pseudomonas putida
SOLVENT: 50 mM Potassium phosphate buffer, pH 7.0
REFERENCE: M. J. Keat and D. J. Hopper, Biochem. J., 175, 649 (1978)

436. Crotapotin

SOURCE: Crotalus terrificus venom
SOLVENT: a. 0.1 M HCl, pH 1.0
b. 0.1 M $NaHCO_3$, pH 8.5
c. 0.1 M NaOH, pH 12.9
REFERENCE: H. Breithaupt, R. Rubsamen, and E. Habermann, Eur. J. Biochem., 49, 333 (1974)

437. Crustacyanin

SOURCE: Lobster (Homarus vulgaris)
SOLVENT: 0.006 M Ammonium acetate
COMMENTS: Spectral changes due to increase in ionic strength
REFERENCE: L. Wahlgren-Brannstrom and M. Baltscheffsky, Acta Chem. Scand., 33B, 613 (1979)

438. α-Crystallin

SOURCE: Bovine lens
REFERENCE: F. A. Bettelheim, Exp. Eye Res., 14, 251 (1972)

439. Crystallin, β_1-, β_2-, β_3-, and γ-

SOURCE: Human lens
REFERENCE: P. J. Ringens, K. N. Lem-The, H. J. Hoenders, and J. Wollensak, Interdiscipl. Topics. Gerontol., 13, 193 (1978)

440. γ-Crystallin

SOURCE: Human lens
SOLVENT: pH 6.5-pH 13.0
REFERENCE: S. Lerman and R. F. Borkman, Interdiscipl. Topics Gerentol., 13, 154 (1978)

441. γ-Crystallin

SOURCE: Human lens
SOLVENT: 0.05 M Phosphate buffer, pH 7.0
COMMENTS: Spectra of ultraviolet-tryptophan treated and ultraviolet-tryptophan untreated protein
REFERENCE: D. Grover and S. Zigman, Exp. Eye Res., 13, 70 (1972)

442. δ-Crystallin

SOURCE: Embryonic chick lens
SOLVENT: 1 mM Tris-HCl, pH 8.3
COMMENTS: Spectra in presence and absence of SDS
REFERENCE: J. Piatigorsky, J. Horowitz, and R. T. Simpson, Biochim. Biophys. Acta, 490, 279 (1977)

443. Cuprein, Cerebro-

SOURCE: Beef, white matter of brain
REFERENCE: S. G. Sharoyan, A. A. Shalijian, R. M. Nalbandyan, and H. C. Buniatian, Biochim. Biophys. Acta, 493, 478 (1977)

444. Cuprodoxin

SOURCE: Yeast
SOLVENT: 10 mM Potassium phosphate buffer, pH 7.5
COMMENTS: Spectra of native and oxidized protein
REFERENCE: R. Prinz and U. Weser, FEBS Lett., 54, 224 (1975)

445. Cusacyanin

SOURCE: Cucumber (Cucumis sativus)
REFERENCE: V. T. Aikazyan, K. A. Markasyan, and R. M. Nalbandyan, Biokhimiya (Biochemistry), 39, 982 (1974)

446. β-Cyanoalanine Synthase

SOURCE: Blue lupine seedlings
SOLVENT: 0.1 M Tris buffer

REFERENCE: T. N. Akopyan, A. E. Braunstein, and E. V. Goryachenkova, Proc. Nat. Acad. Sci., 72, 1617 (1975)

447. Cyclohexanone Oxygenase

SOURCE: Nocardia globerula
SOLVENT: 21 mM Phosphate buffer
REFERENCE: N. A. Donoghue, D. B. Norris, and P. W. Trudgill, Eur. J. Biochem., 63, 175 (1976)

448. Cyclooxygenase, Fatty Acid

SOURCE: Sheep vesicular glands
COMMENTS: Spectra of unsupplemented oxidized enzyme and heme-supplemented enzyme
REFERENCE: M. Hemler, W. E. M. Lands, and W. L. Smith, J. Biol. Chem., 251, 5575 (1976)

449. Cyclopentanone Oxygenase

SOURCE: Pseudomonas NCIB 9872
SOLVENT: 20 mM Phosphate buffer, pH 7.1
COMMENTS: Spectral titration with NADPH
REFERENCE: M. Griffin and P. W. Trudgill, Eur. J. Biochem., 63, 199 (1976)

450. β-Cyclopiazonate Oxidocyclase, I, II, and III

SOURCE: Penicillium cyclopium
SOLVENT: 0.02 M Sodium phosphate buffer, pH 7.6
COMMENTS: Spectra before and after addition of sodium dithionite or β-cyclopiazonic acid
REFERENCE: J. C. Schabort and D. J. J. Potgieter, Biochim. Biophys. Acta, 250, 329 (1971)

451. β-Cyclopiazonate Oxidocyclase

SOURCE: Penicillium cyclopium
SOLVENT: 4:1 Mixture of 0.05 M sodium phosphate buffer, pH 6.8 and ethylene glycol; T = 6°C
COMMENTS:
a. Enzyme reduced with borohydride and allowed to reoxidize by air
b. Aerobic reaction of borohydride-reduced enzyme with β-cyclopiazonate
c. Aerobic reaction of borohydride-reduced enzyme with α-cyclopiazonate

REFERENCE: D. J. Steenkamp and J. C. Schabort, Eur. J. Biochem., 40, 163 (1973)

452. β-Cyclopiazonate Oxidocyclase

SOURCE: Penicillium cyclopium
COMMENTS: Spectral changes seen after addition of β-cyclopiazonic acid or L-tryptophan
REFERENCE: D. J. Steenkamp, J. C. Schabort, and N. P. Ferreira, Biochim. Biophys. Acta, 309, 440 (1973)

453. β-Cyclopiazonate Oxidocyclase

SOURCE: Penicillium cyclopium
SOLVENT: 0.05 M Sodium phosphate buffer, pH 6.8
COMMENTS: Spectra of:
a. oxidized isoenzyme 1.3
b. enzyme-β-cyclopiazonic acid-intermediate complex
c. enzyme reduced with excess substrate
d. enzyme reacted with 4-γ,γ,-di-methylallyl tryptophan
e. carbethoxylated enzyme after reaction with different amounts of β-cyclopiazonic acid
REFERENCE: D. J. Steenkamp, J. C. Schabort, C. W. Holzapfel, and N. P. Ferreira, Biochim. Biophys. Acta, 358, 126 (1976)

454. Cystathionase

SOURCE: Rat liver
SOLVENT: 0.1 M Phosphate buffer, pH 7.4
REFERENCE: K. J. Oh and J. E. Churchich, J. Biol. Chem., 248, 7370 (1973)

455. Cystathionase

SOURCE: Rat liver
COMMENTS: Protein labeled with 1.6 residues of dimethylamino-naphthylsulfonyl cystine per mole
REFERENCE: K. J. Oh and J. E. Churchich, J. Biol. Chem., 249, 4737 (1974)

456. Cystathionase

SOURCE: Rat liver
SOLVENT: 0.1 M Phosphate, pH 7.3
COMMENTS: Changes in visible spectrum seen on addition of amino-oxyacetic acid, canaline, or α,γ-diamino-butyric acid
REFERENCE: T. Beeler and J. E. Churchich, J. Biol. Chem., 251, 5267 (1976)

457. γ-Cystathionase

SOURCE: Rat liver
SOLVENT: 0.2 M Potassium phosphate, pH 7.2 containing 0.02 M EDTA, 0.07 M 2-mercaptoethanol, 0.0005 M pyridoxal 5'-phosphate
COMMENTS: Changes in spectrum after reaction with either propargylglycine, trifluoroalanine, or propargylglycine and trifluoroalanine
REFERENCE: R. B. Silverman and R. H. Abeles, Biochemistry, 16, 5515 (1975)

458. Cystathionine β-Synthase

SOURCE: Human liver
SOLVENT: 25 mM Potassium phosphate buffer, pH 7.8
REFERENCE: J. Kraus, S. Packman, B. Flowler, and L. E. Rosenberg, J. Biol. Chem., 254, 6523 (1978)

459. Cystathionine γ-Synthase

SOURCE: Salmonella typhimurium
COMMENTS: Spectral changes upon reaction with L-propargylglycine (0-120 minutes)
REFERENCE: M. Johnston, D. Jankowski, P. Marcotte, H. Tanaka, N. Esaki, K. Soda, and C. Walsh, Biochemistry, 18, 4690 (1979)

460. Cysteine Desulfhydrase

SOURCE: Salmonella typhimurium
SOLVENT: 0.1 M Tris-HCl, pH 7.6 containing 10 mM 2-mercaptoethanol
REFERENCE: N. M. Kredich, B. S. Keenan, and L. J. Foote, J. Biol. Chem., 247, 7157 (1972)

461. Cysteine Desulfhydrase

SOURCE: Aerobacter aerogenes
SOLVENT: 0.05 M Potassium phosphate buffer, pH 7.0, containing 10^{-5} M pyridoxal phosphate, 10^{-4} M EDTA, 10^{-2} 2-mercaptoethanol, and 20% glycerol
REFERENCE: H. Kumagai, S. Sejima, Y. J. Choi, H. Tanaka, and Y. Yamada, FEBS Lett., 52, 304 (1975)

462. Cysteine Lyase

SOURCE: Chick embryo
SOLVENT: pH 4.0, 7.0, 8.0, 10.0

COMMENTS: Changes in spectrum on addition of either phenylhydrazine, amino-oxyacetic acid, sodium borohydride, or cysteine and sodium borohydride
REFERENCE: E. A. Tolosa, E. V. Goryachinhova, R. M. Khomatov and E. S. Severin in Pyridoxal Catalysis: Enzymes & Model Systems (E. E. Snell, A. E. Braunstein, E. S. Severin and Y. M. Torchinsky, eds.), 525 (1968), Interscience

463. Cysteine Synthase

SOURCE: Rape leaves
SOLVENT: 0.1 M Potassium phosphate buffer, pH 8.0
REFERENCE: M. Masada, K. Fukushima, and G. Tamura, J. Biochem. Tokyo, 77, 1107 (1975)

464. Cytochrome 558

SOURCE: Euglena gracilis
SOLVENT: 0.1 M Phosphate buffer, pH 7.0
COMMENTS: Spectra of reduced and oxidized forms. Spectrum of reduced form at 70°K
REFERENCE: T. E. Meyer and M. A. Cusmanovich, Biochim. Biophys. Acta, 267, 383 (1972)

465. Cytochrome a

SOURCE: Beef heart muscle
SOLVENT: 50 mM Sodium phosphate buffer, pH 7.4, containing (a) 1% cholate or (b) 25% Emasol 1130
COMMENTS:
1. Spectra of reduced and oxidized forms
2. Spectrum of proteinase treated protein
3. Spectra of cyanide complexes of (a) and (b) and CO complexes of (a) and (b)

REFERENCE: T. Yamamoto and K. Okunuki, J. Biochem. Tokyo, 71, 435 (1972)

466. Cytochrome a_1

SOURCE: Nitrosomonas
SOLVENT: 50 mM Sodium/potassium phosphate buffer, pH 7.5
COMMENTS: Spectra of reduced and oxidized forms
REFERENCE: R. H. Erickson, A. B. Hooper, and K. R. Terry, Biochim. Biophys. Acta, 283, 155 (1972)

467. Cytochrome aa_3

SOURCE: Beef heart
SOLVENT: 100 mM Tris sulphate, pH 8.0, containing 0.5% potassium cholate and 0.5% Tween 80

REFERENCE: K. J. H. Van Buren, P. F. Zuurendonk, B. F. Van Gelder, and A. O. Muijsers, Biochim. Biophys. Acta, 256, 243 (1972)

468. Cytochrome aa_3

SOURCE: Beef liver
SOLVENT: a. 100 mM Phosphate, pH 7.4, containing 0.5% Tween 80
b. 100 mM Tris-H_2SO_4 buffer, pH 8.0, containing 5% potassium cholate
COMMENTS: Changes in spectra of reduced and oxidized protein on addition of cyanide
REFERENCE: K. J. H. Van Buren, P. Nicholls, and B. F. Van Gelder, Biochim. Biophys. Acta, 256, 258 (1972)

469. Cytochrome $a(a_3)$

SOURCE: Beef heart
SOLVENT: 0.01 M Sodium phosphate buffer, containing 1% deoxycholate, pH 7.4
COMMENTS: Reduced with sodium dithionite. Spectra taken at 8.5°K-224.0°K
REFERENCE: B. Hagihara, R. Oshino, and T. Iizuka, J. Biochem. Tokyo, 75, 45 (1974)

470. Cytochrome b

SOURCE: Ascaris lumbricoides var Summ (muscle microsomes)
COMMENTS: Spectra of reduced and oxidized forms
REFERENCE: S. Matuda, J. Biochem. Tokyo, 85, 351 (1979)

471. Cytochrome b

SOURCE: Baker's yeast
SOLVENT: 0.05 M Tris/acetate, pH 7.5
COMMENTS: Spectra of reduced and oxidized forms
REFERENCE: L. F. H. Lin and D. S. Beattie, J. Biol. Chem., 253, 2412 (1978)

472. Cytochrome b

SOURCE: (a) Neurospora crassa
(b) Locusta migratoria
COMMENTS: Spectra of reduced and oxidized forms
REFERENCE: B. Lorenz, W. Kleinow and H. Weiss, Hoppe-Seyler's Z. Physiol. Chem., 355, 300 (1974);H. Weiss and B. Ziganke, Eur. J. Biochem., 41, 63 (1974)

473. Cytochrome b_2

SOURCE: Yeast
SOLVENT: 0.15 M Phosphate buffer, pH 7.2, containing 0.15 M NaCl and 20 x 10^{-6} M EDTA
COMMENTS: Changes in spectrum of apoprotein upon addition of heme
REFERENCE: M. Mevel-Ninio, P. Pajot, and F. Labeyrie, Biochimie, 53, 35 (1971)

474. Cytochrome b_2

SOURCE: Yeast
SOLVENT: 0.2 M Phosphate, 1 mM EDTA, pH 7.2
COMMENTS: Changes in spectrum upon addition of oxalate
REFERENCE: D. Thusius, B. Blazy, and A. Baudras, Biochemistry, 15, 250 (1976)

475. Cytochrome b_2

SOURCE: Baker's yeast
SOLVENT: 0.1 M Sodium potassium phosphate buffer, 1 mM EDTA, pH 7
COMMENTS: Changes in spectrum upon addition of sulfite, and spectra of reduced and oxidized forms
REFERENCE: F. Lederer, Eur. J. Biochem., 88, 425 (1978)

476. Cytochrome b_5

SOURCE: Rat liver, outer membrane of mitochondria
SOLVENT: 50 mM Potassium phosphate buffer, pH 7.5, containing 0.02% Tween 20
COMMENTS: Spectra of reduced and oxidized forms
REFERENCE: Y. Nisimoto, F. Takeuchi, and Y. Shibata, J. Biochem. Tokyo, 82, 1257 (1977)

477. Cytochrome b_5 and Cytochrome d-b_5

SOURCE: Rat liver
SOLVENT: 8 M Urea, pH 4
COMMENTS: Changes in spectrum upon addition of N-bromosuccinimide
REFERENCE: J. Poensgen and V. Ullrich, Biochim. Biophys. Acta, 465, 34 (1977)

478. Cytochrome b_5

SOURCE: Calf liver
SOLVENT: 10 mM Phosphate buffer, pH 7.2, containing 0.2 mM EDTA

COMMENTS: Changes in spectrum upon autooxidation, 1-32 minutes
REFERENCE: M. C. Berman, C. M. Adnams, K. M. Ivanetich, and J. E. Kench, Biochem. J., 157, 237 (1976)

479. Cytochrome b_5

SOURCE: Rat liver
SOLVENT: 0.05 M Potassium phosphate buffer, pH 7.25, containing 20% glycerol, 0.35% deoxycholate & 0.15% cholate
COMMENTS: Spectra of reduced and oxidized forms
REFERENCE: Y. Imai, J. Biochem. Tokyo, 80, 267 (1976)

480. Cytochrome b_5, Reconstituted

REFERENCE: F. Adar, Arch. Biochem. Biophys., 170, 644 (1975)

481. Cytochrome b_5

SOURCE: House fly
SOLVENT: 0.1 M Sodium phosphate buffer, pH 7.0
COMMENTS: Spectra of reduced and oxidized forms
REFERENCE: J. Capdevila, N. Ahmad, and M. Agosin, J. Biol. Chem., 250, 1048 (1975)

482. Cytochrome b_5

SOURCE: Yeast
SOLVENT: a. 50 mM Potassium phosphate buffer, pH 7.5
b. 50 mM Potassium phosphate buffer containing 50% glycerol, pH 7.5
COMMENTS: 1. Spectra taken at room temperature and liquid nitrogen temperature
2. Spectra of reduced and oxidized forms
REFERENCE: Y. Yoshida, H. Kumaoka, and R. Sato, J. Biochem. Tokyo, 75, 1211 (1974)

483. Cytochrome b_5

SOURCE: Beef liver
SOLVENT: 10 mM Tris buffer, pH 8.1
COMMENTS: Spectra of reduced and oxidized forms
REFERENCE: J. Ozols, Biochemistry, 13, 426 (1974)

484. Cytochrome b_5

SOURCE: Pig liver
SOLVENT: a. 0.5 M Phosphate buffer, pH 3.5
b. 0.5 M Tris-acetate, pH 4.0, 5.0, 8.0

c. 0.5 M Sodium bicarbonate-sodium carbonate-NaOH, pH 11.0, 12.0
REFERENCE: M. Ikeda, T. Iizuka, H. Takao, and B. Hagihara, Biochim. Biophys. Acta, 336, 15 (1974)

485. Cytochrome b_5 and OM-Cytochrome

SOURCE: Rat liver outer mitochondrial membrane
SOLVENT: 50 mM Potassium phosphate buffer, pH 7.5
COMMENTS: Spectra of reduced and oxidized forms
REFERENCE: K. Fukushima and R. Sato, J. Biochem. Tokyo, 74, 161 (1973)

486. Cytochrome b_5

SOURCE: Human erythrocytes
SOLVENT: 0.04 M Tris-HCl, pH 7.5
REFERENCE: P. G. Passon, D. W. Reed, and D. E. Hultquist, Biochim. Biophys. Acta, 275, 51 (1972)

487. Cytochrome b_5

SOURCE: Human erythrocytes
SOLVENT: 0.1 M Phosphate buffer, pH 7.0
COMMENTS: Spectra of reduced and oxidized forms
REFERENCE: K. Abe and Y. Sugita, Eur. J. Biochem., 101, 423 (1979)

488. Cytochrome b_6-f particle

SOURCE: Lettuce chloroplasts
SOLVENT: a. 0.025 M Tris, pH 8, containing 0.1 M NaCl, and 0.25% sodium cholate
b. 0.1 M Tris-HCl, pH 8
COMMENTS: Spectra of reduced and oxidized forms
REFERENCE: N. Nelson and J. Neumann, J. Biol. Chem., 247, 1817 (1972)

489. Cytochrome b_9

SOURCE: Rat liver
SOLVENT: 0.05 M Tris buffer, pH 7.4
COMMENTS: Spectra of oxidized and dithionite-reduced forms
REFERENCE: I. N. Kim, J. Biol. Chem., 254, 10615 (1979)

490. Cytochrome b_{556}

SOURCE: E. coli K12
SOLVENT: 10 mM Tris-HCl, pH 8.0

COMMENTS: Spectra of reduced and oxidized forms
REFERENCE: K. Kita, I. Yamato, and Y. Anraku, J. Biol. Chem., 253, 8910 (1978)

491. Cytochrome b_{559}

SOURCE: Alga (Bumilleriopsis filiformis)
SOLVENT: 4 M Urea, 2% Triton X-100, 2 mM DTT, 50 mM Tris, pH 8.0
COMMENTS: Spectra of reduced and oxidized forms
REFERENCE: H. J. Lach and P. Boger, Z. Naturforsch., 32C, 75 (1977)

492. Cytochrome b_{561}

SOURCE: Beef adrenal glands: chromaffin granules
SOLVENT: 0.09 M Tris/0.05 M phosphate buffer, pH 6.9
COMMENTS: Spectra of reduced and oxidized forms
REFERENCE: T. Silsand and T. Flatmark, Biochim. Biophys. Acta, 359, 257 (1974)

493. Cytochrome b_{562}

SOURCE: Escherichia coli
SOLVENT: 0.1 M Phosphate, pH 4.02, 6.8, 9.98
REFERENCE: Y. P. Myer and P. A. Bullock, Biochemistry, 17, 3723 (1978)

494. Cytochrome b_{562}

SOURCE: Escherichia coli
SOLVENT: 0.1 M Phosphate, pH 6.8
COMMENTS: Spectra of reduced and oxidized forms
REFERENCE: P. A. Bullock and Y. P. Meyer, Biochemistry, 17, 3084 (1978)

495. Cytochrome b_{562}, Cytochrome C, Cytochrome C_2, and Cytochrome C_3

SOURCE:
a. Escherichia coli (b-562)
b. Horse heart (C)
c. Rhodospirillum rubrum (C_2)
d. C. ethylicum (C_3)

SOLVENT:
a. 0.33 M Phosphate, pH 8.0, 50% (V/V) glycerol
b. 50 mM Phosphate, pH 7.4, 50% (V/V) glycerol
c. 75 mM Tris-Cl, pH 7.3, 50% (V/V) glycerol
d. 5 mM Phosphate, pH 7.8, 0.1 M NaCl, 50% (V/V) glycerol

COMMENTS: Spectra taken at 77°K
REFERENCE: G. C. Wagner and R. J. Hassner, Biochem. Biophys. Res. Comm., 63, 385 (1975)

496. Cytochrome $b_{562.5}$

SOURCE: Green alga (Ulva pertusa)
SOLVENT: a. 0.4 M Potassium phosphate buffer, pH 7.4
b. 50% Glycerol containing 2.5 mM potassium phosphate buffer, pH 7.0
COMMENTS: Spectrum at 77°K in solvent b; spectra of reduced and oxidized forms
REFERENCE: Y. Sugimura and E. Yakushiji, J. Biochem. Tokyo, 85, 641 (1979)

497. Cytochrome $b-c_1$

SOURCE: Yeast (Saccharomyces cerevisiae)
SOLVENT: 50 mM Tris-acetate, pH 7.5, containing 0.5% cholate
COMMENTS: Spectrum of CO complex. Spectra of reduced and oxidized forms
REFERENCE: M. B. Katan, L. Pool, and G. S. P. Groot, Eur. J. Biochem., 65, 95 (1976)

498. Cytochrome $b-c_1$

SOURCE: Beef heart muscle
SOLVENT: 50 mM Phosphate buffer, pH 7.4
COMMENTS: Spectra of reduced and oxidized forms
REFERENCE: C. A. Yu, L. Yu and T. E. King, J. Biol. Chem., 249, 4905 (1974)

499. Cytochrome $b-c_1$ (Complex III)

SOURCE: Baker's yeast
SOLVENT: 0.1 M Potassium phosphate buffer, pH 7.4, containing 0.1% deoxycholate, 0.1% Triton X, and 1 mM EDTA
COMMENTS: Spectra of reduced and oxidized forms
REFERENCE: J.N. Siedow, S. Power, F. F. De La Rosa, and G. Palmer, J. Biol. Chem., 253, 2392 (1978)

500. Cytochrome $b-c_1$

SOURCE: Pigeon heart muscle mitochondria
SOLVENT: 50 mM Phosphate buffer, pH 7.4
COMMENTS: Spectra of cytochrome $b-c_1$ complex in hydrated oriented proteoliposome multilayers taken with polarized light. Spectra of reduced and oxidized forms
REFERENCE: M. Erecinska and D. A. Wilson, Arch. Biochem. Biophys., 192, 80 (1979)

501. Cytochrome c

SOLVENT: 9 M Urea
COMMENTS: Spectrum of carboxymethylated cytochrome c
REFERENCE: J. Babul and E. Stellwagen, Biopolymers, 10, 2359 (1971)

502. Cytochrome c

SOURCE: Beef heart
SOLVENT: Citrate-phosphate buffer, pH 3.05 and pH 12.85
COMMENTS: a. Changes in spectrum due to NO
b. Changes in spectrum due to NADH and nitrate
c. Spectra of reduced and oxidized forms
REFERENCE: Y. Orii and H. Shimada, J. Biochem. Tokyo, 84, 1543 (1978)

503. Cytochrome c

SOURCE: Commercial
SOLVENT: pH 6
COMMENTS: Spectra of oxidized protein and its tannic acid complexes in presence of Tween 80
REFERENCE: A. Zanobini, P. Vanni, and A. M. Firenzuoli, Experientia, 15, 1015 (1967)

504. Cytochrome c

SOURCE: Commercial
SOLVENT: a. 50 mM Sodium cacodylate, pH 7
b. 1% (V/V) buffer in 9% (W/V) AOT/silicone oil
c. 9% (V/V) buffer in 9% (W/V) AOT/silicone oil
(AOT = disodium diisooctyl sulfosuccinate)
REFERENCE: C. Balny and P. Douzou, Biochimie, 61, 445 (1979)

505. Cytochrome c

SOURCE: Methylomonas methanica
COMMENTS: Spectra of the protein, after reduction with dithionite, after reduction with dithionite and CO, and after oxidation with ferricyanide
REFERENCE: R. Patel, C. T. Hou, and A. Felix, J. Gen. Appl. Microbiol., 25, 197 (1979)

506. Cytochrome c

SOURCE: Horse heart
SOLVENT: 0.05 M Phosphate, 0.2 M KCl, pH 7.0
COMMENTS: Spectra of ferricytochrome C in presence and absence of 4.5 M urea and 9 M urea

REFERENCE: Y. P. Meyer, L. H. MacDonald, B. C. Verma, and A. Pande, Biochemistry, 19, 199 (1980); Y. P. Meyer, A. F. Saturno, B. C. Verma, and A. Pande, J. Biol. Chem., 254, 11202 (1979)

507. Cytochrome c, Manganese-

SOURCE: Horse heart
COMMENTS: Spectrum of NO complex; spectra of reduced and oxidized forms
REFERENCE: L. C. Dickinson and J. C. W. Chien, J. Biol. Chem., 252, 6165 (1977)

508. Cytochrome c, Nickel-

SOURCE: Horse heart
SOLVENT: Physiological pH, pH 2.5, and pH 13.0
REFERENCE: M. C. Findlay and J. C. W. Chien, Eur. J. Biochem., 76, 79 (1977)

509. Cytochrome c

SOURCE: Horse heart
SOLVENT: a. 0.05 M Ammonium acetate, pH 7.0
b. pH 1.15, pH 2.51, pH 2.98, pH 3.53, pH 4.51, pH 5.04, and pH 5.40
COMMENTS: Spectrum of methionine sulfoxide ferricytochrome c
REFERENCE: K. M. Ivanetich, J. J. Bradshaw, and L. S. Kaminsky, Biochemistry, 15, 1144 (1976)

510. Cytochrome c

SOURCE: Horse heart
SOLVENT: a. pH 7.0, 0.025 M Phosphate buffer
b. pH 7.0, and pH 9.0, 0.33 M NaCl
COMMENTS: Spectra of maleylated, and amidinated forms
REFERENCE: G. W. Pettigrew, I. Aviram, and A. Schejter, Biochem. Biophys. Res. Comm., 68, 807 (1976)

511. Cytochrome c, Zinc-, Tin-, Porphyrin-

SOURCE: Horse heart
SOLVENT: 10 mM Phosphate, pH 7.2
COMMENTS: Changes in spectrum of zinc-cytochrome upon irradiation with light
REFERENCE: J. M. Vanderkooi, F. Adar, and M. Erecinska, Eur. J. Biochem., 64, 381 (1976)

512. Cytochrome c

SOURCE: Horse heart
SOLVENT: Phosphate buffer, pH 7.2
COMMENTS: Changes in spectrum on addition of dihydroxyphenylalanine
REFERENCE: S. S. Tate, J. Orlando, and A. Meister, Proc. Nat. Acad. Sci., 69, 2505 (1972)

513. Cytochrome c

SOURCE: Horse heart
SOLVENT: 1 mM HCl adjusted to pH 3.8
COMMENTS: Effect of methylation on 695 nm band absorption
REFERENCE: E. Stellwagen and J. Babul, Biochemistry, 14, 5135 (1975)

514. Cytochrome c

SOURCE: Horse heart
SOLVENT: 0.1 M Phosphate buffer, pH 7.0
COMMENTS: Spectra of NBS FIII and NBS FII components in ferric state
REFERENCE: D. J. O'Hern, P. K. Pal, and Y. P. Meyer, Biochemistry, 14, 382 (1975)

515. Cytochrome c, guanidinated

SOURCE: Horse heart
SOLVENT: a. Tris-HCl buffer, pH 7.0
b. Borate buffer, pH 10.5
c. Borate-NaOH, pH 12.3
REFERENCE: T. Brittain and C. Greenwood, Biochem. J., 147, 175 (1975)

516. Cytochrome c

SOURCE: Horse heart
SOLVENT: 0.1 M Phosphate, pH 7.3
COMMENTS: Spectrum of dinitroazidophenyl cytochrome c also given
REFERENCE: M. Erecinska, J. M. Vanderkooi, and D. F. Wilson, Arch. Biochem. Biophys., 171, 108 (1975)

517. Cytochrome c

SOURCE: Horse heart
SOLVENT: 0.1 M Potassium phosphate, pH 7.0
COMMENTS: Spectra of imidazole and azide complexes
REFERENCE: M. Ikeda-Saito and T. Iizuka, Biochim. Biophys. Acta, 393, 335 (1975)

518. Cytochrome c

SOURCE: Horse heart
SOLVENT: 10 mM Phosphate buffer, pH 7.4, containing 0.2 mM EDTA
COMMENTS: Reduction by ferrocytochrome b_5
REFERENCE: M. C. Berman, C. M. Adnams, K. M. Ivaetich, and J. E. Kench, Biochem. J., 157, 237 (1976)

519. Cytochrome c, Copper-

SOURCE: Horse heart
SOLVENT: Neutral pH
REFERENCE: M. Erecinska and J. M. Vanderkooi, Meth. Enzymol., 53D, 165 (1978)

520. Cytochrome c

SOURCE: Horse heart
SOLVENT: 20 mM Acetate buffer, pH 6.0
COMMENTS: Changes in spectrum due to increasing amounts of methanol and spectrum at pH 1.5 without methanol
REFERENCE: H. R. Drew and R. E. Dickerson, J. Biol. Chem., 253, 8420 (1978)

521. Cytochrome c, Cobalt-

SOURCE: Horse heart
SOLVENT: 0.02 M Sodium phosphate, pH 8.0
COMMENTS: Spectra of reduced and oxidized forms
REFERENCE: L. C. Dickinson and J. C. W. Chien, Biochem. Biophys. Res. Comm., 58, 238 (1974)

522. Cytochrome c

SOURCE: Horse heart
SOLVENT: 8 M Urea-0.1 M phosphate buffer, pH 7.0
COMMENTS: Spectrum of cytochrome c-phospholipid complex and effect of CO on spectrum
REFERENCE: K. M. Ivanetich, J. J. Henderson, and L. S. Kaminsky, Biochemistry, 13, 1469 (1974)

523. Cytochrome c

SOURCE: Horse heart
SOLVENT: Water:ethylene glycol (50:50)
COMMENTS: Spectra of reduced and oxidized forms; spectra taken at room temperature and 77°K
REFERENCE: L. A. Blumenfeld, R. M. Davydov, N. S. Fel, S. N. Magonov, and R. O. Vilu, FEBS Lett., 45, 256 (1974)

524. Cytochrome c

SOURCE:	Horse heart
SOLVENT:	pH 7, Various buffers
COMMENTS:	Spectra of reduced and oxidized forms; spectra of ferricyanide-treated protein at 16°C, 25°C, and 35°C
REFERENCE:	R. Margalit and A. Schejter, Eur. J. Biochem., 32, 492 (1973)

525. Cytochrome c

SOURCE:	Horse heart
SOLVENT:	25 mM Sodium tetraborate plus 0.30 M NaCl, pH 10.5 and pH 7.5
COMMENTS:	Spectra of reduced and oxidized forms
REFERENCE:	D. O. Lambeth, K. L. Campbell, R. Zand, and G. Palmer, J. Biol. Chem., 248, 8130 (1973)

526. Cytochrome c

SOURCE:	Horse heart
SOLVENT:	0.1 N Acetic acid
COMMENTS:	Spectra of oxidized apocytochrome c and of fragments 1-59 and 60-104
REFERENCE:	A. Fontana, C. Vita, and C. Toniolo, FEBS Lett., 32, 139 (1973)

527. Cytochrome c

SOURCE:	Horse heart (commercial)
SOLVENT:	Neutral pH and pH 1.8
COMMENTS:	Effect of sodium perchlorate on spectrum of ferricytochrome c
REFERENCE:	I. Aviram, J. Biol. Chem., 248, 1894 (1973)

528. Cytochrome c

SOURCE:	Horse heart
COMMENTS:	Spectra of reduced and oxidized forms in the presence of heme fragment 1-65
REFERENCE:	W. R. Fisher, H. Taniuchi, and C. B. Anfinsen, J. Biol. Chem., 248, 3188 (1973)

529. Cytochrome c

SOURCE:	Horse heart
SOLVENT:	10% Acetic acid
COMMENTS:	Spectra of cytochrome c and 2-nitrophenylsulfenyl-cytochrome c

REFERENCE: A. Fontana, F. M. Veronese, and E. Bocca, FEBS Lett., 32, 135 (1973)

530. Cytochrome c

SOURCE: Horse heart
SOLVENT: pH 5.5
COMMENTS: Spectra of unmodified protein, methionine-80 sulfoxide-cytochrome c, histidine-18-modified cytochrome c, and histidine-18-modified plus methionine-80-modified cytochrome c
REFERENCE: M. Folin, A. Azzi, A. M. Tamburro, and G. Jori, Biochim. Biophys. Acta, 285, 337 (1972)

531. Cytochrome c

SOURCE: Horse heart
SOLVENT: pH 2.0, 4.2, 4.5, 6.8, 7.2, 10.0 and 10.1
COMMENTS: Spectra of types I, II, III, and IV cytochrome c, reduced and oxidized after treatment with N-bromosuccinimide
REFERENCE: Y. P. Meyer and P. K. Pal, Biochemistry, 11, 4209 (1972)

532. Cytochrome c

SOURCE: Horse heart
SOLVENT: 0.1 M Tris-HCl, pH 9
COMMENTS: Changes in spectrum due to bleaching by lipoxidase-linoleate system; effect of erythrocuprein upon system
REFERENCE: A. Finazzi Agro, C. Giovagnoli, P. De Sole, L. Calabrese, G. Rotilo, and B. Mondovi, FEBS Lett., 21, 183 (1972)

533. Cytochrome c

SOURCE: Horse heart
SOLVENT: 0.1 M Acetate buffer, pH 5, 15.6 mole % and 24 mole % ethanol
COMMENTS: Effect of carbon monoxide on spectrum
REFERENCE: L. S. Kaminsky, P. E. Burger, A. J. Davison, and D. Helfet, Biochemistry, 11, 3702 (1972)

534. Cytochrome c

SOURCE: Horse heart
SOLVENT: 10 mM Phosphate buffer, pH 2.16–pH 7.10
COMMENTS: Spectra of native and $(carboxymethyl)_2$ cytochrome c

REFERENCE: J. Babul and E. Stellwagen, Biochemistry, 11, 1195 (1972)

535. Cytochrome c

SOURCE: Horse heart
SOLVENT: 0.05 M Sodium borate, pH 9.0
COMMENTS: Spectra of protein and protein-gladiolic acid complexes
REFERENCE: G. A. White and W. B. Elliott, Biochem. Biophys. Res. Comm., 47, 1186 (1972)

536. Cytochrome c

SOURCE: Horse heart
SOLVENT: 1-Propanol; 17 mole %, 8.5 mole %, 2.5 mole %
COMMENTS: Spectra of reduced cytochrome c under anaerobic, and anaerobic plus CO conditions
REFERENCE: L. S. Kaminsky, F. C. Yong, and T. E. King, J. Biol. Chem., 247, 1354 (1972)

537. Cytochrome c

SOURCE: Horse heart
COMMENTS: Changes in spectrum upon addition of malonaldehyde
REFERENCE: T. W. Kwon and H. S. Olcott, Biochim. Biophys. Acta, 130, 528 (1968)

538. Cytochrome c

SOURCE: Horse heart
SOLVENT: 0.1 M Sodium phosphate buffer, pH 7.0
COMMENTS: Spectra of reduced and oxidized forms
REFERENCE: L. C. Dickinson and J. C. W. Chien, Biochemistry, 14, 3526 (1975)

539. Cytochrome c, Cobalt-

SOURCE: Horse heart
SOLVENT: 0.1 M Sodium-phosphate buffer, pH 7.0
COMMENTS: Spectra of reduced and oxidized forms
REFERENCE: L. C. Dickinson and J. C. W. Chien, Biochemistry, 14, 3526 (1975)

540. Cytochrome c, carboxymethyl-

SOURCE: Horse heart
SOLVENT: a. 2% Sodium borate buffer, pH 9.2
b. 0.1 M Potassium phosphate buffer, pH 6.5

COMMENTS: Spectra of reduced cytochrome c and its CO complex
REFERENCE: M. T. Wilson, M. Brunori, C. C. Rotilo, and E. Antonini, J. Biol. Chem., 248, 8162 (1973)

541. Cytochrome c, formylated

SOURCE: Horse heart
SOLVENT: 0.1 M Potassium phosphate buffer, pH 7.0
COMMENT: Spectra of reduced and oxidized forms
REFERENCE: T. Brittain and C. Greenwood, Biochem. J., 149, 713 (1975)

542. Cytochrome c, formylated

SOURCE: Horse heart
SOLVENT: Mixtures of 0.1 M citric acid and 0.2 M potassium phosphate at pH 2.2, 4.0, 6.1, and 7.0
REFERENCE: M. Erecinska, Arch. Biochem. Biophys., 169, 199 (1975)

543. Cytochrome c, nitrotyrosyl-

SOURCE: Horse heart
SOLVENT: 0.1 M Sodium acetate, pH 4.5
COMMENTS: Changes in spectrum due to 6 M guanidine HCl; spectra of reduced and oxidized forms
REFERENCE: A. J. Do Nascimento, Biochem. J., 155, 589 (1976)

544. Cytochrome c

SOURCE: Soybean
COMMENTS: Spectra of reduced and oxidized forms
REFERENCE: C. Fridman, H. Lis, N. Sharon, and E. Katchalski, Arch. Biochem. Biophys., 126, 299 (1968)

545. Cytochrome c

SOURCE: Buckwheat
SOLVENT: 0.05 M Phosphate buffer, pH 8.0
COMMENTS: Spectra of reduced and oxidized forms
REFERENCE: M. Richardson, M. V. Laycock, J. A. M. Ramshaw, E. W. Thompson, and D. Boulter, Phytochemistry, 9, 2271 (1970)

546. Cytochrome c

SOURCE: Pigeon breast muscle
SOLVENT: 0.1 M Phosphate buffer, pH 7.4
COMMENTS: Spectra of reduced and oxidized forms at several temperatures (room temperature down to 4°K)

REFERENCE: B. Hagihara, R. Oshino, and T. Iizuka, J. Biochem. Tokyo, 75, 45 (1974)

547. Cytochrome c

SOURCE: Chlorella
SOLVENT: Water:Ethylene glycol (50:50)
COMMENTS: Spectra of reduced and oxidized forms at 77°K
REFERENCE: L. A. Blumenfeld, R. M. Davydov, N. S. Fel, S. N. Magonov, and R. O. Vilu, FEBS Lett., 45, 256 (1974)

548. Cytochrome c

SOURCE: Dictyostelium discoideum
SOLVENT: 0.1 M Phosphate buffer, pH 6.5
COMMENTS: Spectra of reduced and oxidized forms
REFERENCE: A. I. Al-Ayash and M. T. Wilson, Comp. Biochem. Physiol., 56B, 147 (1977)

549. Cytochrome c

SOURCE: Methylosinus trichosporium OB36
SOLVENT: 20 mM Sodium phosphate buffer, pH 7.0
COMMENTS: Spectra of reduced and oxidized forms
REFERENCE: G. M. Tonge, D. E. F. Harrison, and I. J. Higgins, Biochem. J., 161, 333 (1977)

550. Cytochrome c

SOURCE: Rice embryo (Oryzae sativa L.)
COMMENTS: Spectra of reduced and oxidized forms
REFERENCE: Y. Morita and S. Ida, Agric. Biol. Chem., 32, 441 (1968)

551. Cytochrome c

SOURCE: Rice bran (Oryzae sativa L.)
COMMENTS: Spectra of reduced and oxidized forms
REFERENCE: S. Ida and Y. Morita, Agric. Biol. Chem., 33, 10

552. Cytochrome c

SOURCE: Yeast (Candida krusei)
SOLVENT: 0.1 M Phosphate buffer, pH 7.4
COMMENTS: Spectra of reduced form at various temperatures (room temperature-4°K)
REFERENCE: B. Hagihara, R. Oshino, and T. Iizuka, J. Biochem. Tokyo, 75, 45 (1974)

553. Cytochrome c and iso-Cytochrome c

SOURCE: Yeast (Schizosaccharomyces pombe)
Yeast (Saccharomyces cerevisiae)
SOLVENT: 0.1 M Sodium-potassium phosphate buffer, pH 7.0
COMMENTS: a. Spectra of reduced and oxidized forms
b. Low temperature spectra of yeast and horse heart cytochromes c, and yeast iso-cytochrome c (reduced forms)
REFERENCE: M. L. Claisse and A. M. Simon-Becam, Eur. J. Biochem., 86, 399 (1978)

554. Cytochrome c

SOURCE: Rhizopus oryzae
COMMENTS: Spectra of reduced and oxidized forms
REFERENCE: A. Obayashi, H. Yorifuji, T. Yamagata, T. Ijichi, and M. Kanie, Agric. Biol. Chem., 30, 717 (1966)

555. Cytochrome c

SOURCE: Thiocapsa roseopersicina
COMMENTS: Spectra of reduced and oxidized forms
REFERENCE: I. N. Gogotov and T. V. Laurinavichene, Biokhimiya (Biochemistry), 42, 1002 (1977)

556. Cytochrome c

SOURCE: Blue-green alga (Synechoccus lividus)
SOLVENT: Neutral phosphate buffer
COMMENTS: Spectra of reduced and oxidized forms
REFERENCE: H. L. Crespi, V. Smith, L. Gajda, T. Tisue, and R. M. Ammeraal, Biochim. Biophys. Acta, 256, 611 (1972)

557. Cytochrome c-I/Cytochrome c-II

SOURCE: Pseudomonas K62 and Pseudomonas CF
COMMENTS: Spectra of reduced and oxidized forms
REFERENCE: K. Furukawa and K. Tonomura, Biochim. Biophys. Acta, 325, 413 (1973)

558. Cytochrome c, crypto-

SOURCE: Pseudomonas denitrificans
SOLVENT: pH 7
COMMENTS: Spectra of reduced, oxidized, CO-reduced, and NO-reduced forms; effect of sodium dodecylbenzenesulfonate and 4 M guanidine on spectra

REFERENCE: H. Iwasaki and S. Shidara, Plant Cell Physiol., 10, 291 (1969)

559. Cytochrome c'

SOURCE: Purple sulfur photosynthetic bacterium (Chromatium)
SOLVENT: 0.1 M Potassium phosphate, pH 7.4
COMMENTS: Spectra of reduced, CO-reduced, and oxidized forms
REFERENCE: M. Cusanovichi and Q. H. Gibson, J. Biol. Chem., 248, 822 (1973)

560. Cytochrome c'

SOURCE: Rhodopseudomonas palustris
SOLVENT: pH 7.0, 10.6, 12.0
COMMENTS: Spectra of reduced and oxidized forms
REFERENCE: T. C. Strekas and I. G. Spiro, Biochim. Biophys. Acta, 351, 237 (1974)

561. Cytochrome c'

SOURCE: Chromatium
COMMENTS: Spectra of oxidized form at 293°K and 77°K
REFERENCE: M. M. Maltempo, Biochim. Biophys. Acta, 434, 513 (1976)

562. Cytochrome c'

SOURCE: Thiocapsa roseopersicina
SOLVENT: 20 mM Tris-HCl, pH 7.0 and pH 8.0
COMMENTS: Spectra of oxidized enzyme and dithionite-reduced, dithionite-reduced-CO, and sulfide-reduced forms
REFERENCE: U. Fischer and H. G. Truper, Current. Microbiol., 3, 41 (1979)

563. Cytochrome c'

SOURCE: Rhodopseudomonas spheroides, iron deficient
SOLVENT: 0.02 M Tris, pH 7.5
COMMENTS: Spectra of reduced and oxidized forms
REFERENCE: I. Agalidis, E. Janeau, and F. Reiss-Husson, Eur. J. Biochem., 47, 573 (1974)

564. Cytochrome cc'

SOURCE: Rhodospirillum rubrum
SOLVENT: Potassium phosphate buffer, pH 7.0; sodium carbonate buffer, pH 10.0; disodium hydrogen phosphate plus NaOH, pH 12.5

COMMENTS: Spectra of reduced and oxidized forms
REFERENCE: Y. Imai, K. Imai, R. Sato, and T. Horio, J. Biochem. Tokyo, 65, 225 (1969)

565. Cytochrome c_1

SOURCE: Beef heart
SOLVENT: 50 mM Phosphate buffer, pH 3.5-6.95
COMMENTS: Spectra of reduced form at 77°K at different pH values
REFERENCE: L. S. Kaminsky, Y. L. Chiang, and T. E. King, J. Biol. Chem., 250, 7280 (1975)

566. Cytochrome c_1

SOURCE: Beef heart
SOLVENT: 50 mM Phosphate buffer, pH 7.4
COMMENTS: Protein denatured by alcohol or sodium dodecylsulfate; spectra of reduced and oxidized forms before and after treatment with cyanide
REFERENCE: L. S. Kaminsky, Y. L. Chiang, and T. E. King, J. Biol. Chem., 250, 7280 (1975)

567. Cytochrome c_1

SOURCE: Beef heart
SOLVENT: 50 mM Oxygen-free phosphate buffer, pH 7.4
COMMENTS: Spectra of photoreduction of protein after different times of illumination
REFERENCE: C. A. Yu, Y. L. Chiang, L. Yu, and T. E. King, J. Biol. Chem., 250, 6218 (1975)

568. Cytochrome c_1

SOURCE: Beef heart
SOLVENT: 0.1 M Phosphate buffer
COMMENTS: Low temperature spectrum
REFERENCE: C. A. Yu, L. Yu, and T. E. King, Biochim. Biophys. Acta, 267, 300 (1972)

569. Cytochrome c_1

SOURCE: Beef heart
SOLVENT: 0.1 M and 0.05 M Phosphate buffer, pH 7.4
COMMENTS: Spectra of reduced and oxidized forms at 23°C and -160°C
REFERENCE: C. A. Yu, L. Yu, and T. E. King, J. Biol. Chem., 247, 1012 (1972)

570. Cytochrome c_1

SOURCE: Beef cardiac muscle
COMMENTS: Spectra of ferricyanide-oxidized and dithionite-reduced forms
REFERENCE: R. Holmquist and H. Moise, Biochimie, 61, 697 (1979)

571. Cytochrome c_1, tryptic-resistant heme peptide

SOURCE: Beef heart
SOLVENT: 25 mM Ammonium phosphate, pH 7.4
COMMENTS: Spectra of reduced and oxidized forms and of cyanide complexes; spectrum of reduced form at -155°C
REFERENCE: L. Yu, Y. L. Chiang, C. Yu, and T. E. King, Biochim. Biophys. Acta, 379, 33 (1975)

572. Cytochrome c_1

SOURCE: Beef heart
SOLVENT: 0.1 M Phosphate buffer, pH 7.4
COMMENTS: Spectra of reduced form at various temperatures (room temperature-7°K)
REFERENCE: B. Hagihara, R. Oshina, and T. Iizuka, J. Biochem. Tokyo, 75, 45 (1974)

573. Cytochrome c_1

SOURCE: Yeast (Saccharomyces cerevisiae)
SOLVENT: 0.1 M Sodium phosphate-0.5 mM EDTA buffer, pH 7.5, with or without glycerol
COMMENTS: Spectra of reduced and oxidized forms; low temperature spectrum of reduced form
REFERENCE: E. Ross and G. Schatz, J. Biol. Chem., 251, 1991 (1976)

574. Cytochrome c_3

SOURCE: Desulfovibrio vulgaris
SOLVENT: Potassium phosphate, 40 mM and 100 mM, pH 6.0-11.0
COMMENTS: Spectra of reduced form in presence of histidine and imidazole at 77°K
REFERENCE: D. D. Dervartanian and J. LeGall, Biochim. Biophys. Acta, 502, 458 (1978)

575. Cytochrome c_3

SOURCE: Desulfovibrio vulgaris, Miyazaki
SOLVENT: Dry film

COMMENTS: Spectra of time course of oxidation of reduced film
REFERENCE: K. Kimura, A. Suzuki, H. Inokuchi, and T. Yagi, Biochim. Biophys. Acta, 567, 96 (1979)

576. Cytochrome c_4

SOURCE: Pseudomonas denitrificans
SOLVENT: 0.1 M Phosphate buffer, pH 6.8
COMMENTS: Spectra of oxidized and dithionite-reduced forms
REFERENCE: A. R. W. Coulson and R. I. C. Oliver, Biochem. J., 181, 159 (1979)

577. Cytochrome c-550

SOURCE: Bacterium PS3
SOLVENT: 50 mM Phosphate buffer, pH 7.0
COMMENTS: Spectra of oxidized and dithionite-reduced protein
REFERENCE: K. Hon-Nami, T. Oshima, H. Kihara, and Y. Kagawa, Biochem. Biophys. Res. Comm., 87, 1066 (1979)

578. Cytochrome c-550

SOURCE: Rhodopseudomonas spheroides, iron deficient
SOLVENT: 0.02 M Tris, 0.1 M NaCl, pH 7.5 , and 0.02 M Tris, pH 7.5
COMMENTS: Spectra of reduced and oxidized forms
REFERENCE: I. Agalidis, E. Ja neau, and F. Reiss-Husson, Eur. J. Biochem., 47, 573 (1974)

579. Cytochrome c-550

SOURCE: Tapeworm (Moniezia expansa)
COMMENTS: Spectra of reduced and oxidized forms; spectrum of reduced form at -196°C
REFERENCE: K. S. Cheah, Comp. Biochem. Physiol., 51B, 41 (1975)

580. Cytochrome c-550

SOURCE: Thiocapsa roseopersicina
SOLVENT: 20 mM Tris-HCl, pH 8.0
COMMENTS: Spectra of reduced and oxidized forms; changes in spectrum of oxidized form by reduction with sulfide under aerobic conditions
REFERENCE: U. Fischer and H. G. Truper, FEBS Lett., 1, 87 (1977)

581. Cytochrome c-550

SOURCE: Hemispora stellata
COMMENTS: Spectra of reduced and oxidized forms
REFERENCE: F. J. Sala, T. Burgos, and J. Burgos, Arch. Microbiol., 88, 11 (1973)

582. Cytochrome c-550 and c-553

SOURCE: Marine diatom (Phaeodactylum tricornutum)
SOLVENT: 0.5 M Tris-HCl, pH 8.0; 50 mM and 100 mM phosphate buffer, pH 7.0
COMMENTS: Spectra of reduced and oxidized forms with and without cyanide, and reduced CO-complex
REFERENCE: K. Shimazaki, K. Takamiya, and M. Nishimura, J. Biochem. Tokyo, 83, 1631 (1978)

583. Cytochrome c-551

SOURCE: Pseudomonas aeruginosa
COMMENTS: Spectra of reduced and oxidized forms
REFERENCE: F. W. Adair, S. G. Geftic, and J. Gelzer, Trans. N. Y. Acad. Sci., 33, (II) 799 (1971)

584. Cytochrome c-551

SOURCE: Pseudomonas aeruginosa
SOLVENT: Sodium phosphate buffer, pH 6.0
COMMENTS: Spectra of ferrous and ferric forms
REFERENCE: M. Ronnenberg, N. Ellfolk, and R. Soininen, Biochim. Biophys. Acta, 578, 392 (1979)

585. Cytochrome c-551

SOURCE: Pseudomonas aeruginosa
SOLVENT: pH 8.7, pH 11.0, and pH 11.8
REFERENCE: Y. Y. H. Chao, R. Bersohn, and P. Aisen, Biochemistry, 18, 774 (1979)

586. Cytochrome c-551 and c-555

SOURCE: Azotobacter vinelandii
SOLVENT: 0.05 M Sodium phosphate buffer, pH 7.4
COMMENTS: Spectra of reduced and oxidized forms
REFERENCE: H. Campbell, W. H. Orme-Johnson, and R. H. Burris, Biochem. J., 135, 617 (1973)

587. Cytochrome c-551.5

SOURCE: Chloropseudomonas ethylica 2K
SOLVENT: 0.05 M Tris-HCl buffer, pH 7.0
COMMENTS: Spectra of reduced and oxidized forms; spectrum of reduced form at 77°K
REFERENCE: Y. Shioi, K. Takamiya, and M. Nishimura, J. Biochem. Tokyo, 71, 285 (1972)

588. Cytochrome c-551.5

SOURCE: Desulfuromonas acetoxidans
SOLVENT: 0.1 M Tris-HCl buffer, pH 7.6
COMMENTS: Spectra of reduced and oxidized forms
REFERENCE: I. Probst, M. Bruschi, N. Pfennig, and J. LeGall, Biochim. Biophys. Acta, 460, 58 (1977)

589. Cytochrome c-552

SOURCE: Euglena gracilis
SOLVENT: Tris-cacodylate buffer, pH 7
COMMENTS: Spectra of reduced and oxidized forms
REFERENCE: T. Goldkorn and A. Schejter, Arch. Biochem. Biophys., 177, 39 (1976)

590. Cytochrome c-552

SOURCE: Chromatium vinosum
SOLVENT: 0.1 M Tris-HCl buffer, pH 8.5, containing 6 M urea
COMMENTS: Spectra of oxidized and dithionite-reduced forms
REFERENCE: Y. Fukumori and T. Yamanaka, J. Biochem. Tokyo, 85, 1405 (1979)

591. Cytochrome c-552, and Cytochrome c-559

SOURCE: Rhodopseudomonas spheroides y cells
SOLVENT: 0.02 M Tris buffer, 0.35 M NaCl, pH 7.5
COMMENTS: Spectra of reduced and oxidized forms
REFERENCE: I. Agalidis, E. Janeau, and F. Reiss-Husson, Eur. J. Biochem., 47, 573 (1974)

592. Cytochrome c-552

SOURCE: Thiobacillus thioxidans
SOLVENT: 0.05 M Phosphate buffer, pH 7.0
COMMENTS: Spectra of reduced and oxidized forms
REFERENCE: S. Takakwa, J. Biochem. Tokyo, 78, 181 (1975)

593. Cytochrome c-552

SOURCE: Euglena gracilis
SOLVENT: a. pH 6
b. pH 11.2
c. pH 11.2 in 0.4 M NaCl
REFERENCE: I. Aviram, G. W. Pettigrew, and A. Schejter, Biochemistry, 15, 635 (1976)

594. Cytochrome c-552

SOURCE: Euglena gracilis
SOLVENT: pH 5.5-pH 12.0
COMMENTS: Spectra of acetylated and maleylated forms
REFERENCE: I. Aviram, Arch. Biochem. Biophys., 181, 199 (1977)

595. Cytochrome c-552

SOURCE: Euglena gracilis
COMMENTS: Spectra of reduced and oxidized forms
REFERENCE: A. Mitsui and K. Tsushima in Structure and Function of Cytochromes (M. Okunuki, M. D. Kamen, and I. Sekuzu, eds.), 459 (1968), University Park Press, Baltimore, Md.

596. Cytochrome c-552 and Cytochrome c-554

SOURCE: Nitrosomonas europaea
SOLVENT: 0.05 M Phosphate buffer, pH 7.0
COMMENTS: Spectra of reduced and oxidized forms
REFERENCE: T. Yamanaka and M. Shinra, J. Biochem. Tokyo, 75, 1265 (1974)

597. Cytochrome c-552

SOURCE: Horse heart
SOLVENT: pH 7
COMMENTS: Spectra of reduced and oxidized forms
REFERENCE: H. Kihara, H. Nakatani, K. Hiromi, K. Hon-Nami, and T. Oshima, Biochim. Biophys. Acta, 460, 480 (1977)

598. Cytochrome c-552

SOURCE: Thermus thermophilus HB8
SOLVENT: 0.01 M Phosphate buffer, pH 7.0
COMMENTS: Spectra of reduced (dithionite) and oxidized forms
REFERENCE: K. Hon-Nami and T. Oshima, J. Biochem. Tokyo, 82, 769 (1977)

599. Cytochrome c-552

SOURCE: Thermus thermophilus HB8
SOLVENT: a. pH 7.5
b. pH 0.1
c. 6.4 M Guanidine-HCl
d. 6.4 M Guanidine-HCl, pH 7.0
e. 6.4 M Guanidine-HCl, pH 2.4
REFERENCE: K. Hon-Nami and T. Oshima, Biochemistry, 18, 5693 (1979)

600. Cytochrome c-553 (550)

SOURCE: Desulfovibrio desulfuricans norway
SOLVENT: 50 mM Tris-HCl, pH 7.6
COMMENTS: Spectra of reduced (ascorbate or dithionite) and oxidized forms
REFERENCE: G. Fauque, M. Bruschi, and J. LeGall, Biochem. Biophys. Res. Comm., 86, 1020 (1979)

601. Cytochrome c-553

SOURCE: Chlorobium thiosulfatophilum
SOLVENT: 0.2 M Phosphate buffer, pH 7.0
COMMENTS: Spectra of reduced and oxidized forms
REFERENCE: T. Yamanaka, J. Biochem. Tokyo, 79, 655 (1976)

602. Cytochrome c-553

SOURCE: Chlorobium thiosulfatophilum
SOLVENT: 0.1 M Tris-HCl buffer, pH 7.4
COMMENTS: Spectra of oxidized, oxidized plus cyanide, and reduced forms
REFERENCE: A. Kusai and T. Yamanaka, FEBS Lett., 34, 235 (1973)

603. Cytochrome c-553

SOURCE: Alga (Petalonia fascia)
COMMENTS: Spectra of reduced and oxidized forms
REFERENCE: Y. Sugimura, F. Toda, T. Murata, and E. Yakushiji in Structure and Function of Cytochromes (K. Okunuki, M. D. Kamen, and I. Sekuzu, eds.), 452 (1968), University Park Press, Balitmore, Md.

604. Cytochrome c-554 and Cytochrome c-549

SOURCE: Thermus thermophilus HB8
SOLVENT: 0.01 M Phosphate buffer, pH 7.0

COMMENTS: Spectra of reduced and oxidized forms
REFERENCE: K. Hon-Nami and T. Oshima, J. Biochem. Tokyo, 83, 629 (1978)

605. Cytochrome c-554.5

SOURCE: Alga (Codium latum)
COMMENTS: Spectra of reduced and oxidized forms
REFERENCE: Y. Sugimura, F. Toda, T. Murata, and E. Yakushiji in Structure and Function of Cytochromes (K. Okunuki, M. D. Kamen, and I. Sekuzu, eds.), 452 (1968), University Park Press, Baltimore, Md.

606. Cytochrome c-555

SOURCE: Prosthecochloris aestuarii
SOLVENT: 0.2 M Potassium phosphate buffer, pH 7.0
COMMENTS: Spectra of reduced and oxidized forms
REFERENCE: M. D. Fiechtner and R. J. Kassner, Biochemistry, 17, 1028 (1978)

607. Cytochrome c-555

SOURCE: Protozoan (Crithidia fasciculata)
SOLVENT: 50 mM Sodium potassium phosphate buffer, pH 7.0
COMMENTS: Spectra of reduced and oxidized forms
REFERENCE: J. B. Kusel, J. R. Suriano, and M. M. Weber, Arch. Biochem. Biophys., 133, 293 (1969)

608. Cytochrome 555 (550)

SOURCE: Chloropseudomonas ethylica 2K
SOLVENT: 0.05 M Tris-HCl buffer, pH 7.0
COMMENTS: Spectra of reduced and oxidized forms; spectr reduced form at liquid nitrogen temperature
REFERENCE: Y. Shioi, K. Takamiya, and M. Nishimura, J. Biochem. Tokyo, 71, 285 (1972)

609. Cytochrome c-556

SOURCE: Agrobacterium tumefaciens B2a
COMMENTS: Spectra of reduced (2-mercaptoethanol) and oxidized forms
REFERENCE: C. Van Den Branden, J. Van Beeumen, J. De Lay, and A. Vande Meroop, Hoppe-Seyler's Z. Physiol. Chem., 356, 1251 (1975)

610. Cytochrome c-556

SOURCE: Pseudomonas aeruginosa
COMMENTS: Spectra of reduced and oxidized forms; spectrum of reduced form in 6M urea
REFERENCE: J. Singh and D. C. Wharton, Biochim. Biophys. Acta, 292, 391 (1973)

611. Cytochrome c-557

SOURCE: Crithid oncopelti
SOLVENT: 0.1 M Potassium phosphate buffer, pH 7.0
COMMENTS: a. Change in spectrum with temperature
b. Spectrum of ferrocytochrome cyanide adduct
REFERENCE: G. W. Pettigrew, I. Aviram, and A. Schejter, Biochem. J., 149, 155 (1975)

612. Cytochrome c-558

SOURCE: Euglena gracilis
SOLVENT: 0.1 M Potassium phosphate buffer, pH 7.0
COMMENTS: a. Change in spectrum with temperature
b. Spectrum of ferrocytochrome cyanide adduct
REFERENCE: G. W. Pettigrew, I. Aviram, and A. Schejter, Biochem. J., 149, 155 (1975)

613. Cytochrome c-558

SOURCE: Euglena gracilis
SOLVENT: pH 8.10, pH 11.17 in 0.38 M NaCl
REFERENCE: E. Stellwagen and R. Cass, Biochem. Biophys. Res. Comm., 60, 371 (1974)

614. Cytochrome d

SOURCE: Tetrahymena pyriformis
SOLVENT: 0.25 M Sucrose-0.01 M phosphate buffer, pH 7.2
COMMENTS: Spectra of oxidized, dithionite-reduced, dithionite-reduced-CO, oxidized-cyanide, dithionite-reduced-cyanide, and dithionite-reduced-NO forms
REFERENCE: L. Kilpatrick and M. Erecinska, Arch. Biochem. Biophys., 197, 1 (1979)

615. Cytochrome f

SOURCE: Green alga (Kirchneriella obesa)
COMMENTS: Spectrum of reduced protein
REFERENCE: R. M. Nalbadyan, Biokhimiya (Biochemistry), 37, 970 (1972)

616. Cytochrome f

SOURCE: Dinoflagellate (Cachonina niei)
COMMENTS: Spectra of reduced and oxidized (ferricyanide) forms
REFERENCE: C. W. Mehard, B. L. Prezelin, and F. T. Haxo, Phytochemistry, 14, 2379 (1979)

617. Cytochrome f

SOURCE: Spinach chloroplasts
SOLVENT: 0.05 M N-Tris (hydroxymethyl)-methylglycine, pH 8
REFERENCE: N. Nelson and E. Racker, J. Biol. Chem., 247, 3848 (1972)

618. Cytochrome f

SOURCE: Charlock (Sinapis arvensis L.)
SOLVENT: 50 mM Sodium phosphate, pH 7.5
COMMENTS: Spectra of reduced and oxidized forms
REFERENCE: J. C. Gray, Eur. J. Biochem., 82, 133 (1978)

619. Cytochrome f

SOURCE: Chlamydomonas
SOLVENT: 20 mM Phosphate plus 60 mM NaCl, pH 7.0
COMMENTS: Spectrum of reduced form
REFERENCE: P. M. Wood, Eur. J. Biochem., 72, 605 (1977)

620. Cytochrome f

SOURCE: Euglena
SOLVENT: Tris-HCl buffer, pH 7.0 and carbonate buffer, pH 10.0
COMMENTS: Spectra of reduced and oxidized forms
REFERENCE: G. Ben-Hayyim and A. Schejter, Eur. J. Biochem., 46, 569 (1974)

621. Cytochrome f

SOURCE: Japanese radish (Raphanus sativus L. var. acanthioformis Makino)
SOLVENT: 50 mM Potassium phosphate, pH 7.8
COMMENTS: Spectra of reduced and oxidized forms at room temperature and low temperature (77°K)
REFERENCE: K. Tanaka, M. Takahashi, and K. Asada, J. Biol. Chem., 253, 7397 (1978)

622. Cytochrome f

SOURCE: a. Spirula maxima
b. Spinach Chloroplasts
SOLVENT: 50 mM Sodium phosphate buffer, pH 7.5
COMMENTS: Spectra of oxidized and reduced forms of cytochrome from (a)
REFERENCE: K. K. Ho and D. W. Krogman, J. Biol. Chem., 255, 3855 (1980)

623. Cytochrome h

SOURCE: Land snail (Helix pomatia)
SOLVENT: 0.1 M Phosphate buffer, pH 7.25
COMMENTS: Spectra of reduced and oxidized forms at room temperature and low temperature
REFERENCE: J. Keilin in Structure and Function of Cytochromes (K. Okunuki, M. D. Kamen, and I. Sekuzu, eds.) 691 (1967), University Park Press, Baltimore, Md.

624. Cytochrome o

SOURCE: Azotobacter vinelandii membrane
COMMENTS: Spectra of reduced, oxidized, and CO-reduced forms at room temperatures and low temperature (77°K)
REFERENCE: T. Y. Yang and P. Jurtshuk, Jr., Biochem. Biophys. Res. Comm., 81, 1032 (1978)

625. Cytochrome o

SOURCE: Vitreoscilla
SOLVENT: 0.1 M Phosphate buffer, pH 7.5
COMMENTS: Spectra of reduced, CO-reduced, oxidized, CO-oxygenated, cyanide treated, and hydrogen peroxide treated forms
REFERENCE: C. Y. Liu and D. A. Webster, J. Biol. Chem., 249, 4261 (1974)

626. Cytochrome o

SOURCE: Vitreoscilla
SOLVENT: 0.1 M Sodium phosphate, pH 7.0
COMMENTS: Spectra of reduced, oxygenated, CO-oxygenated forms and high molecular weight fraction at 23°C and 0°C
REFERENCE: B. Tyree and D. A. Webster, J. Biol. Chem., 254, 176 (1979)

627. Cytochrome o

SOURCE: Vitreoscilla

SOLVENT: 0.1 M Sodium phosphate, pH 7.0
COMMENTS: a. Spectra during anaerobic titration with dithionite
b. Spectra during equilibrium photochemical titration with toluylene blue and indigo carmine
REFERENCE: B. Tyree and D. A. Webster, in Cytochrome Oxidase (T. E. King, Y. Orii, B. Chance and K. Okunuki, eds.), 21 (1979), Elsevier, New York

628. Cytochromes, B-Types

SOURCE: Marine diatom (Phaeodactylum tricornutum)
SOLVENT: 50 mM Phosphate buffer containing 0.6 M NaCl, pH 7.0
COMMENTS: Spectra of reduced and oxidized forms at room temperature and low temperature
REFERENCE: K. Shimazaki, K. Takamiya, and M. Nishimura, J. Biochem. Tokyo, 83, 1631 (1978)

629. Cytochrome P-420

SOURCE: Rabbit liver (phenobarbital-induced)
SOLVENT: 0.1 M Potassium phosphate buffer, pH 9.25, containing 20% glycerol
COMMENTS: Spectrum of ethyl isocyanide compound of reduced P-420 made by treatment of P-450 with potassium isothiocyanate
REFERENCE: Y. Imai and R. Sato, J. Biochem. Tokyo, 75, 689 (1974)

630. Cytochrome P-420

SOURCE: Pseudomonas putida (camphor grown)
SOLVENT: 50 mM Potassium phosphate, pH 7.0
COMMENTS: Spectra of reduced, CO-reduced, and oxidized forms
REFERENCE: C. A. Yu and I. C. Gunsalas, J. Biol. Chem., 249, 102 (1974)

631. Cytochrome P-420

SOURCE: Rabbit liver (phenobarbital-induced)
SOLVENT: 100 mM Potassium phosphate buffer, pH 7.5, containing 20% glycerol
COMMENTS: Spectrum of reduced-ethylisocyanide complex at 0°C
REFERENCE: T. Shimizu, T. Nozawa, M. Hatano, Y. Imai, and R. Sato, Biochemistry, 14, 4172 (1975)

632. Cytochrome P-420 LM

SOURCE: Rabbit liver microsomes

COMMENTS: Spectral changes associated with anaerobic titration of P-420 LM with dithionite
REFERENCE: F. P. Guengerich, D. P. Ballou, and M. J. Coon, J. Biol. Chem., 250, 7405 (1975)

633. Cytochrome P-420 LM2

SOURCE: Rabbit liver
COMMENTS: Spectra of reduced, CO-reduced, and oxidized forms
REFERENCE: J. H. Dawson, J. R. Trudell, R. E. Linder, G. Barth, E. Bunnenberg, and C. Djerassi, Biochemistry, 17, 33 (1978)

634. Cytochrome P-420, Reconstituted

SOURCE: Pseudomonas putida (grown in camphor)
SOLVENT: 50 mM Potassium phosphate buffer, pH 7.0
COMMENTS: Spectra of reduced, CO-reduced, oxidized, and apoprotein forms
REFERENCE: C. A. Yu and I. C. Gunsalus, J. Biol. Chem., 249, 107 (1974)

635. Cytochrome P-448

SOURCE: Rat liver (methylcholanthrene treated)
SOLVENT: 0.1 M Potassium phosphate buffer, pH 7.4
COMMENTS: Spectra of reduced, CO-reduced, and oxidized forms
REFERENCE: W. Levin, A. Y. Lu, D. Ryan, S. West, R. Kuntzman, and A. H. Conney, Arch. Biochem. Biophys., 153, 543 (1972); A. Y. Lu, S. B. West, D. Ryan, and W. Levin, Drug Metab. and Dispos., 1, 29 (1973); W. Levin, A. Y. Lu, D. Ryan, S. West, R. Kuntzman, and A. H. Conney, Adv. Exp. Med. Biol., 58, 1 (1975)

636. Cytochrome P-450 (P-448)

SOURCE: Aroclor 1254-treated male rats
SOLVENT: 0.1 M Potassium phosphate, pH 7.4
COMMENTS: Spectra of reduced, CO-reduced, and oxidized forms
REFERENCE: A. P. Alvares and A. Kappas, J. Biol. Chem., 252, 6373 (1971)

637. Cytochrome P-450 and Cytochrome P-448

SOURCE: Phenobarbital (PB) and 3-methylcholanthrene (MC)-induced rabbit liver microsomes
COMMENTS: 1. Spectra of PB P-450, PB P-448, and MC-P-448 with 3-MC

2. Spectra of dithionite-reduced PB P-450, and dithionite-reduced MC P-448 with 3-MC
3. Spectra of dithionite-reduced-CO complexes of PB P-450 and MC P-448 with 3-MC
4. Spectra of oxidized forms of PB P-450 and MC P-448 with 3-MC

REFERENCE: T. Shimizu, T. Nozawa, M. Hatano, H. Satake, Y. Imai, C. Hashimoto, and R. Sato, Biochim. Biophys. Acta, 579, 122 (1979)

638. Cytochrome P-450

SOLVENT: 6% (V/V) of 200 mM Tris-HCl, pH 8.4, in 10% AOT-n-heptane; AOT is sodium diisooctyl sulfosuccinate
COMMENTS: Spectrum of oxyferroprotein at 0°C
REFERENCE: C. Balny and P. Douzou, Biochimie, 61, 445 (1979)

639. Cytochrome P-450

SOURCE: Rabbit lung microsomes
SOLVENT: 0.1 M Potassium phosphate, pH 7.7, 0.1 mM EDTA, 30% glycerol
COMMENTS: Spectra of reduced, CO-reduced, and oxidized forms
REFERENCE: E. Arinc and R. M. Philpot, J. Biol. Chem., 251, 3212 (1976)

640. Cytochrome P-450

SOURCE: Liver microsomes from phenobarbital-treated rabbits
SOLVENT: a. 0.1 M Tris-HCl buffer, pH 8.3, 30% glycerol
b. 0.1 M Potassium phosphate buffer, pH 6.9, containing 30% glycerol
COMMENTS: Spectrum of the ethylisocyanide compound of reduced form
REFERENCE: R. Sato, H. Satake, and Y. Imai, Drug. Metab. and Dispos., 1, 6 (1973)

641. Cytochrome P-450

SOURCE: Rabbit liver microsomes
COMMENTS: Spectra during dithionite titration
REFERENCE: M. J. Coon, T. A. Van Der Hoeven, D. A. Haugen, F. P. Guengerich, J. L. Vermilion, and D. P. Ballou, Adv. Exp. Med. Biol., 58, 25 (1975); D. P. Ballou, C. Veeger, T. A. Van Der Hoeven, and M. J. Coon, FEBS Lett., 38, 337 (1974)

642. Cytochrome P-450

SOURCE: Rabbit liver microsomes
SOLVENT: 0.2 M Potassium phosphate, pH 6.0, 7.0, 8.0, containing 20% glycerol
COMMENTS: Spectra of ethylisocyanide compound of reduced protein at several pH values
REFERENCE: Y. Imai and R. Sato, J. Biochem. Tokyo, 75, 689 (1974)

643. Cytochrome P-450

SOURCE: Rabbit liver (phenobarbital-treated)
SOLVENT: a. 0.05 M Tris-acetate buffer, pH 7.4, containing 20% glycerol and 1.0 mM EDTA
b. 0.05 M Phosphate buffer, pH 7.4, containing 20% glycerol and 1.0 mM EDTA
COMMENTS: Spectra of reduced, CO-reduced, and oxidized forms
REFERENCE: T. A. Van Der Hoeven and M. J. Coon, J. Biol. Chem., 249, 6302 (1974); T. A. Van Der Hoeven and D. A. Haugen, Biochem. Biophys. Res. Comm., 60, 569 (1974)

644. Cytochrome P-450

SOURCE: Rabbit liver microsomes (phenobarbital-treated)
SOLVENT: a. 0.1 M Potassium phosphate buffer, pH 7.25, containing 20% glycerol and 0.2% Emulgen 913
b. Same without Emulgen 913
COMMENTS: Spectra of reduced, CO-reduced, and oxidized forms at room temperature and 0°C
REFERENCE: Y. Imai and R. Sato, Biochem. Biophys. Res. Comm., 60, 8 (1974); T. Shimizu, T. Nozawa, M. Hatano, Y. Imai, and R. Sato, Biochemistry, 14, 4172 (1975)

645. Cytochrome P-450

SOURCE: 3-Methylcholanthrene treated rabbits
SOLVENT: 300 mM Potassium phosphate buffer, pH 7.25, containing 20% glycerol
COMMENTS: Spectra of reduced, CO-reduced, and oxidized forms
REFERENCE: C. Hashimoto and Y. Imai, Biochem. Biophys. Res. Comm., 68, 821 (1976)

646. Cytochrome P-450LM

SOURCE: Rabbit liver microsomes
SOLVENT: 0.05 M Tris-acetate buffer, pH 7.5, 20% glycerol, 1 mM EDTA, 0.75 mM benzphetamine, NADPH

COMMENTS: Spectra during reduction by cytochrome P-450 reductase under aerobic conditions
REFERENCE: F. P. Guengerich, D. P. Ballou, and M. J. Coon, J. Biol. Chem., 250, 7405 (1975)

647. Cytochrome P-450

SOURCE: Rabbit liver microsomes
SOLVENT: 0.1 M Phosphate buffer, pH 7.7, containing 20% glycerol
COMMENTS: Spectra of reduced, CO-reduced, and oxidized forms
REFERENCE: R. M. Philpot and E. Arinc, Molec. Pharmacol., 12, 883 (1976)

648. Cytochrome P-450, LM_2, and LM_4

SOURCE: Phenobarbital-treated rabbit liver microsomes; β-naphthoflavone-treated rabbit liver microsomes; untreated rabbit liver microsomes
SOLVENT: 0.1 M Potassium phosphate buffer, pH 7.4, containing 20% glycerol
COMMENTS: Spectra of reduced, CO-reduced, and oxidized forms. Spectral changes due to Renex 690 and octylamine
REFERENCE: D. A. Haugen and M. J. Coon, J. Biol. Chem., 251, 7929 (1976); M. J. Coon, R. E. White, G. D. Nordblom, D. P. Ballou, and F. P. Guengerich, Croat. Chem. Acta, 49, 163 (1977)

649. Cytochrome P-450 LM

SOURCE: Liver microsomes from phenobarbital-treated rabbits
SOLVENT: Tris-acetate buffer, pH 7.5
COMMENTS: Spectra of reduced, oxidized, and oxygenated complexes in presence of benzphetamine
REFERENCE: F. P. Guengerich, D. P. Ballou, and M. J. Coon, Biochem. Biophys. Res. Comm., 70, 951 (1976)

650. Cytochrome P-450 LM_2

SOURCE: Liver microsomes from phenobarbital-treated rabbits
SOLVENT: 0.1 M Potassium phosphate buffer, pH 7.4, containing 20% glycerol, 0.1 M EDTA
COMMENTS: Spectral changes due to sodium dithionite addition under CO atmosphere in presence of all of the following: methyl viologen, benzphetamine, dilauroylglyceryl-3-phosphorylcholine, glucose, catalase and glucose oxidase
REFERENCE: J. A. Peterson, R. E. White, Y. Yasukochi, M. L. Coomes, D. H. O'Keeffe, R. E. Ebel, B. S. S. Masters, D. P. Ballou, and M. J. Coon, J. Biol. Chem., 252, 4431 (1977)

651. Cytochrome P-450 LM

SOURCE: Rabbit liver microsomes
COMMENTS: Spectra of protein in presence of benzphetamine, hexobarbital, aniline, imidazole, or cyanide ion
REFERENCE: H. Rein, O. Ristan, J. Friedrich, G.-R. Janig, and K. Ruckpaul, FEBS Lett., 75, 19 (1977)

652. Cytochrome P-450 LM_2 and LM_4

SOURCE: Phenobarbital-treated rabbit liver microsomes
SOLVENT: 0.1 M Potassium phosphate buffer, pH 7.4 containing 20% glycerol
COMMENTS: Spectra of oxidized, reduced and reduced-CO forms
REFERENCE: R. Hansson and K. Wikvall, J. Biol. Chem., 255, 1643 (1980)

653. Cytochrome P-450 LM

SOURCE: Liver microsomes from phenobarbital-treated male rabbits
COMMENTS: Spectra of reduced and oxidized forms in the presence of isopropyl cyanide
REFERENCE: K. Ruckpaul, H. Rein, G.-R. Janig, W. Winkler, and O. Ristau, Croat. Chem. Acta, 49, 339 (1977)

654. Cytochrome P-450 LM_2, Cytochrome P-450 LM 3:4, and Cytochrome P-450 LM4

SOURCE: Liver microsomes from phenobarbital-treated male New Zealand rabbits
COMMENTS: Spectra of reduced, CO-reduced, and oxidized forms
REFERENCE: J. H. Dawson, J. R. Trudell, R. H. Linder, G. Barth, E. Bunnenberg, and C. Djerassi, Biochemistry, 17, 33 (1978)

655. Cytochrome P-450c

SOURCE: Liver microsomes from 2, 3, 7, 8-tetrachlordibenz-p-dioxin-treated rabbits
SOLVENT: 0.1 M Potassium phosphate buffer, pH 7.4, containing 20% glycerol
COMMENTS: Spectra of reduced, reduced-carbonyl complex, and oxidized forms
REFERENCE: E. F. Johnson and U. Muller-Eberhard, Biochem. Biophys. Res. Comm., 76, 652 (1977)

656. Cytochrome P-450

SOURCE: Liver microsomes from phenobarbital-treated rats
SOLVENT: 0.1 M Potassium phosphate buffer, pH 7.4
COMMENTS: Spectra of reduced, CO-reduced, and oxidized forms
REFERENCE: W. Levin, D. Ryan, S. West, and A. Y. H. Lu, J. Biol. Chem., 249, 1747 (1974)

657. Cytochrome P-450a, P-450b, and P-450c

SOURCE: Liver microsomes from rats treated with Aroclor 1254
SOLVENT: a. 0.1 M Potassium phosphate buffer, pH 7.4, containing 20% glycerol
b. 0.1 M Potassium phosphate buffer, pH 7.4, containing 20% glycerol, 0.1 M EDTA, 0.5% sodium cholate and 0.2% Emulgen 911
COMMENTS: Spectra of reduced, CO-reduced, and oxidized forms
REFERENCE: D. E. Ryan, P. E. Thomas, D. Korzeniowski, and W. Levin, J. Biol. Chem., 254, 1365 (1979)

658. Cytochrome P-450

SOURCE: Rat liver microsomes
SOLVENT: a. 0.05 M Phosphate buffer, pH 7.4, containing 25% glycerol
b. 0.1 M Tris buffer, pH 7.4
COMMENTS: Spectral changes due to temperature changes: 4.5°C, 24.5°C, 40°C, in (a); 6.5°C, 16°C, 26°C, 37°C, 42°C in (b)
REFERENCE: D. L. Cinti, S. G. Sligar, S. G. Gibson, and J. B. Schenkman, Biochemistry, 18, 36 (1979)

659. Cytochrome P-450

SOURCE: Liver microsomes from phenobarbital-treated mice
SOLVENT: 0.1 M Potassium phosphate, pH 7.4
COMMENTS: Spectra of reduced, CO-reduced, and oxidized forms
REFERENCE: M.-T. Huang, S. B. West, and A. Y. H. Lu, J. Biol. Chem., 251, 4659 (1976)

660. Cytochrome P-450 and Cytochrome P-450-I

SOURCE: Housefly, strain Fc
SOLVENT: 0.025 M Sodium phosphate, pH 7.0, containing 0.1 mM EDTA, 0.1 mM DTT, and 20% glycerol
COMMENTS: Spectra of reduced, CO-reduced, and oxidized forms
REFERENCE: J. Capdevila, N. Ahmad, and M. Agosin, J. Biol. Chem., 250, 1048 (1975)

661. Cytochrome P-450

SOURCE: Tobacco hornworm (Manduca sexta J.) midgut
SOLVENT: 0.5 M Potassium phosphate, pH 7.5, containing 0.1 M KCl
COMMENTS: Spectra of reduced, CO-reduced, and oxidized forms
REFERENCE: A. P. Kulkarni and E. Hodgson, Insect Biochem., 6, 385 (1976)

662. Cytochrome P-450

SOURCE: Yeast (Saccharomyces cerevisiae)
SOLVENT: 0.1 M Potassium phosphate, pH 7.0, containing 20% glycerol and 0.2% Emulgen 913
COMMENTS: Spectra of reduced, CO-reduced, and oxidized forms; spectral changes during conversion of P-450 to P-420 by the addition of KSCN
REFERENCE: Y. Yoshida, Y. Aoyama, H. Kumaoka, and S. Kubota, Biochem. Biophys. Res. Comm., 78, 1005 (1977)

663. Cytochrome P-450

SOURCE: Yeast (Saccharomyces cerevisiae)
SOLVENT:
a. 50 mM Potassium phosphate buffer, pH 7.5, containing 0.25% Triton X-100
b. 0.3% Triton X-100
c. 2.0 M KCl and 0.05% Triton X-100

COMMENTS: Spectra of reduced, CO-reduced, and oxidized forms in solvent a; oxidized form in solvents b and c
REFERENCE: Y. Yoshida and H. Kumaoka, J. Biochem. Tokyo, 78, 785 (1975)

664. Cytochrome P-450

SOURCE: Beef adrenal cortex
SOLVENT: 50 mM Phosphate buffer, pH 7.0
COMMENTS: Spectra of reduced and oxidized forms
REFERENCE: H. Schleyer, D. Y. Cooper, and O. Rosenthal, J. Biol. Chem., 247, 6103 (1972)

665. Cytochrome P-450 scc

SOURCE: Beef adrenal cortex
SOLVENT: 10 mM Potassium phosphate buffer, pH 7.4, containing 0.1 mM EDTA

COMMENTS: 1. Spectral changes due to the addition of the following substances:
a. Emulgen 913 (0-0.5 mg/ml)
b. Ethylene glycol (0-50%)
c. Dimyristoyl-L-α-phosphatidylcholine (0-88.8 mM)
d. Amphenone B (0-0.25 mM)
e. Aminoglutethimide (0-1.25 mM)
f. 20α-OH-cholesterol (0-2.34μM)
g. Adrenodoxin in the presence and absence of Emulgen 913
h. Metyrapone (0-1.0 mM)
2. Spectral changes due to temperature (20°C-45°C) and due to pH (pH 7.0-pH 10.5)
NOTE: Aminoglutethimide is α-ethyl-α-(p-aminophenyl) glutarimide; Amphenone B is 3,3-bis (p-aminophenyl)-2-butanone; metyrapone is 2-methyl-1,2-di-3-pyridyl-1-propanone

REFERENCE: T. Kido, M. Arakawa, and T. Kimura, J. Biol. Chem., 254, 8377 (1979)

666. Cytochrome P-450 scc

SOURCE: Beef adrenal cortex mitochondria
COMMENTS: Spectra of cytochrome P-450 scc with egg phosphatidylcholine vesicles or cholesterol vesicles or alone
REFERENCE: D. W. Seybert, J. R. Lancaster, Jr., J. D. Lambeth, and H. Kamin, J. Biol. Chem., 254, 12088 (1979)

667. Cytochrome P-450 (High spin type)

SOURCE: Beef adrenal cortex
SOLVENT: 0.1 M Phosphate buffer, pH 7.0
COMMENTS: a. Spectra of reduced, CO-reduced, and oxidized forms
b. Oxidized form plus 20 α-hydroxycholesterol
c. Spectral change on standing at 0°C for 24 hours and 48 hours
REFERENCE: N. Ando and S. Horie, J. Biochem. Tokyo, 72, 583 (1972)

668. Cytochrome P-450

SOURCE: Beef adrenal mitochondria
SOLVENT: 100 mM Potassium phosphate buffer, pH 7.0, containing 5 mM cysteine and 1 mM EDTA
REFERENCE: M. Shikita and P. F. Hall, J. Biol. Chem., 248, 5598 (1973)

669. Cytochrome P-450

SOURCE:	Beef adrenal cortex
SOLVENT:	50 mM Potassium phosphate buffer containing 100 mM DTT and 100 mM EDTA
COMMENTS:	Spectra of deoxycorticosterone treated, reduced, and CO-reduced
REFERENCE:	M. Katagiri and S. Takemori, Molec. Cell Biochem., 2, 137 (1973)

670. Cytochrome P-450

SOURCE:	Beef adrenal mitochondria
SOLVENT:	50 mM Potassium phosphate, pH 7.0, containing 0.1 mM EDTA, 0.1 mM DTT, and 0.01% cholate
COMMENTS:	Spectra of reduced, CO-reduced, and oxidized forms
REFERENCE:	S. Takemori, K. Suhara, S. Hashimoto, M. Hashimoto, H. Sato, T. Gomi, and M. Katagiri, Biochem. Biophys. Res. Comm., 63, 588 (1975)

671. Cytochrome P-450

SOURCE:	Beef adrenal mitochondria
SOLVENT:	10 mM Sodium phosphate buffer, pH 7.4, containing 0.1 mM EDTA and 20% glycerol
COMMENTS:	Spectra of reduced, CO-reduced, and oxidized forms
REFERENCE:	H -P. Wang and T. Kimura, J. Biol. Chem., 251, 6068 (1978)

672. Cytochrome P-450 (Large)

SOURCE:	Beef adrenals
SOLVENT:	50 mM Potassium phosphate buffer, pH 7.0, containing 0.1 mM EDTA
COMMENTS:	Spectra of reduced, CO-reduced, and oxidized forms
REFERENCE:	B. E. Tilley, M. Watanuki, and P. F. Hall, Biochim. Biophys. Acta, 493, 260 (1977)

673. Cytochrome P-450

SOURCE:	Beef adrenocortical microsomes
SOLVENT:	100 mM Potassium phosphate buffer, pH 7.2, containing 0.1 mM EDTA, 0.1 mM DTT, and 20% glycerol
COMMENTS:	Spectra of reduced, CO-reduced, oxidized, and in presence of 17 α-hydroxyprogesterone
REFERENCE:	S. Kominami, S. Mori, and S. Takemori, FEBS Lett., 89, 215 (1978)

674. Cytochrome P-450 scc

SOURCE: Beef adrenal cortex mitochondria
SOLVENT: 10 mM Potassium phosphate buffer, pH 7.4, 100 μM Emulgen 913, EDTA
COMMENTS: Spectral changes upon addition of adrenodoxin
REFERENCE: T. Kido and T. Kimura, J. Biol. Chem., 254, 11806 (1979)

675. Cytochrome P-450 scc

SOURCE: Beef adrenal cortex mitochondria
SOLVENT: 50 mM Potassium phosphate, pH 7.4, containing 0.1 mM EDTA, 0.1 mM DTT and 0.01% sodium cholate
COMMENTS: Spectra of reduced, CO-reduced, oxidized, and cholesterol complexed
REFERENCE: M. Katagiri, S. Takemori, E. Itagaki, K. Suhara, T. Gomi, and H. Sato, Adv. Exp. Med. Biol., 74, 281 (1976)

676. Cytochrome P-450-11β

SOURCE: Beef adrenal cortex mitochondria
SOLVENT: 50 mM Potassium phosphate, pH 7.3, containing 0.1 mM EDTA, 0.1 mM DTT, 0.3% sodium cholate, and 0.3% Tween 20
COMMENTS: Spectra of oxidized form in absence of deoxycorticosterone, reduced, CO-reduced, and oxidized in presence of deoxycorticosterone, and in presence of metyrapone
REFERENCE: S. Takemori, H. Sato, T. Gomi, K. Suhara, and M. Katagiri, Biochem. Biophys. Res. Comm., 67, 1151 (1975); H. Sato, N. Ashida, K. Suhara, E. Itagaki, S. Takemori, and M. Katagiri, Arch. Biochem. Biophys., 190, 307 (1978)

677. Cytochrome P-450

SOURCE: Pseudomonas putida (ATCC 17453)
SOLVENT: 50 mM Potassium phosphate buffer, pH 7.4, containing 0.1 M KCl and 5 mM camphor
COMMENTS: Spectra of reduced, CO-reduced and oxidized forms, and NO-complexes of reduced and oxidized forms
REFERENCE: J. A. Peterson and B. W. Griffin, Arch. Biochem. Biophys., 151, 427 (1972); R. E. Ebel, D. H. O'Keeffe, and J. A. Peterson, FEBS Lett., 55, 198 (1975)

678. Cytochrome P-450 (oxygenated)

SOURCE: Pseudomonas putida (ATCC 17453)
SOLVENT: 50 mM Potassium phosphate buffer, pH 7.4, containing 0.1 M KCl
COMMENTS: Spectrum of oxygenated form and spectra of interconversions to CO-form and back to oxygenated form
REFERENCE: J. A. Peterson, Y. Ishimura, and B. W. Griffin, Arch. Biochem. Biophys., 149, 197 (1972)

679. Cytochrome P-450

SOURCE: Pseudomonas putida (ATCC 17453)
SOLVENT: 20 mM Potassium phosphate buffer, pH 7.4, containing 0.1 M KCl
COMMENTS: Spectra of pyridine complexes of reduced and oxidized forms
REFERENCE: B. W. Griffin and J. A. Peterson, Biochemistry, 11, 4740 (1972)

680. Cytochrome P-450-cam

SOURCE: Pseudomonas putida, strain PpG 786
SOLVENT: 50 mM Potassium phosphate buffer, pH 7.4, containing 0.300 mM D(+) camphor
COMMENTS: Spectra of reduced, oxidized, and oxygenated forms
REFERENCE: C. A. Tyson, J. D. Lipscomb, and I. C. Gunsalus, J. Biol. Chem., 247, 5777 (1972)

681. Cytochrome P-450 cam-CO

SOURCE: Pseudomonas putida (ATCC 17453)
COMMENTS: Spectral changes due to photo-dissociation in presence and absence of camphor
REFERENCE: U. Shimada, T. Izuka, R. Ueno, and Y. Ishimura, FEBS Lett., 98, 290 (1979); T. Iizuka, H. Shimada, R. Ueno, and Y. Ishimura in Cytochrome Oxidase, 1978 (T. E. King, Y. Orii, B. Chance, and K. Okunuki, eds.), 9 (1979), Elsevier, New York

682. Cytochrome P-450-cam

SOURCE: Pseudomonas putida
SOLVENT: 50 mM Phosphate buffer, pH 7
COMMENTS: Spectra of reduced, CO-reduced, and oxidized in the presence of camphor; P-450-cam in the presence of phenylimidazole or mercaptoethanol

REFERENCE: C. A. Yu, I. C. Gunsalus, M. Katagiri, K. Suhara, and S. Takemori, J. Biol. Chem., 249, 94 (1974)

683. Cytochrome P-450

SOURCE: Pseudomonas putida
SOLVENT: 0.1 M Morpholinopropane sulfonate-Tris, pH 7.4, containing 0.1 M KCl
COMMENTS: Spectra of complex of 1-oxyl-2,2,6,6-tetramethyl-4-piperidinyl nicotinate (RNO) with oxidized cytochrome, in presence and absence of camphor, of reduced cytochrome-RNO complex, and CO-reduced cytochrome-RNO complex
REFERENCE: B. W. Griffin, S. M. Smith, and J. A. Peterson, Arch. Biochem. Biophys., 160, 323 (1974)

684. Cytochrome P-450

SOURCE: Pseudomonas putida
COMMENTS: Spectral changes on addition of N-ethylmaleimide to ferric-cytochrome P-450-camphor complex
REFERENCE: K. Dus, Adv. Exp. Biol. Med., 58, 287 (1975)

685. Cytochrome P-450-cam

SOURCE: Pseudomonas putida
SOLVENT: 0.1 M Phosphate buffer, pH 7, plus ethylene glycol (1:1 V/V)
COMMENTS: Spectra of oxy-complexes at -40°C and -20°C in the presence and absence of substrate
REFERENCE: L. Eisenstein, P. Debey, and P. Douzou, Biochem. Biophys. Res. Comm., 77, 1377 (1977)

686. Cytochrome P-450

SOURCE: Bacillus megaterium
SOLVENT: 50 mM Phosphate buffer, pH 7.4
COMMENTS: Spectra of reduced, CO-reduced, and oxidized forms
REFERENCE: A. Berg, J. A. Gustafsson, M. Ingelman-Sundberg, and K. Carlstrom, J. Biol. Chem., 251, 2831 (1976); A. Berg, M. Ingelman-Sundberg, and J. A. Gustafsson, J. Biol. Chem., 254, 5264 (1979)

687. Cytochrome P-450 meg

SOURCE: Bacillus megaterium
SOLVENT: 50 mM Phosphate buffer, pH 7.4
COMMENTS: Spectra of oxidized, dithionite-reduced, and dithionite-reduced-CO forms

REFERENCE: A. Berg, M. Ingleman-Sundberg, and J. A. Gustafsson, Acta Biol. Med. Germ., 38, 333 (1979)

688. Cytochrome P-450

SOURCE: Bacterial
SOLVENT: 1:1 Mixture of 100 mM sodium/potassium phosphate buffer, pH 7.3, and ethylene glycol containing camphor
COMMENTS: Spectral changes due to photoreduction at 7°C and -29°C
REFERENCE: C. Bonfils, J. L. Saldana, P. Derby, P. Maurel, C. Balny, and P. Douzou, Biochimie, 61, 681 (1979)

689. Cytochrome P-450

SOURCE: Bacterial
SOLVENT: Mixture of phosphate buffer, KCl, camphor, corn oil and Span 65
COMMENTS: Spectra of reduced, CO-reduced, and oxidized forms at -20°C
REFERENCE: P. Douzou, P. Debey, and F. Franks, Biochim. Biophys. Acta, 523, 1 (1978)

690. Cytochrome P-450

SOURCE: Rhizobium japonica Strain 505
SOLVENT: 50 mM Potassium phosphate, pH 6.8
COMMENTS:
1. Spectra of oxidized components a, b, and c
2. Spectra of oxidized P-450 b from bacteroids and oxidized P-450 b_1 from anaerobic/nitrate-grown rhizobia
3. Spectra of dithionite-reduced, and dithionite-reduced-CO components a, b, and c

REFERENCE: C. A. Appleby and R. M. Daniel, in International Symposium on Oxidases and Related Redox Systems (T. E. King, H. S. Mason, and M. Morrison, eds.) 516 (1973), University Park Press, Baltimore, Md.

691. Cytochrome P-450

SOURCE: Human liver
SOLVENT: 5 mM Potassium phosphate buffer, pH 7.7, containing 20% glycerol, 0.1 mM EDTA, and 0.2% Emulgen 913
COMMENTS: Spectra of oxidized, dithionite-reduced, and dithionite-reduced-CO forms
REFERENCE: P. Wang, P. S. Mason, and F. P. Guengerich, Arch. Biochem. Biophys., 199, 206 (1980)

692. Cytochrome P-450

SOURCE: Human liver
SOLVENT: 20 mM Potassium phosphate, pH 7.25, and 20% glycerol
COMMENTS: Spectra of oxidized, reduced, and reduced-CO forms
REFERENCE: T. Kamataki, M. Sugira, Y. Yamazoe, and R. Kato, Biochem. Pharm., 28, 1993 (1979)

693. Cytochrome P-450

SOURCE: Human liver
SOLVENT: 100 mM Phosphate buffer, pH 7.4, containing 20% (V/V) glycerol
COMMENTS: Spectra of Fe (III) form, Fe (III) form plus ethylpyridine, and Fe (II)-CO form
REFERENCE: P. Beaune, P. Dansette, J. P. Flinois, S. Columelli, D. Mansuy, and J. P. Leroux, Biochem. Biophys. Res. Comm., 88, 826 (1979)

694. Cytochrome P-460

SOURCE: Nitrosomonas europaea
SOLVENT: 50 mM Sodium-potassium phosphate buffer, pH 7.5
COMMENTS: Spectra of reduced, CO-reduced, and oxidized forms in the presence and absence of hydrazine or hydroxylamine; spectral changes on autooxidation of reduced form; spectral changes on conversion of P-460 form to 423nm absorbing form
REFERENCE: R. H. Erickson and A. B. Hooper, Biochim. Biophys. Acta, 275, 231 (1972)

695. Cytochrome c Oxidase

SOLVENT: 100 mM Potassium phosphate, pH 7.4, and 1% Tween
COMMENTS:
1. Spectra of partially reduced cytochrome C oxidase in the presence of formate and after addition of CO
2. Spectral changes after illumination of partially-reduced cytochrome C oxidase-CO

REFERENCE: R. Boelens and R. Wever, Biochim. Biophys. Acta, 547, 296 (1979)

696. Cytochrome c Oxidase

SOURCE: Beef heart
COMMENTS: Spectra of oxidized and reduced (dithionite) starting material, reduced (dithionite) and aged (14 days) material, and oxidized and aged (14 days) material

REFERENCE: M. Morrison in Oxidases and Related Redox Systems (T. E. King, H. S. Mason, and M. Morrison, eds.), 607 (1965), J. Wiley and Sons, Inc., New York

697. Cytochrome Oxidase

SOURCE: Beef heart
COMMENTS:
1. Spectra of protein and sonicated protein (50 minutes): reduced, CO-reduced, and oxidized
2. Effects of cyanide on spectrum of oxidized protein
3. Effect of high pH (pH 11.6) on spectrum

REFERENCE: T. E. King, M. Kuboyama, and S. Takemori, in Oxidases and Related Redox Systems (T. E. King, H. S. Mason, and M. Morrison, eds.), 607 (1965), J. Wiley and Sons, Inc., New York

698. Cytochrome Oxidase

SOURCE: Beef heart
SOLVENT: 0.1 M Sodium phosphate, pH 7.4, and 0.25% Emasol-1130
COMMENTS: Spectra of reduced, CO-reduced, and oxidized forms and cyanide complexes
REFERENCE: M. Kuboyama, F. C. Yong, and T. E. King, J. Biol. Chem., 247, 6375 (1972); F. C. Yong and T. E. King, J. Biol. Chem., 247, 384 (1972)

699. Cytochrome Oxidase

SOURCE: Beef heart
SOLVENT: pH 7.33-12.79
COMMENTS: Low temperature spectra (86°K) of reduced protein
REFERENCE: Y. Orii and T. Iizuka, Biochem. Biophys. Res. Comm., 48, 885 (1972)

700. Cytochrome Oxidase

SOURCE: Beef heart mitochondria
COMMENTS: Spectra of N-bromosuccinimide modified oxidized and reduced protein
REFERENCE: M. London and R. J. Mayer, FEBS Lett., 27, 153 (1972)

701. Cytochrome c Oxidase

SOURCE: Beef heart mitochondria
COMMENTS: Spectra of reduced and oxidized forms
REFERENCE: R. A. Capaldi and H. Hayashi, FEBS Lett., 26, 261 (1972)

702. Cytochrome Oxidase

SOURCE: Beef heart
SOLVENT: 0.3 M Phosphate buffer, pH 7.0, 0.1 M NaCl, 0.1% Tween 20
COMMENTS: Spectra during titration with methylviologen and O_2
REFERENCE: W. R. Heineman, T. Kuwana, and E. R. Hartzell, Biochem. Biophys. Res. Comm., 49, 1 (1972)

703. Cytochrome Oxidase

SOURCE: Beef heart muscle
SOLVENT: 0.5 M Phosphate buffer, pH 7.4-1% cholate
COMMENTS: Spectra of oxidized and reduced forms
REFERENCE: K. A. Markosyan and R. M. Nalbadyan, Biokhimiya (Biochemistry), 43, 899 (1978)

704. Cytochrome c Oxidase

SOURCE: Beef heart
SOLVENT: 0.1 M Potassium phosphate buffer, pH 7.2, 1.0% Tween 80
COMMENTS: Effect of azide on spectrum
REFERENCE: R. Wever, A. O. Muijsers, B. F. Van Gelder, E. P. Bakker, and K. J. H. Van Buren, Biochim. Biophys. Acta, 325,1 (1973)

705. Cytochrome Oxidase

SOLVENT: Beef heart
SOLVENT: 0.1 M Potassium phosphate buffer, pH 7.4, 1% Tween 80
COMMENTS:
1. Spectral changes associated with addition of CO to aged enzyme
2. Spectra of oxidized, mixed valence-CO form, and "oxygenated" form

REFERENCE: C. Greenwood, M. T. Wilson, and M. Brunori, Biochem. J., 137, 205 (1974)

706. Cytochrome Oxidase

SOURCE: Beef heart
SOLVENT: 0.01 M Phosphate buffer, pH 7.4, with and without 0.75% Tween 20
COMMENTS: Spectra of reduced and oxidized forms
REFERENCE: J. A. Volpe and W. S. Caughey, Biochem. Biophys. Res. Comm., 61, 502 (1974)

707. Cytochrome Oxidase

SOURCE: Beef heart
COMMENTS: Reaction mixtures contained various amounts of ascorbate, cyanide, and oxidase
REFERENCE: S. Yoshikawa and Y. Orii, J. Biochem. Tokyo, 76, 271 (1974)

708. Cytochrome Oxidase

SOURCE: Beef heart
SOLVENT: 50 mM Phosphate buffer, pH 7.4, 1.5% sodium cholate
COMMENTS: Spectra of phospholipid-depleted, oxidized, reduced and reduced-CO forms
REFERENCE: C. A. Yu, L. Yu, and T. E. King, J. Biol. Chem., 250, 1383 (1975)

709. Cytochrome c Oxidase

SOURCE: Beef heart
SOLVENT: 0.1 M Potassium phosphate, pH 7.4, 1% cholate
COMMENTS: Spectra of reduced, oxidized, and reduced CO-forms
REFERENCE: G. Steffens and G. Buse, Hoppe-Seyler's Z. Physiol. Chem., 357, 1125 (1976)

710. Cytochrome c Oxidase

SOURCE: Beef heart
SOLVENT: 0.05 M Phosphate buffer, pH 7.5
COMMENTS: Spectra made during anaerobic titration with CO of dithionite-reduced protein
REFERENCE: D. C. Wharton and Q. H. Gibson, J. Biol. Chem., 251, 2861 (1976)

711. Cytochrome Oxidase

SOURCE: Beef heart
COMMENTS:
1. Spectrum of oxygenated compound prepared by NADH method in presence of phenazine methosulfate
2. Spectrum of decay of compound I

REFERENCE: Y. Orii and T. E. King, J. Biol. Chem., 251, 7487

712. Cytochrome Oxidase

SOURCE: Beef heart
SOLVENT: 0.1 M Phosphate buffer, pH 7.4, 1% Tween 80

COMMENTS: Changes in spectrum caused by phosphine, dithionite, EDTA, fluoride, phosphite, hypophosphite, and time after addition of reagent to oxidized protein
REFERENCE: K. P. Kashi and W. Chefurka, Pesticide Biochem. Physiol., 6, 350 (1976)

713. Cytochrome Oxidase

SOURCE: Beef heart
SOLVENT: 0.1 M Potassium phosphate buffer, pH 7.4, containing 0.5% Tween 80
COMMENTS: Spectra of reduced, oxidized, and partially reduced cyanide and formate complexes
REFERENCE: G. T. Babcock, L. E. Vickery, and G. Palmer, J. Biol. Chem., 251, 7907 (1976)

714. Cytochrome Oxidase

SOURCE: Beef heart
SOLVENT: 50 mM Phosphate buffer, pH 7.4, containing 1% sodium cholate and 0.1 M ammonium sulfate
COMMENTS: Spectra of reduced (dithionite), oxidized, ferrocyanide, and ferrocyanide plus ascorbate treated protein
REFERENCE: C. A. Yu and L. Yu, Biochem. Biophys. Res. Comm., 70, 1115 (1976)

715. Cytochrome Oxidase

SOURCE: Beef heart mitochondria
SOLVENT: 0.1 M Potassium phosphate, pH 6.4, 0.5% Emasol 4130
COMMENTS: Changes in spectrum after incubation with hexacyanoferrate (II)/(III); spectra of reduced and oxidized forms
REFERENCE: B. Lanne, B. G. Malmstrom, and T. Vanngard, FEBS Lett., 77, 146 (1977)

716. Cytochrome c Oxidase

SOURCE: Beef heart
COMMENTS: Spectra of reduced (cyanide form), reduced form, mixed-valence CO-form, reduced (formate form), partially reduced (cyanide form) and oxidized (fluoride form)
REFERENCE: A. J. Thompson, T. Brittain, C. Greenwood, and J. P. Springall, Biochem. J., 165, 327 (1977)

717. Cytochrome Oxidase

SOURCE: Beef heart
SOLVENT: 0.075 M Sodium phosphate, pH 7.4, containing 0.2 M ascorbic acid and 0.25% Tween 20
COMMENTS: Effect of cyanide on the Soret band
REFERENCE: S. Yoshikawa, T. Ueno, and T. Sai, J. Biochem. Tokyo, 82, 1361 (1977)

718. Cytochrome Oxidase

SOURCE: Beef heart mitochondria
SOLVENT: 0.1 M Bicarbonate buffer, pH 8.0, containing 0.5% deoxycholate
COMMENTS: Spectra of reduced, CO-reduced, and oxidized forms
REFERENCE: T. Ozawa, M. Tada, and H. Suzuki, in Cytochrome Oxidase (T. E. King, Y. Orii, B. Chance, and K. Okunuki, eds.), 39 (1979), Elsevier/North Holland Biomed. Press

719. Cytochrome Oxidase

SOURCE: Beef heart
COMMENTS: Spectra of oxidized, oxidized-formate complex, oxidized-cyanide complex, and reduced forms in Soret region
REFERENCE: G. T. Babcock and I. Salmeen, Biochemistry, 18, 2493 (1979)

720. Cytochrome Oxidase

SOURCE: Locust (Locusta migratoria) flight muscle
COMMENTS: Spectra of oxidized and reduced forms
REFERENCE: H. Weiss, B. Lorenz, and W. Kleinow, FEBS Lett., 25, 49 (1972)

721. Cytochrome c Oxidase

SOURCE: House fly (Musca domestica)
COMMENTS: Spectrum of reduced form
REFERENCE: B. Sacktor, J. Gen Physiol., 35, 397 (1952)

722. Cytochrome Oxidase

SOURCE: Pig leukocytes
COMMENTS: Spectra of reduced, CO-reduced, and oxidized forms
REFERENCE: T. Odajima and I. Yamazaki, Biochim. Biophys. Acta, 284, 368 (1972)

723. Cytochrome c Oxidase

SOURCE:	Rat liver mitochondria
SOLVENT:	0.1 M Potassium phosphate, pH 7.4, containing 0.5% Tween 20
COMMENTS:	Spectra of reduced, reduced-CO, and oxidized forms
REFERENCE:	L. Hochli and C. R. Hackenbrock, Biochemistry, <u>17</u>, 3712 (1978); R. J. Rascati and P. Parsons, J. Biol. Chem., <u>254</u>, 1586 (1979); I. Z. Ades and J. Cascarano, J. Bioenerg. Biomem., <u>9</u>, 237 (1977)

724. Cytochrome c Oxidase

SOURCE:	Rat liver
SOLVENT:	50 mM Sodium phosphate buffer, pH 7.5
COMMENTS:	Spectra of air-oxidized enzyme and dithionite-reduced enzyme
REFERENCE:	T. Nagasawa, H. Nagasawa-Fujimori, and P. C. Heinrich, Eur. J. Biochem., <u>94</u>, 31 (1979)

725. Cytochrome c Oxidase

SOURCE:	Paracoccus denitrificans
SOLVENT:	50 mM Potassium phosphate, pH 7.6, 0.1% Triton X-100
COMMENTS:	Spectra of oxidized and dithionite-reduced-CO form
REFERENCE:	B. Ludwig and G. Schatz, Proc. Natl. Acad. Sci., <u>77</u>, 196 (1980)

726. Cytochrome Oxidase

SOURCE:	Pseudomonas aeruginosa
SOLVENT:	a. 0.25 M Phosphate, pH 7.0, containing 6 M urea b. 0.05 M Tris-HCl, pH 8 c. 0.1 M Phosphate buffer, pH 8.0, and pH 6.5
COMMENTS:	a. Spectra of reduced and oxidized forms b. Spectra of reduced and oxidized forms reconstituted with heme d_1 c. Spectra of reduced and oxidized forms reconstituted with heme a
REFERENCE:	K. E. Hill, and D. C. Wharton, J. Biol. Chem., <u>253</u>, 489 (1978); D. C. Wharton, Methods Enzymol., <u>53D</u>, 646 (1978)

727. Cytochrome Oxidase

SOURCE:	Pseudomonas aeruginosa
SOLVENT:	0.04 M Potassium phosphate buffer, pH 6.9

COMMENTS: Spectra of oxidized, ascorbate-reduced, dithionite-reduced, and CO-reduced forms
REFERENCE: D. Barber, S. R. Parr, and C. Greenwood, Biochem. J., 157, 431 (1976)

728. Cytochrome c-551 Oxidase

SOURCE: Pseudomonas aeruginosa
SOLVENT: 0.1 M Potassium phosphate buffer adjusted to pH 7.0, pH 9.4 and pH 10
COMMENTS: a. Effect of cyanide on spectrum of oxidized apoprotein
b. Changes in spectrum of oxidized-cyanide form caused by ascorbate
c. Spectra after addition of different amounts of cyanide to reduced form
REFERENCE: D. Barber, S. R. Parr, and C. Greenwood, Biochem. J., 175, 239 (1978); T. Walsh, M. K. Johnson, C. Greenwood, D. Barber, J. P. Springall, and A. J. Thomson, Biochem. J., 177, 29 (1979)

729. Cytochrome Oxidase

SOURCE: Pseudomonas aeruginosa
SOLVENT: 0.04 M Potassium phosphate buffer, pH 7.4
COMMENTS: Spectra of reduced form in presence and absence of CO
REFERENCE: S. R. Parr, M. T. Wilson, and C. Greenwood, Biochem. J., 151, 51 (1975)

730. Cytochrome c Oxidase

SOURCE: Human heart mitochondria; mouse heart mitochondria
COMMENTS: Spectra of reduced and oxidized forms
REFERENCE: A. J. Jeffreys and I. W. Craig, FEBS Lett., 77, 151 (1977)

731. Cytochrome c Oxidase

SOURCE: Sweet potato (Ipomoea batatas, Kokei No. 14) roots
SOLVENT: 50 mM Tris-acetate buffer, pH 7.5, containing 0.1% Triton X-100 and 0.1 M KCl
COMMENTS: Spectra of reduced and oxidized forms
REFERENCE: M. Maeshima and T. Asahi, Arch. Biochem. Biophys., 187, 423 (1978)

732. Cytochrome Oxidase

SOURCE: Thiobacillus novellus
SOLVENT: 10 mM Tris-HCl buffer, pH 8.0, containing 0.5% Triton X-100

COMMENTS: Spectra of reduced and oxidized forms
REFERENCE: T. Yamanaka, Y. Fukumori, and K. Fujii in Cytochrome Oxidase (T. E. King, Y. Orii, B. Chance, and K. Okunuki, eds.), 399 (1979), Elsevier/North Holland

733. Cytochrome Oxidase

SOURCE: Neurospora crassa
SOLVENT: a. 0.1 M Phosphate buffer, pH 7.4
b. 0.05 M Tris-HCl, pH 8.0
COMMENTS: Spectra of reduced and oxidized forms
REFERENCE: W. Sebald, W. Machleidt, and J. Otto, Eur. J. Biochem., 38, 311 (1973); S. Werner, Eur. J. Biochem., 79, 103 (1977)

734. Cytochrome c Oxidase

SOURCE: Neurospora crassa
COMMENTS: Spectra of air-oxidized and dithionite-reduced forms
REFERENCE: H. Weiss and H. J. Kolb, Eur. J. Biochem., 99, 139 (1979)

735. Cytochrome c Oxidase

SOURCE: Yeast (Candida utilis)
COMMENTS: Spectra of reduced and oxidized forms
REFERENCE: J. Keyhani and E. Keyhani, Arch. Biochem. Biophys., 167, 588 (1975)

736. Cytochrome Oxidase

SOURCE: Yeast (Saccharomyces cerevisiae)
SOLVENT: Phosphate-Tween 20
COMMENTS: Spectra of different resolved forms in the reduced, reduced-CO, oxidized, and reduced and oxidized treated with cyanide
REFERENCE: S. H. Phan and J. R. Mahler, J. Biol. Chem., 251, 257 (1976)

737. Cytochrome c Peroxidase

SOURCE: Baker's yeast
SOLVENT: pH 4
COMMENTS: Spectra after adding various amounts of hydrogen peroxide
REFERENCE: J. E. Erman and T. Yonetani, Biochim. Biophys. Acta, 393, 343 (1975)

738. Cytochrome c Peroxidase

SOURCE: Baker's yeast
SOLVENT: pH 7
COMMENTS: Spectra of reduced, reduced-NO, oxidized, oxidized-NO and oxidized-nitrite forms; spectra of manganese-protoporphyrin IX-peroxidase with manganese in the +2, +3 states and as the Mn-NO complex
REFERENCE: T. Yonetani, H. Yamamoto, J. E. Erman, J. S. Leigh, Jr., and G. H. Reed, J. Biol. Chem., 247, 2447 (1972)

739. Cytochrome c Peroxidase

SOURCE: Baker's yeast
SOLVENT: pH 6.0, pH 8.0
COMMENTS: Changes in spectrum upon addition of ferrocyanide, and ferrocyanide-hydrogen peroxide
REFERENCE: H. G. Jordi and J. E. Erman, Biochemistry, 13, 3734 (1974); H. G. Jordi and J. E. Erman, Biochemistry, 13, 3741 (1974)

740. Cytochrome c Peroxidase

SOURCE: Baker's yeast
SOLVENT: 0.1 M Phosphate buffer, pH 7.0
COMMENTS: Changes in the spectrum upon addition of bromine, O-benzoylhydroxylamine, and O-benzoylhydroxylamine and ferrocyanide
REFERENCE: F. W. Coulson and T. Yonetani, Biochemistry, 14, 2389 (1975)

741. Cytochrome c Peroxidase

SOURCE: Baker's yeast (Saccharomyces cerevisiae)
SOLVENT: Sodium phosphate buffer, pH 6.0
COMMENTS: Spectra of reduced, reduced-CO, oxidized, oxidized-cyanide complex, and oxidized-fluoride complex forms
REFERENCE: G. Sievers, Biochim. Biophys. Acta, 536, 212 (1978)

742. Cytochrome c Peroxidase

SOURCE: Baker's yeast (Saccharomyces cerevisiae)
SOLVENT:
a. 0.1 M Sodium acetate buffer, pH 5.3
b. 0.2 M Potassium phosphate buffer, pH 8.1
c. 0.2 M Potassium phosphate buffer, in deuterium oxide, pH = 8.1

COMMENTS: Spectra of oxidized forms and fluoride complex of 645 nm and 620 nm species

REFERENCE: R. A. Mathews and J. B. Wittenberg, J. Biol. Chem., 254, 5991 (1979)

743. Cytochrome c Peroxidase

SOURCE: Thiobacillus novellus (Starkey)
SOLVENT: 70 mM Tris-HCl buffer, pH 8.5
COMMENTS: Spectra of reduced, reduced-CO, reduced-cyanide, oxidized, and oxidized-cyanide forms
REFERENCE: T. Yamanaka, Biochim. Biophys. Acta, 275, 74 (1972)

744. Cytochrome c Peroxidase

SOURCE: Pseudomonas aeruginosa
SOLVENT: a. Sodium phosphate buffer, pH 6.0
b. 0.1 N NaOH
c. Tris-HCl pH 7.2
d. pH 2.5
COMMENTS: Spectra of reduced and oxidized forms; spectral changes of reduced and oxidized forms due to pH
REFERENCE: R. Soininen and N. Ellfolk, Acta Chem. Scand., 27, 35 (1973); R. Soininen and N. Kalkkinen, Acta Chem. Scand., 31B, 604 (1977)

745. Cytochrome c Peroxidase

SOURCE: Pseudomonas aeruginosa
SOLVENT: Sodium phosphate buffer, pH 6.0
COMMENTS: Spectra of ferrous, ferrous-carbonyl, and ferric forms
REFERENCE: M. Ronnberg, N. Ellfolk, and R. Soininen, 578, 392 (1979)

746. Cytochrome c Peroxidase

SOURCE: Pseudomonas denitrificans
SOLVENT: 0.1 M Phosphate buffer, pH 6.8
COMMENTS: Spectra of oxidized and dithionite-reduced forms
REFERENCE: A. F. W. Coulson and R. I. C. Oliver, Biochem. J., 181, 159 (1979)

747. Cytochrome c Peroxidase (CCP)

SOLVENT: 0.05 M Sodium acetate, pH 5.0
COMMENTS: Spectrum of peroxide compound of CCP formed on addition of indole 3-acetic acid
REFERENCE: H. Yamazaki and I. Yamazaki, Arch. Biochem. Biophys., 154, 147 (1973)

748. Cytochrome c Peroxidase

SOLVENT: 0.1 M Potassium phosphate buffer, pH 6.0
COMMENTS: Spectra of protoporphyrin monomethyl ester-peroxidase complex, protohememonomethylester complex and its reduced, reduced-CO and oxidized forms, and its fluoride, azide, and cyanide derivatives
REFERENCE: T. Asakura and T. Yonetani, J. Biol. Chem., 247, 2278 (1972)

749. Cytochrome b5 Reductase

SOURCE: Calf liver
SOLVENT: 0.1 M Tris-acetate buffer, pH 8.1
COMMENTS: Spectra of reductase and trypsin-modified reductase at 2°C
REFERENCE: P. Strittmatter, R. E. Barry, and D. Corcoran, J. Biol. Chem., 247, 2768 (1972)

750. Cytochrome b5 Reductase

SOURCE: Calf liver
SOLVENT: 0.02 M Tris-acetate, pH 8.1, 0.2 mM EDTA and 0.4% deoxycholate
COMMENTS: Spectra of oxidized and reduced forms of lysosomal-extracted and detergent-extracted forms
REFERENCE: L. Spatz and P. Strittmatter, J. Biol. Chem., 248, 793 (1973)

751. Cytochrome b5 Reductase (NADH)

SOURCE: Rabbit liver microsomes
SOLVENT: 50 mM Tris-HCl buffer, pH 8.0, containing 1 mM EDTA and 0.1 M KCl
COMMENTS: Spectra of detergent-soluble form and lysosome-soluble form
REFERENCE: K. Mihara and R. Sato, J. Biochem. Tokyo, 78, 1057 (1975)

752. Cytochrome b5 Reductase (NADPH)

SOURCE: Pig liver microsomes
SOLVENT: 0.05 M Potassium phosphate, pH 7.7, containing 0.8 M KCl and 0.1 mM EDTA
COMMENTS: Spectra of oxidized, reduced, and half-reduced forms
REFERENCE: R. A. Prough and B. S. S. Masters, Arch. Biochem. Biophys., 165, 263 (1974)

753. Cytochrome c Reductase (NADH)

SOURCE: Pseudomonas arvilla C-1
SOLVENT: 50 mM Tris-HCl buffer, pH 6.8
REFERENCE: M. Yamaguch and H. Fujisawa, J. Biol. Chem., 253, 8848 (1978); H. Fujisawa, M. Yamaguchi, and T. Yamauchi, Adv. Exp. Med. Biol., 74, 118 (1976)

754. Cytochrome c Reductase (hydroxylamine)

SOURCE: Nitrosomonas europaea
SOLVENT: 0.1 M Phosphate buffer, pH 7.0
COMMENTS: Spectra of reduced and oxidized forms
REFERENCE: T. Yamanaka, M. Shinra, and K. Kimura, Biosystems, 9, 155 (1977)

755. Cytochrome c Reductase (NADPH)

SOURCE: Rat liver microsomes
SOLVENT: 0.05 M Potassium phosphate, pH 7.7, containing 0.1 mM EDTA and 0.02% sodium azide
REFERENCE: B. S. S. Masters, R. Prough, and H. Kamin, Biochemistry, 14, 607 (1975)

756. Cytochrome c Reductase (NADPH)

SOURCE: Rabbit liver microsomes
SOLVENT: 0.1 M Potassium phosphate buffer, pH 7.7, containing 1 mM EDTA
COMMENTS: Spectra of trypsin solubilized reduced, half-reduced, and oxidized forms
REFERENCE: T. Iyanagi and H. S. Mason, Biochemistry, 12, 2297 (1973)

757. Cytochrome c Reductase (NADPH)

SOURCE: Vitamin C-3 deficient chick kidney mitochondria
SOLVENT: 10 mM 4-morpholine propane sulfonic acid, pH 7.4
COMMENTS: Spectra of reduced and oxidized forms
REFERENCE: J. A. Kulkoski, J. L. Weber, and J. G. Ghazarian, Arch. Biochem. Biophys., 192, 539 (1979)

758. Cytochrome c Reductase (succinate)

SOLVENT: 0.1 M Phosphate buffer, containing 0.5% cholate
COMMENTS: Spectra of reduced (ascorbate, succinate, and dithionite) and oxidized forms at 23°C and -160°C
REFERENCE: C. A. Yu, L. Yu, and T. E. King, Biochim. Biophys. Acta, 267, 300 (1972)

759. Cytochrome c Reductase, NADPH-

SOURCE: Ale yeast (Saccharomyces cerevisiae)
SOLVENT: 50 mM Potassium phosphate, pH 7.6, 0.1 mM EDTA
COMMENTS: Spectra of enzyme, enzyme plus ferrocyanide, enzyme plus NADPH in air, and dithionite-reduced
REFERENCE: E. Tryon, M. C. Cress, M. Hamada, and S. A. Kuby, Arch. Biochem. Biophys., 197, 104 (1979)

760. Cytochrome c Reductase

SOURCE: Neurospora crassa
COMMENTS: Spectra of air-oxidized and dithionite-reduced forms (room temperature and low temperature)
REFERENCE: H. Weiss and H. J. Kalb, Eur. J. Biochem., 99, 139 (1979)

761. Cytochrome c Reductase, NADPH-

SOURCE: Southern armyworm (Spodoptera eridania)
SOLVENT: 0.05 M Tris-HCl, pH 8, containing 1 mM EDTA and 20% glycerol
COMMENTS: Spectra of oxidized and NADPH-reduced forms
REFERENCE: D. L. Crankshaw, K. Hetnarski, and C. F. Wilkinson, Biochem. J., 181, 593 (1979)

762. Cytochrome o Reductase, NADH-

SOURCE: Vitreoscilla
SOLVENT: a. 0.02 M Sodium phosphate buffer, pH 7.5
b. 0.1 M Sodium phosphate buffer, pH 7.0
COMMENTS: 1. Spectra of oxidized and NADH-reduced forms (solvent a, under argon)
2. Spectral changes during ferricyanide titration of dithionite-reduced form in solvent b under argon
3. Spectral changes during NADH titration of oxidized form in solvent b under argon
REFERENCE: V. Gonzales-Prevatt and D. A. Webster, J. Biol. Chem., 255, 1478 (1980)

763. Cytochrome P-450 Reductase (NADPH)

SOURCE: Rabbit liver microsomes
SOLVENT: 0.1 M Potassium phosphate buffer, pH 7.7
COMMENTS: Spectral changes caused by ferricyanide, NADPH, and dithionite
REFERENCE: T. Iyanagi, N. Makino, and H. S. Mason, Biochemistry, 13, 1701 (1974)

764. Cytochrome P-450 Reductase (NADPH)

SOURCE: Rabbit liver microsomes
SOLVENT: 0.1 M Potassium phosphate buffer, pH 7.0, containing 20% glycerol
COMMENTS: Spectra of detergent-solubilized, reduced, half reduced, and oxidized forms; spectral changes caused by ferricyanide
REFERENCE: T. Iyanagi, F. K. Anan, Y. Imai, and H. S. Mason, Biochemistry, 17, 2224 (1978)

765. Cytochrome P-450 Reductase (NADPH)

SOURCE: Rabbit liver microsomes
SOLVENT: 0.05 mM Tris buffer, pH 7.7, containing 10% glycerol and 5.0 mM EDTA
COMMENTS: Spectral changes upon photo-reduction and reoxidation with NADP and O_2, and upon denaturation with 4 M guanidine
REFERENCE: J. S. French and M. J. Coon, Arch. Biochem. Biophys., 195, 565 (1979)

766. Cytochrome P-450 Reductase (NADPH)

SOURCE: Rat liver microsomes
SOLVENT: 0.3 M Potassium phosphate buffer, pH 7.7
COMMENTS: Spectral changes due to ferricyanide and NADPH
REFERENCE: J. M. Vermilion and M. J. Coon, Biochem. Biophys. Res. Comm., 60, 1315 (1974); J. L. Vermilion and M. J. Coon, J. Biol. Chem., 253, 2694 (1978)

767. Cytochrome P-450 Reductase (NADPH)

SOURCE: Rat liver microsomes
SOLVENT:
a. 0.1 M Phosphate buffer, pH 7.7, containing 10% glycerol, 0.1 mM EDTA, and 0.1% deoxycholate
b. 0.15 M Phosphate buffer, pH 7.4, containing 10% glycerol, 3.0 mM EDTA, and 0.1% deoxycholate

COMMENTS: Spectra of photochemical reduction of FMN-depleted reductase and reoxidation by NADP under anaerobic conditions, and spectral changes during transfer of electrons from FMN-depleted enzyme to FMN under anaerobic conditions
REFERENCE: J. L. Vermilion and M. J. Coon, J. Biol. Chem., 253, 8812 (1978)

768. Cytochrome P-450 Reductase (NADPH)

SOURCE: Rat liver
SOLVENT: 0.05 M Tris-HCl, pH 7.7, 20% glycerol, 1 mM EDTA, 0.1 mM DTT, and 0.1% deoxycholate
COMMENTS: Changes in spectrum due to aerobic titration with NADPH and anaerobic titration with NADPH
REFERENCE: J. D. Dignam and H. W. Strobel, Biochem. Biophys. Res. Comm., 63, 845 (1975); J. D. Dignam and H. W. Strobel, Biochemistry, 16, 1116 (1977)

769. Cytochrome P-450 Reductase, NADPH-

SOURCE: Beef adrenal microsomes
SOLVENT: 0.1 M Potassium phosphate buffer, pH 7.4
COMMENTS: Spectra of oxidized and dithionite-reduced forms
REFERENCE: A. Hiwatashi and Y. Ichikawa, Biochim. Biophys. Acta, 580, 44 (1979)

770. Cytochrome c (P-450) Reductase, NADPH-

SOURCE: Human liver microsomes
SOLVENT: 10 mM Potassium phosphate, pH 7.7, containing NADP and 20% glycerol
REFERENCE: T. Kamataki, M. Sugira, Y. Yamazoe, and R. Kato, Biochem. Pharm., 28, 1993 (1979)

771. Cytochrome c (P-450) Reductase, NADPH-

SOURCE: Pig liver
SOLVENT: 0.1 M Potassium phosphate buffer, pH 7.7, containing 20% glycerol, 0.2 mM EDTA
COMMENTS: Spectral changes during anaerobic titration of oxidized enzyme with dithionite
REFERENCE: Y. Yasukochi, J. A. Peterson, and B. S. S. Masters, J. Biol. Chem., 254, 7047 (1979)

772. Cytochrome c (P-450) Reductase (NADPH)

SOURCE: Pig liver
SOLVENT: 0.05 mM Potassium phosphate buffer, pH 7.7, containing 0.1 mM EDTA, 20% glycerol, and 0.1% sodium deoxycholate
COMMENTS: Changes in spectrum upon aerobic titration with potassium ferricyanide
REFERENCE: Y. Yasukochi and B. S. S. Masters, J. Biol. Chem., 251, 5337 (1976)

773. Cytochrome c (P-450) Reductase

SOURCE: Microsomes from diazinon resistant house flies
SOLVENT: 0.01 M Potassium phosphate buffer, pH 7.7, 0.1 mM EDTA, 20% glycerol, and 0.1% Emulgen 913
COMMENTS: Spectra of oxidized and air stable semiquinone forms
REFERENCE: R. T. Mayer and J. L. Durant, J. Biol. Chem., 254, 756 (1979)

774. Cytochrome Deaminase

SOURCE: Pseudomonas aureofaciens
SOLVENT: 0.01 M Potassium phosphate buffer, pH 8.0
REFERENCE: T. Sakai, T. Yu, K. Taniguchi, and S. Omata, Agr. Biol. Chem., 39, 2015 (1975)

775. Denitrifying Enzyme

SOURCE: Pseudomonas denitrificans
SOLVENT: pH 7
COMMENTS: Spectra of oxidized and reduced forms
REFERENCE: H. Iwasaki and T. Matsubara, J. Biochem. Tokyo, 71, 645 (1972)

776. Deoxynucleic Acid Ligase

SOURCE: Escherichia coli
SOLVENT: 0.01 Potassium phosphate, pH 7.5, containing 1 mM EDTA and 1 mM 2-mercaptoethanol
COMMENTS: Spectra of enzyme and enzyme-AMP complex
REFERENCE: P. Modrich, Y. Anraku, and I. R. Lehman, J. Biol. Chem., 248, 7495 (1973)

777. Deoxynucleic Acid Photoreactivating Enzyme

SOURCE: Streptomyces griseus
REFERENCE: A. P. M. Eker and M. J. Fichtinger-Shepman, Biochim. Biophys. Acta, 378, 54 (1975)

778. Desulforedoxin

SOURCE: Desulfovibrio gigas
SOLVENT: 0.01 M Tris-HCl, pH 7.6
COMMENTS: Spectrum of oxidized form
REFERENCE: I. Moura, M. Bruschi, J. LeGall, G. Moura, and A. V. Xavier, Biochem. Biophys. Res. Comm., 75, 1037 (1977)

779. Desulforubidin

SOURCE: Desulfovibrio desulfuricans, Norway
COMMENTS: Spectra of protein and its CO complex
REFERENCE: M. J. Murphy, L. M. Siegel, H. Kamins, D. V. Dervartanian, J. P. Lee, J. LeGall, and H. D. Peck, Biochem. Biophys. Res. Comm., 54, 82 (1973)

780. Desulfoviridin

SOURCE: Desulfovibrio gigas
REFERENCE: M. J. Murphy, L. M. Siegel, H. Kamin, D. V. Dervartanian, J. P. Lee, J. LeGall, and H. D. Peck, Biochem. Biophys. Res. Comm., 54, 82 (1973)

781. Desulfoviridin I and II

SOURCE: Desulfovibrio vulgaris
SOLVENT: 0.05 M Phosphate buffer, pH 7.0
REFERENCE: Y. Seki, K. Koboyashi, and M. Ishimoto, J. Biochem. Tokyo, 85, 705 (1979)

782. Diamine Oxidase

SOURCE: Pig kidney
SOLVENT: 0.1 M Potassium phosphate buffer, pH 7.4
COMMENTS: Spectra of enzyme, enzyme plus cadaverine (anaerobic) and then plus oxygen; effect of phenylhydrazine or diethyldithiocarbamate, and effect of phenylhydrazine and diethyldithiocarbamate, and then pyridoxal phosphate added
REFERENCE: B. Mondovi, M. T. Costa, A. Finazzi Agro, and G. Rotilio in Pyridoxal Catalysis: Enzymes and Model Systems (E. E. Snell, A. E. Braunstein, E. S. Severin, and Y. M. Torchinsky, eds.), 403 (1968), Interscience, New York

783. Diamine Oxidase

SOURCE: Pig kidney
COMMENTS: Spectra of enzyme, enzyme plus putrescine (anaerobically), and then aeration; enzyme plus phenylhydrazine, and enzyme plus hydroxylamine
REFERENCE: E. V. Goryachenkova, L. I. Shcherbatiuk, and C. I. Zamaraev in Pyridoxal Catalysis: Enzymes and Model Systems (E. E. Snell, A. E. Braunstein, E. S. Severin, and Y. M. Torchinsky, eds.), 391 (1968), Interscience, New York

784. Diaphorase

SOURCE: Pig heart
SOLVENT: 0.03 M Phosphate, pH 7.6, containing 0.0004 M EDTA
COMMENTS: Spectra of oxidized, DPNH reduced, dithionite reduced, and reoxidized forms, and same in 6.5 M urea
REFERENCE: V. Massey, J. Biol. Chem., 235, P47 (1961)

785. Diaphorase

SOURCE: Bluewhite dolphin (Stenella caerulo-alba)
REFERENCE: T. Matsui, C. Shimizu, and F. Matsuura, Bull. Jap. Soc. Sci. Fish., 41, 761 (1975)

786. Dihydrodipicolinate Reductase

SOURCE: Bacillus subtilis PCI 219
SOLVENT: 20 mM Tris-HCl buffer, pH 7.4
COMMENTS: Spectra of reduced and oxidized forms
REFERENCE: K. Kimura, J. Biochem. Tokyo, 77, 405 (1975)

787. Dihydrodipicolinic Acid Reductase

SOURCE: Escherichia coli
SOLVENT: 0.01 M Tris-chloride, pH 8.1
REFERENCE: H. Tamir and C. Gilvarg, J. Biol. Chem., 249, 3034 (1974)

788. Dihydrofolate Reductase I and II

SOURCE: Lactobacillus casei (Amethopterin resistant)
SOLVENT: 0.1 M Potassium phosphate, pH 7.0
REFERENCE: R. B. Dunlap, L. E. Gundersen, and F. M. Huennekens, Biochem. Biophys. Res. Comm., 42, 772 (1971); L. E. Gundersen, R. B. Dunlap, N. G. L. Harding, J. H. Freisheim, F. Otting, and F. M. Huennekens, Biochemistry, 11, 1018 (1972)

789. Dihydrofolate Reductase

SOURCE: Lactobacillus casei (Amethopterin resistant)
SOLVENT: 0.1 M Potassium phosphate buffer, pH 7.0
COMMENTS: Spectra of enzyme and dimethyl-aminonaphthylsulfonyl derivative
REFERENCE: G. A. Vehar, A. V. Reddy, and J. H. Freisheim, Biochemistry, 15, 2512 (1976)

790. Dihydrofolate Reductase II

SOURCE: Lactobacillus casei
COMMENTS: Effect of N-bromosuccinimide on spectrum of protein
REFERENCE: J. K. Liu and R. B. Dunlap, Biochemistry, 13, 1807 (1974)

791. Dihydrofolate Reductase

SOURCE: Escherichia coli B strain MB1428
SOLVENT: 0.1 M NaCl, pH 7.2
REFERENCE: N. J. Greenfield, M. N. Williams, M. Poe, and K. Hoogsteen, Biochemistry, 11, 4706 (1972)

792. Dihydrofolate Reductase

SOURCE: Escherichia coli B MB1428
SOLVENT: 0.05 M NaCl, 0.05 M Tris-HCl, pH 7.20
COMMENTS: Spectra of 1:1 and 2:1 complexes with TPNH
REFERENCE: M. Poe, N. J. Greenfield, and M. N. Williams, J. Biol. Chem., 249, 2710 (1974)

793. Dihydrofolate Reductase

SOURCE: Escherichia coli DR
SOLVENT: 0.04 M Potassium phosphate buffer, pH 7.0
REFERENCE: J. S. Erickson and C. K. Mathews, Biochemistry, 12, 372 (1973)

794. Dihydrofolate Reductase

SOURCE: Phage T4 DR
SOLVENT: 0.04 M Potassium phosphate, pH 7.0
REFERENCE: J. S. Erickson and C. K. Mathews, Biochemistry, 12, 372 (1973)

795. Dihydrofolate Reductase

SOURCE: Streptococcus faecium
SOLVENT: 0.05 M Potassium phosphate, pH 7.5
COMMENTS: Spectra of protein and derivative from reaction with 5,5'-dithiobis (2-nitrobenzoic acid)
REFERENCE: P. E. Warwick and J. H. Freisheim, Biochemistry, 14, 664 (1975)

796. Dihydrofolate Reductase

SOURCE: Mouse leukemia cells (L1210)
SOLVENT: pH 7.0

COMMENTS: Spectra of protein and its complexes with NADPH, $NADP^+$, acid-modified NADPH, dihydrofolate, and amethopterin
REFERENCE: V. G. Neef and F. M. Huennekens, Arch. Biochem. Biophys., 171, 435 (1975)

797. Dihydrofolate Reductase

SOURCE: Chicken liver
SOLVENT: 0.2 M Potassium phosphate, pH 7.8
COMMENTS: Spectra of pH 7.4 and pH 8.4 fractions
REFERENCE: B. T. Kaufman and V. F. Kemerer, Arch. Biochem. Biophys., 179, 420 (1977)

798. Dihydrolipoyl Transacetylase

SOURCE: Escherichia coli
SOLVENT: pH 11.2
COMMENTS: Spectra of protein and iodinated derivatives
REFERENCE: E. R. Schwartz and L. J. Reed, J. Biol. Chem., 243, 639 (1968)

799. Dihydroorotate Dehydrogenase

SOURCE: Rat liver mitochondria
SOLVENT:
a. 0.05 M Sodium phosphate buffer, pH 7.5, 0.05% Triton X-100
b. 0.1% Sodium dodecylsulfate
c. 0.02% Triton X-100
REFERENCE: H. J. Forman and J. Kennedy, Arch. Biochem. Biophys., 191, 23 (1978)

800. Dihydroorotate Oxidase

SOURCE: Escherichia coli K12
REFERENCE: D. Karibian and P. Couchoud, Biochim. Biophys. Acta, 364, 218 (1974)

801. Dihydropteridine Reductase

SOURCE: Beef liver
SOLVENT: 0.1 M Potassium phosphate, pH 7.3
COMMENTS: Spectra of protein and NADH complex
REFERENCE: H. Hasegawa, J. Biochem. Tokyo, 81, 169 (1977)

802. 3,4-Dihydroxyphenylacetate 2,3-Dioxygenase

SOURCE: Pseudomonas ovalis
SOLVENT: 25 mM Tris-HCl buffer, pH 7.5
REFERENCE: M. Ono-Kamimoto, J. Biochem. Tokyo, 74, 1049 (1973)

803. Dihydroxyphenylalanine Decarboxylase

SOURCE: Pig kidney
SOLVENT: a. 0.1 M Potassium phosphate buffer, pH 6.8
b. 0.1 N NaOH
COMMENTS: Spectra of enzyme in solvents a and b, and after treatment with sodium borohydride in solvents a and b, and after reaction with hydroxylamine, and α-methyldihydroxyphenylalanine
REFERENCE: C. Borri-Voltattorni, A. Minelli, and C. Turano, FEBS Lett., 17, 231 (1975)

804. 3,4-Dihydroxyphenylalanine Decarboxylase

SOURCE: Pig kidney
SOLVENT: a. Potassium phosphate buffer, pH 6
b. Triethanolamine-HCl buffer, pH 8.6
REFERENCE: C. B. Voltattorni, A. Minelli, P. Vecchini, A. Fiori, and C. Turano, Eur. J. Biochem., 93, 181 (1979)

805. Dihydroxyphenylalanine Decarboxylase

SOURCE: Pig kidney
SOLVENT: Tris-HCl buffer, pH 8.5
COMMENTS: Changes in spectrum of enzyme upon addition of 5-hydroxy-L-tryptophan, 5-hydroxy-tryptamine, and DL-m-tyrosine
REFERENCE: A. Fiori, C. Turano, C. Borri-Voltattorni, A. Minelli, and M. Codini, FEBS Lett., 54, 122 (1975)

806. Dihydroxyphenylalanine Decarboxylase

SOURCE: Pig kidney
SOLVENT: Potassium phosphate buffer, pH 6.8
COMMENTS: Spectral changes due to the addition of L-α-methyl-α-hydrazino-3,4-dihydroxyphenyl propionic acid (HMD), and pyridoxal phosphate plus HMD
REFERENCE: C. Borri-Voltattorni, A. Minelli, and P. Borri, FEBS Lett., 75, 277 (1977)

807. Dihydroxyphenylalanine Decarboxylase

SOURCE: Guinea pig kidneys
SOLVENT: 0.1 M Potassium phosphate buffer, pH 7.1
REFERENCE: K. Srinivasan and J. Awapara, Biochim. Biophys. Acta, 526, 597 (1978)

808. Dihydroxyphenylalanine Decarboxylase

SOLVENT: 0.05 M Tris-HCl, pH 5.75
COMMENTS: Spectral changes upon reaction with amino-oxyacetate
REFERENCE: R. A. John, A. Charteris, and L. J. Fowler, Biochem. J., 171, 771 (1978)

809. Dipeptidase, Cobalt

SOURCE: Pig kidney
REFERENCE: D. J. Armstrong, S. K. Mukhopadhyay, and B. J. Campbell, Biochemistry, 13, 1745 (1974)

810. Diphosphopyridine Nucleotide (reduced) Peroxidase

SOURCE: Streptococcus faecalis, 10Cl
SOLVENT: 0.05 M Potassium phosphate buffer, pH 7.0, containing 0.3 mM EDTA
COMMENTS:
1. Spectral changes during anaerobic titration with dithionite
2. Spectral changes during anaerobic titration with DPNH and then air-reoxidation
3. Spectral changes on reaction with phenylmercuric acetate

REFERENCE: M. I. Dolin, J. Biol. Chem., 250, 310 (1975)

811. Dopamine β-Monooxygenase

SOURCE: Beef adrenal medulla
SOLVENT: 20 mM Potassium phosphate, pH 7.0
COMMENTS: Spectral changes upon addition of bathocuproine disulfonate
REFERENCE: T. Skotland and T. L. Jones, Eur. J. Biochem., 94, 145 (1979)

812. Egg Albumin

SOURCE: Chicken
COMMENTS: Effect of N-bromosuccinimide, urea, and urea and N-bromosuccinimide on spectra of native and performic acid-oxidized protein
REFERENCE: F. V. Shmakova and E. D. Kaverzneva, Biokhimiya (Biochemistry), 34, 587 (1969)

813. Ekkrinosiderophilin

SOURCE: Human milk
REFERENCE: A. L. Schade, C. Pallavicini, and U. Weismann, Protides of Biol. Fluids, 16, 619 (1969)

814. Elastase

SOLVENT: Water and 20% glycerol
REFERENCE: L. Visser and E. R. Blout, Biochemistry, 10, 741 (1971)

815. Elastase II

SOURCE: Pig pancreas
SOLVENT: 0.15 M Borate buffer, pH 8.7; 0.2 M NaOH; 0.05 M HCl
REFERENCE: W. Ardelt, Biochim. Biophys. Acta, 341, 318 (1974)

816. Elastase

SOURCE: Pseudomonas aeruginosa
SOLVENT: Glycinate buffer, pH 10
REFERENCE: K. Morihara, H. Tsuzuki, T. Oka, H. Inoue, and M. Ebata, J. Biol. Chem., 240, 3295 (1965)

817. Electron-transferring Flavoprotein

SOURCE: Peptostreptococcus elsdenii
SOLVENT: 0.05 M Potassium phosphate, pH 6.0
COMMENTS: Change in spectrum of flavoprotein by bleaching with reduced D-lactate dehydrogenase or NADH
REFERENCE: H. L. Brochman and W. A. Wood, J. Bacteriol., 124, 1447 (1975)

818. Electron-transfer Flavoprotein

SOURCE: Pig liver mitochondria
SOLVENT: 0.02 M Potassium phosphate buffer, pH 7.6
REFERENCE: C. L. Hall and H. Kamin, J. Biol. Chem., 250, 3476 (1975)

819. Endonuclease

SOURCE: Acrocyclindrium sp.
REFERENCE: I. Suhara and M. Yoneda, J. Biochem. Tokyo, 73, 647 (1973)

820. Endotoxin

SOURCE: Microcystis aeruginosa NRC-1
COMMENTS: Spectra before and after treatment with 1.5 M HCl
REFERENCE: P. Rabin and A. Darbe, Biochem. Soc. Trans., 3, 428 (1975)

821. Enkephalin, Leu-

SOURCE: Commercial
SOLVENT: 1 mM Tris-HCl, pH 7.5, with and without DTT
COMMENTS: Effect of storage for 50 days on spectrum
REFERENCE: Z. Vogel, T. Miron, M. Altstein, and M. Wilchek, Biochem. Biophys. Res. Comm., 85, 226 (1978)

822. Enkephalin, Met^5-

SOURCE: Synthetic
SOLVENT: pH 11.5, pH 5.75, and trifluoroethylene
REFERENCE: M. A. Sprites, R. W. Schwarts, W. L. Mattice, and D. H. Coy, Biochem. Biophys. Res. Comm., 81, 602 (1978)

823. (D-Ala^2-met^5) Enkephalin-tyr-N-(2-nitro-4-azidophenyl) ethylenediamine

SOLVENT: 50% ethanol
COMMENTS: Spectra before and after photolysis
REFERENCE: T. T. Lee, R. E. Williams, and C. F. Fox, J. Biol. Chem., 254, 11787 (1979)

824. 2-Enoate Reductase

SOURCE: Clostridium kluyveri
COMMENTS: Spectra of oxidized and NADH reduced forms
REFERENCE: W. Tischer, J. Bader, and H. Simon, Eur. J. Biochem., 97, 103 (1979)

825. Entner-Doudoroff Enzyme

SOURCE: Pseudomonas fluorescens
SOLVENT: 10 mM Potassium phosphate buffer, pH 7.0
REFERENCE: D. Lebmann, K. L. Schimz, and G. Kurz, Eur. J. Biochem., 59, 545 (1975)

826. Enzyme, Mercuric Ion Reducing

SOURCE: Escherichia coli W 2252
REFERENCE: K. Izaki, Y. Tashiro, and T. Funaba, J. Biochem. Tokyo, 75, 591 (1974)

827. Enzyme, Milk Clotting

SOURCE: Irpex lacteus Fr.
REFERENCE: M. Kawai, Agr. Biol. Chem., 35, 1517 (1971)

828. Enzyme, Transfer of Amino Acid

SOURCE: Rat liver
SOLVENT: 0.1 M KCl - 0.02 M Tris, pH 7.6
REFERENCE: M. Takanami, Biochim. Biophys. Acta, 51, 85 (1961)

829. Epoxide Hydrase

SOURCE: Rat liver
SOLVENT: 5 mM Sodium phosphate buffer, pH 7.0
REFERENCE: P. Bentley, F. Oesch, and A. Tsugita, FEBS Lett., 59, 296 (1975)

830. Erythrocruorin

SOURCE: Annelid worm (Cirraformia grandis)
COMMENTS: Spectra of oxy-, deoxy-, and CO forms
REFERENCE: J. B. Swaney and I. M. Katz, Arch. Biochem. Biophys., 147, 475 (1971)

831. Erythrocruorin, 10 S Subunit

SOURCE: Earthworm (Lumbricus terrestris)
SOLVENT: 0.1 M Phosphate, pH 6.8
COMMENTS: Spectra of oxy- and deoxy- forms
REFERENCE: M. M. David and E. Daniel, FEBS Lett., 32, 293 (1973)

832. Erythrocruorin, 60 S

SOURCE: Earthworm (Lumbridae sp)
SOLVENT: 0.1 M Phosphate buffer, pH 6.8
COMMENTS: Spectrum of oxy-form
REFERENCE: M. M. David and E. Daniel, J. Mol. Biol., 87, 89 (1974)

833. Erythrocruorin

SOURCE: Planorbis corneus (snail)
COMMENTS: Spectra of oxy-, deoxy-, and CO forms
REFERENCE: E. J. Wood and L. J. Mosby, Biochem. J., 149, 437 (1975)

834. Erythrocruorin, Ferric

SOURCE: Earthworm (Lumbricus terrestris)
SOLVENT: a. Phosphate buffer, pH 7.0
b. Glycine-NaOH buffer, pH 10.9

COMMENTS: In a: Spectra of met- and hemichrome derivatives
In b: Spectrum of hemichrome derivative
REFERENCE: F. Ascoli, M. R. Rossi-Fanelli, E. Chiancone, P. Vecchini, and E. Antonini, J. Mol. Biol., 119, 191 (1978)

835. Erythrocuprein

COMMENTS: Spectra before and after heating to 75° and 100°
REFERENCE: M. A. Simonyah and R. M. Nalbandyan, Biokhimiya (Biochemistry), 40, 616 (1975)

836. Erythrocuprein

SOURCE: Beef
SOLVENT: 5-10 mM Potassium phosphate buffer, pH 7.2
COMMENTS: Spectra of protein (at room temperature and 77°K), apo-protein, Cu^{2+}-apo-protein chelates, Zn^{2+}-apo-protein chelates, reconstituted protein, and apo-protein in guanidine-HCl at pH 11.7
REFERENCE: U. Weser, G. Barth, C. Djerassi, H. J. Hartmann, P. Krauss, G. Voelcker, W. Voelter, and W. Voetsch, Biochim. Biophys. Acta, 278, 28 (1972); U. Weser, Str. and Bonding, 17, 1 (1973)

837. Erythrocuprein (cupro-zinc protein)

SOURCE: Human erythrocytes
SOLVENT: Water
COMMENTS: Spectra of protein and apo-protein
REFERENCE: W. H. Bannister, B. G. Dalgleish, J. V. Bannister, and E. S. Wood, Int. J. Biochem., 3, 560 (1972); U. Weser, Str. and Bonding, 17, 1 (1973)

838. Erythrocytes

SOURCE: Human blood
COMMENTS: Spectra of inclusion in situ and bulk inclusions
REFERENCE: P. Fessas, D. Loukopoulos, and B. Thorell, Blood, 25, 105 (1965)

839. Erythrocytes

SOURCE: Human blood
COMMENTS: Spectra of thalassemic red cells and ghosts with and without inclusions
REFERENCE: E. A. Rachmilewitz and B. Thorell, Blood, 39, 794 (1972)

840. Erythropoietin

SOURCE: Human urine
REFERENCE: J. Espada, A. A. Longton, and M. Dorado, Biochim. Biophys. Acta, 285, 427 (1972)

841. Esterase

SOURCE: Clay loam
COMMENTS: Spectra of untreated enzyme, acid hydrolyzed enzyme, and hyaluronidase-treated enzyme
REFERENCE: T. Satyanarayana and L. W. Getzin, Biochemistry, 12, 1566 (1970)

842. Esterase

SOURCE: Pig liver
SOLVENT: 5 mM Phosphate buffer, pH 8.0
COMMENTS: Spectra of native and maleylated protein
REFERENCE: W. Junge, K. Krisch, and H. Hollandt, Eur. J. Biochem., 43, 379 (1974)

843. Extract

SOURCE: Soybean nodules (Glycine max var Kanrich) from Rhizobium japonica infected plants
SOLVENT: 0.01 M Phosphate buffer, pH 6.8
COMMENTS: Spectra of nodules prepared under CO, argon or air, with polyclar AT, from plants 16-75 days old
REFERENCE: D. T. Nash and H. M. Schulman, Biochem. Biophys. Res. Comm., 68, 781 (1976)

844. Extract

SOURCE: Trema cannabina nodules
SOLVENT: 0.1 M Potassium phosphate, pH 6.8
COMMENTS: Spectra of air-oxidized and dithionite-reduced forms
REFERENCE: D. R. Coventry, M. J. Trinick, and C. A. Appleby, Biochim. Biophys. Acta, 420, 105 (1976)

845. Extract and Fractions

SOURCE: Wheat germ
SOLVENT: 0.1 M Tris-HCl, pH 7.9
COMMENTS: Spectra of extracts and Polymin p precipates
REFERENCE: J. J. Jendrisak and R. R. Burgess, Biochemistry, 14, 4639 (1975)

846. Extract

SOURCE: Rhodopseudomonas spheroides mutants
SOLVENT: a. Ether
b. 0.1 M Tris-Cl buffer, pH 6.8
COMMENTS: Spectra of bacteriochlorophyll extract and same extract plus pigment-protein complex before and after gel filtration
REFERENCE: W. R. Richards, R. B. Wallace, M. S. Tsao, and E. Ho, Biochemistry, 14, 5554 (1975)

847. α-Factor

SOURCE: Saccharomyces cerevisiae
SOLVENT: a. 0.05 M Ammonium acetate, pH 5.5
b. pH 4.7
c. pH 13, 0.1 N NaOH
REFERENCE: W. Duntze, D. Stotzler, E. Bucking-Throm, and S. Kalbitzer, Eur. J. Biochem., 35, 357 (1973)

848. Fl (ATPase)

SOURCE: Beef heart mitochondria
COMMENTS: Spectra after reaction with 7-chloro-4-nitrobenz-2-oxa-1,3-diazole, after ammonium sulfate precipitation and through Sephadex G25, with DTT, and after precipitation and through Sephadex G25
REFERENCE: S. J. Ferguson, W. J. Lloyd, and G. K. Radda, FEBS Lett., 38, 234 (1974)

849. Factor III^{lac}

SOURCE: Staphylococcus aureus
SOLVENT: a. 10 mM Phosphate buffer, pH 7.4, containing 0.1 mM EDTA and 0.5 mM DTT
b. 10 mM Phosphate buffer, pH 6.5, containing 5.7 M guanidinium hydrochloride
REFERENCE: J. B. Hays, R. D. Simoni, and S. Roseman, J. Biol. Chem., 248, 941 (1973)

850. Factor F_{420}

SOURCE: Methanobacterium str. M.O.H.
SOLVENT: a. 0.1 M NaOH
b. 0.1 M HCl
c. 0.05 M Phosphate buffer, pH 7.3
COMMENTS: Spectra before and after reduction with sodium borohydride
REFERENCE: P. Cheeseman, A. Toms-Wood, and R. S. Wolfe, J. Bacteriol., 112, 527 (1972)

851. Factor, Calcium-sensitizing and Factor, Inhibitory

SOURCE: Rabbit skeletal muscle
SOLVENT: 40 mM Tris-HCl buffer, pH 7.4
REFERENCE: M. C. Schaum, S. V. Perry, and W. Hacker, Biochem. J., 126, 237 (1972)

852. Factor, Conversion

SOURCE: Human lung
REFERENCE: H. Nishihara, S. Ishikawa, K. Shinkai, and H. Akedo, Biochim. Biophys. Acta, 302, 429 (1973)

853. Factor, Direct Hemolytic

SOURCE: Central Asian cobra venom
SOLVENT: Water and 7 M urea
REFERENCE: L. Y. Yukelson, E. Sadykov, and V. M. Sorokin, Biokhimiya (Biochemistry), 39, 681 (1974)

854. Factor, Initiation, M1

SOURCE: Rabbit reticulocytes
SOLVENT: 100 mM Tris-HCl, pH 8.0
REFERENCE: W. C. Merrick and W. F. Anderson, J. Biol. Chem., 250, 1197 (1975)

855. Factor, Intrinsic and Nonintrinsic

SOURCE: Pig gastric juice
SOLVENT: 0.05 M Potassium phosphate, pH 7.5, containing 0.075 M NaCl
REFERENCE: R. H. Allen and C. S. Mehlman, J. Biol. Chem., 248, 3670 (1973)

856. Factor, Intrinsic

SOURCE: Human gastric juice
SOLVENT: 0.05 M Potassium phosphate, pH 7.5, containing 0.75 M NaCl
REFERENCE: R. H. Allen and C. S. Mehlman, J. Biol. Chem., 248, 3660 (1973)

857. Factor, Intrinsic

SOURCE: Human gastric juice
SOLVENT: Water and NaOH solutions
REFERENCE: K. Visuri and R. Grasbeck, Biochim. Biophys. Acta, 310, 508 (1973)

858. Factor, Intrinsic

SOURCE: Human gastric juice
COMMENTS: Spectrum of protein-cyanocobalamin complex
REFERENCE: J. M. Christensen, E. Hipps, H. Oleson, M. Rye, E. Haber, L. Lee, and J. Thomsen, Biochim. Biophys. Acta, 303, 319 (1973)

859. Factor, Mating

SOURCE: Saccharomyces cerevisiae
SOLVENT: a. 0.05 M Sodium carbonate buffer, pH 10
b. 0.05 M Acetate buffer, pH 4.0
REFERENCE: T. Tanaka, H. Kita, T. Murakami, and K. Narita, J. Biochem. Tokyo, 82, 1681 (1977)

860. Factor, Permeability; Factor, Permeability Increasing

SOURCE: Guinea pig skin, lymph node cells and kidneys
REFERENCE: T. Inderbitzen, A. Keel, and G. Blumenthal, Int. Arch. Allergy, 29, 417 (1966)

861. Factor, Transcription

SOURCE: Bacillus subtilis, SP01-infected
SOLVENT: 1 mM Sodium phosphate, pH 6.3
REFERENCE: G. G. Johnson and E. P. Guiduschek, J. Biol. Chem., 247, 3571 (1972)

862. Fatty Acyl Coenzyme A Dehydrogenase

SOURCE: Pig liver mitochondria
COMMENTS: Spectra of "general" fatty acyl Coenzyme A dehydrogenase and the acetoacetyl-coenzyme A complex
REFERENCE: M. Benecky, T. Y. Li, J. Schmidt, F. Freeman, K. L. Watters, and J. McFarland, Biochemistry, 18, 3471 (1979)

863. Fatty Acid Synthetase

SOURCE: Pig liver
SOLVENT: 0.2 M Potassium phosphate buffer, pH 7.4, containing 1 mM EDTA
REFERENCE: I. C. Kim, C. J. Unkefer, and W. C. Deal, Arch. Biochem. Biophys., 178, 475 (1977)

864. Ferredoxin

SOURCE: Beef adrenals
SOLVENT: 50 mM Potassium phosphate buffer, pH 7.4

COMMENTS: Spectra of reconstituted and apo-protein
REFERENCE: K. Suhara, K. Kanayama, S. Takemori, and M. Katagiri, Biochim. Biophys. Acta, 336, 309 (1974)

865. Ferredoxin

SOURCE: Beef adrenals
SOLVENT: pH 7.4
COMMENTS: Spectra of oxidized and reduced selenium-replaced protein
REFERENCE: K. Mukai, J. J. Huang, and T. Kimura, Biochim. Biophys. Acta, 336, 427 (1974)

866. Ferredoxin

SOURCE: Beef adrenals
COMMENTS: Spectra of reconstituted, plus DTT, in the presence of DL-dihydrolipoate, and in presence of reduced glutathione
REFERENCE: T. Taniguchi and T. Kimura, Biochim. Biophys. Acta, 364, 284 (1974)

867. Ferredoxin

SOURCE: Beef adrenals
SOLVENT: 50 mM Sodium potassium phosphate buffer, pH 7.4
COMMENTS: Changes in spectrum due to 0.5% sodium dodecylsulphate after 2 minutes and 6 hours
REFERENCE: S. S. Mardanyan, N. A. Grigoryan, and R. M. Nalbandyan, Biochem. J., 153, 725 (1976)

868. Ferredoxin

SOURCE: Rat liver mitochondria
COMMENTS: Spectra of reduced and oxidized forms
REFERENCE: Y. Atsuta and K. Okuda, J. Biol. Chem., 253, 4653 (1978)

869. Ferredoxin

SOURCE: Beef brain mitochondria
SOLVENT: 50 mM Tris, pH 7.5, containing 0.5 M KCl and 0.5 mM DTT
COMMENTS: Spectra of reduced and oxidized forms
REFERENCE: H. Oftebro, F. C. Stormer, and J. I. Pedersen, J. Biol. Chem., 254, 4331 (1979)

870. Ferredoxin

SOURCE: Pigweed (Amaranthus edulis)
SOLVENT: 0.3 M Tris-HCl buffer, pH 7.3, containing 0.54 M NaCl
REFERENCE: P. Schurmann, B. B. Buchanan, and H. Matsubara, Biochim. Biophys. Acta, 223, 450 (1972)

871. Ferredoxin

SOURCE: Cucumber (Cucumis sativus) leaves
COMMENTS: Spectra of reduced and oxidized forms
REFERENCE: E. N. Mukhin and V. K. Gins, Biokhimiya (Biochemistry), 37, 846 (1972)

872. Ferredoxin

SOURCE: Fern (Plystichum munitum)
SOLVENT: 0.3 M Tris-HCl buffer, pH 7.3, containing 0.54 M NaCl
REFERENCE: P. Schurmann, B. B. Buchanan, and H. Matsubara, Biochim. Biophys. Acta, 223, 450 (1973)

873. Ferredoxin

SOURCE: Parsley
COMMENTS: Spectra of reduced and oxidized forms
REFERENCE: K. Garbett, R. D. Gillard, P. F. Knowles, and J. E. Stangroom, Nature, 215, 824 (1967)

874. Ferredoxin

SOURCE: Pea
COMMENTS: Spectral changes during reduction with NADP reductase, 0-1 hour
REFERENCE: I. N. Gogotov and T. V. Laurinavichene, Biokhimiya (Biochemistry), 42, 1285 (1977)

875. Ferredoxins I and II

SOURCE: Pokewood (Phytolacca americana)
SOLVENT: 0.02 M Phosphate buffer, pH 7.0
REFERENCE: S. Takaichi, K. Makino, A. Iwasaki, and K. Suzuki, J. Biochem. Tokyo, 83, 1151 (1978)

876. Ferredoxin

SOURCE: Spinach
COMMENTS: Changes in spectrum due to 8 M urea, and pH 11 with time (0-730 minutes)

REFERENCE: K. Garbett, R. D. Gillard, P. F. Knowles, and J. E. Stangroom, Nature, 215, 824 (1967)

877. Ferredoxin

SOURCE: Spinach
SOLVENT: 0.15 M Phosphate buffer, pH 7.3
COMMENTS: Spectrum of dithionite reduced form
REFERENCE: J. C. Sutherland, I. Salmeen, A. S. K. Sun, and M. P. Klein, Biochim. Biophys. Acta, 263, 550 (1972)

878. Ferredoxin

SOURCE: Spinach
SOLVENT: 1:1 D_2O:ethylene glycol, 77°K
COMMENTS: Spectra of reduced and oxidized forms
REFERENCE: J. Rawlins, O. Siiman, and H. B. Gray, Proc. Nat. Acad. Sci., 71, 125 (1974)

879. Ferredoxin

SOURCE: Spinach
SOLVENT: 5 mM Phosphate buffer, pH 7.5, with or without 1 M NaCL
COMMENTS: Spectra of acetylated and reversed (native) forms
REFERENCE: H. Matsubara, K. Wada, and R. Masaki, Adv. Exp. Med. Biol., 74, 1 (1976)

880. Ferredoxin

SOURCE: Spinach
REFERENCE: R. Thauer, H. Schirrmacher, W. Schymanski, and P. Schonheit, Z. Naturforsch., 33c, 495 (1978)

881. Ferredoxin

SOURCE: Spinach
SOLVENT: 10 mM Tris-HCl buffer, pH 7.4
COMMENTS: Spectral changes during denaturation at room temperature in absence of 1 M NaCl
REFERENCE: H. Hasumi, S. Nakamura, K. Koga, and H. Yoshizumi, Biochem. Biophys. Res. Comm., 87, 1095 (1979)

882. Ferredoxin

SOURCE: Spinach
SOLVENT: 10 mM Tris-HCl, pH 7.4, containing 1 mM sodium pyrophosphate and 2 mM NaCl

COMMENTS: Changes in the spectrum of the protein caused by p-mercuribenzoate
REFERENCE: H. Hasumi and S. Nakamura, J. Biochem. Tokyo, 84, 707 (1978)

883. Ferredoxin

SOURCE: Spinach
COMMENTS: Spectral changes due to X-irradiation
REFERENCE: V. Krasnobajeu, Biophysik, 10, 213 (1973)

884. Ferredoxin

SOURCE: Spinach
SOLVENT: 5 mM Sodium phosphate buffer, pH 7.5
COMMENTS: Changes in the spectrum due to changes in NaCl concentration
REFERENCE: R. Masaki, K. Wada, and H. Matsubara, J. Biochem. Tokyo, 81, 1 (1977)

885. Ferredoxin

SOURCE: Wheat (Triticum aestivum)
SOLVENT: 0.02 M Tris-HCl buffer, pH 7.5, containing 0.33 M NaCl
REFERENCE: M. Shin, Z. Yokoyama, A. Abe, and H. Fukasawa, J. Biochem. Tokyo, 85, 1075 (1979)

886. Ferredoxin

SOURCE: Anacystis nidulans
REFERENCE: R. M. Smillie, Plant. Physiol., 40, 1124 (1965)

887. Ferredoxin

SOURCE: Aphanothece sacrum (Saringar) okada
SOLVENT: 0.1 M Tris-HCl buffer, pH 7.5
REFERENCE: K. Wada, H. Kagamiyama, M. Shin, and H. Matsubara, J. Biochem. Tokyo, 76, 1217 (1974)

888. Ferredoxin

SOURCE: Aphanothece sacrum
SOLVENT: 0.1 M Tris-HCl buffer, pH 7.5
COMMENTS: Spectra of major and minor components
REFERENCE: T. Hase, K. Wada, and H. Matsubara, J. Biochem. Tokyo, 78, 605 (1975)

889. Ferredoxin

SOURCE: Azotobacter vinelandii
SOLVENT: 0.02 M N-2-Hydroxyethyl piperazine-N'-2-ethanesulfonic acid
COMMENTS: Spectra of reduced and oxidized forms
REFERENCE: D. C. Yoch, J. R. Benemann, R. C. Valentine, and D. I. Arnon, Proc. Nat. Acad. Sci., 64, 1404 (1969)

890. Ferredoxins I and II

SOURCE: Azotobacter vinelandii
SOLVENT: 0.02 M and 0.05 M Phosphate buffer, pH 7.3
REFERENCE: D. C. Yoch and D. I. Arnon, J. Biol. Chem., 247, 4514 (1972)

891. Ferredoxin

SOURCE: Azotobacter vinelandii
SOLVENT: 0.1 M Tris-HCl buffer, pH 7.41
COMMENTS: Spectra of oxidized (ferricyanide) and reduced (dithionite) forms
REFERENCE: W. V. Sweeney, J. C. Rabinowitz, and D. C. Yoch, J. Biol. Chem., 250, 7842 (1975)

892. Ferredoxin I

SOURCE: Azotobacter vinelandii
COMMENTS: Spectra of oxidized, photoreduced, and dithionite reduced forms
REFERENCE: G. Scherings, H. Haaker, and C. Veeger, Eur. J. Biochem., 77, 621 (1977)

893. Ferredoxin I

SOURCE: Azotobacter vinelandii
SOLVENT: 0.1 M Tris-HCl buffer, pH 8.73
COMMENTS: Spectra of isolated form and hydrogen/hydrogenase form
REFERENCE: W. V. Sweeney, J. C. Rabinowitz, and D. C. Yoch, J. Biol. Chem., 250, 7842 (1975)

894. Ferredoxins I and II

SOURCE: Bacillus polymyxa
SOLVENT: 50 mM Potassium phosphate buffer, pH 7.0
REFERENCE: N. A. Stombaugh, R. H. Burris, and W. H. Orme-Johnson, J. Biol. Chem., 248, 7951 (1973)

895. Ferredoxins I and II

SOURCE: Bacillus polymyxa
SOLVENT: 0.05 M Tris-HCl buffer, pH 7.4
REFERENCE: D. C. Yoch, Arch. Biochem. Biophys., 158, 633 (1973)

896. Ferredoxin

SOURCE: Bacillus stearothermophilus
COMMENTS: Spectra of reduced and oxidized forms
REFERENCE: P. J. Stephens, A. J. Thomson, J. B. R. Dunn, T. A. Keiderling, J. Rawlings, K. K. Rao, and D. O. Hall, Biochemistry, 17, 4770 (1978); R. N. Mullinger, R. Cammack, K. K. Rao, D. O. Hall, D. P. E. Dickson, C. E. Johnson, J. D. Rush, and A. Simopoulos, Biochem. J., 151, 75 (1975)

897. Ferredoxin

SOURCE: Bacillus stearothermophilus
COMMENTS: Spectra of oxidized and dithionite-reduced forms
REFERENCE: R. Cammack, D. O. Hall, R. N. Mullinger, and K. K. Rao, in Enzymes and Proteins from Thermophilic Microorganisms (H. Zuber, ed.), 307 (1976), Birkhauser Verlag, Basel.

898. Ferredoxin

SOURCE: Ceratophyllum demersum
REFERENCE: E. P. Oganezova and R. M. Nalbandyan, Biokhimiya (Biochemistry), 41, 651 (1976)

899. Ferredoxin

SOURCE: Chlorobium thiosulfatophilum
SOLVENT: 0.30 M Tris buffer, pH 7.3, containing 0.54 M NaCl
REFERENCE: B. B. Buchanan, H. Matsubara, and M. Evans, Biochim. Biophys. Acta, 189, 46 (1969)

900. Ferredoxin

SOURCE: Clostridium acidi-urici
COMMENTS: Spectra of native and potassium ferricyanide modified
REFERENCE: W. V. Sweeney, A. J. Bearden, and J. C. Rabinowitz, Biochem. Biophys. Res. Comm., 59, 188 (1974)

901. Ferredoxin and (LEU^2)-Ferredoxin

SOURCE: Clostridium acidi-urici and synthesis
SOLVENT: 0.01 M Tris-HCl buffer, pH 7.4
REFERENCE: E. T. Lode, C. L. Murray, W. V. Sweeney, and J. C. Rabinowitz, Proc. Nat. Acad. Sci., 71, 1361 (1974)

902. Ferredoxin, (LEU^2)-Ferredoxin, (LYS^1)-Ferredoxin and (TRP^2)-Ferredoxin

SOURCE: Clostridium acidi-urici and synthesis
SOLVENT: 0.01 M Tris-Chloride buffer, pH 7.4
REFERENCE: E. T. Lode, C. L. Murray, and J. C. Rabinowitz, J. Biol. Chem., 251, 1675 (1976)

903. Ferredoxin

SOURCE: Clostridium acidi-urici, 9α Type; Clostridium cylindrosporum, HCl type
REFERENCE: A. B. Champion and J. C. Rabinowitz, J. Bacteriol., 132, 1003 (1977)

904. Ferredoxin , azo-

SOURCE: Clostridium pasteurianum W5
SOLVENT: 0.05 M Tris-HCl buffer, pH 7.5
COMMENTS: Spectra of reduced, physiologically oxidized (ATP), and oxygen oxidized forms
REFERENCE: M. Walker and L. E. Mortenson, Biochem. Biophys. Res. Comm., 54, 669 (1973)

905. Ferredoxin

SOURCE: Clostridium pasteurianum
SOLVENT: 0.05 M Tris-HCl, pH 7.5, 0.2 M NaCl
COMMENTS: Spectra of reduced and oxidized forms
REFERENCE: W. G. Zumft and L. E. Mortenson, Eur. J. Biochem., 35, 401 (1973)

906. Ferredoxin

SOURCE: Clostridium pasteurianum
SOLVENT: 0.15 M Tris-HCl, pH 7.3
REFERENCE: P. W. Johnson and E. Canale-Parola, Arch. Mikrobiol., 89, 341 (1973)

907. Ferredoxin

SOURCE: Clostridium pasteurianum
COMMENTS: Spectra of reduced and oxidized forms

REFERENCE: P. J. Stephens, A. J. Thomson, J. B. R. Dunn, T. A. Keiderling, J. Rawlings, K. K. Rao, and D. O. Hall, Biochemistry, 17, 4770 (1978)

908. Ferredoxin

SOURCE: Clostridium pasteurianum
REFERENCE: P. Schonheit, C. Wascher, and R. K. Thauer, FEBS Lett., 89, 219 (1978)

909. Ferredoxin

SOURCE: Clostridium perfringens
COMMENTS: Spectra of reduced and oxidized forms
REFERENCE: S. Seki, M. Hagiwara, K. Kudo, and M. Ishimoto, J. Biochem. Tokyo, 85, 833 (1979)

910. Ferredoxin

SOURCE: Clostridium thermoaceticum
SOLVENT: a. 0.1 M Tris-HCl, pH 7.6, and pH 8.4
b. 80% Hexamethyl phosphoric triamide
COMMENTS: Spectra in (a) of reduced and oxidized forms; spectral changes in (b) upon addition of benzene-thiol
REFERENCE: S. S. Yang, L. G. Jungdahl, and J. LeGall, J. Bacteriol., 130, 1048 (1977)

911. Ferredoxins I and II

SOURCE: Corynebacterium autotrophicum
COMMENTS: Spectra of reduced and oxidized forms
REFERENCE: H. Berndt, D. J. Lowe, and M. G. Yates, Eur. J. Biochem., 86, 133 (1978)

912. Ferredoxin

SOURCE: Desulfuromonas acetoxidans
SOLVENT: 50 mM Tris-HCl, pH 7.6
COMMENTS: Spectra of reduced, oxidized, denatured, and reconstituted forms
REFERENCE: I. Probst, J. J. G. Moura, I. Moura, M. Bruschi, and J. LeGall, Biochim. Biophys. Acta, 502, 38 (1978)

913. Ferredoxins I, I', and II

SOURCE: Desulfovibrio gigas
COMMENTS: Spectra of reduced and oxidized forms

REFERENCE: M. Bruschi, E. C. Hatchikian, J. LeGall, J. J. G. Moura, and A. V. Xavier, Biochim. Biophys. Acta, 449, 275 (1976)

914. Ferredoxins I and II

SOURCE: Equisetum arvense
SOLVENT: 0.05 M Tris-HCl buffer, pH 7.5, containing 0.35 M NaCl
REFERENCE: T. Hase, K. Wada, and H. Matsubara, J. Biochem. Tokyo, 82, 277 (1977)

915. Ferredoxin

SOURCE: Escherichia coli
SOLVENT: pH 7.2
COMMENTS: Spectra of reduced and oxidized forms
REFERENCE: H. E. Knoell and J. Knappe, Eur. J. Biochem., 50, 245 (1974)

916. Ferredoxin

SOURCE: Halobacter halobium
COMMENTS: Spectra of reduced and oxidized forms
REFERENCE: L. Kerscher and D. Oesterhelt, FEBS Lett., 67, 320 (1976); L. Kerscher, D. Oesterhelt, R. Cammack, and D. O. Hall, Eur. J. Biochem., 71, 101 (1976)

917. Ferredoxin

SOURCE: Halobacterium
SOLVENT: 10 mM Sodium phosphate, pH 7.3, 4.3 M NaCl
COMMENTS: Spectra of holoferredoxin, apoferredoxin, dithionite-reduced ferredoxin, and reoxidized ferredoxin
REFERENCE: A. Gafni and M. M. Werber, Arch. Biochem. Biophys., 196, 363 (1979)

918. Ferredoxin

SOURCE: Nicotiana glutinosa leaves
SOLVENT: 0.05 M Tris-HCl, pH 7.5
COMMENTS: Spectra of reduced and oxidized forms
REFERENCE: P. Kwanyuen and S. G. Wildman, Biochim. Biophys. Acta, 405, 167 (1975)

919. Ferredoxins I and II

SOURCE: Nostoc verrucosum
SOLVENT: 0.02 M Tris-HCl buffer, pH 7.5, containing 0.33 M NaCl
REFERENCE: M. Shin, M. Sukenobu, R. Oshino, and Y. Kitazume, Biochim. Biophys. Acta, 460, 85 (1977)

920. Ferredoxin

SOURCE: Peptococcus aerogenes
SOLVENT: 0.1 M Potassium phosphate buffer, pH 7.9
COMMENTS: Spectra of reduced and oxidized forms
REFERENCE: N. A. Stombaugh, J. E. Sundquist, R. H. Burris, and W. H. Orme-Johnson, Biochemistry, 15, 2633 (1976)

921. Ferredoxins I and II

SOURCE: Rhodospirillum rubrum, light and dark grown cells
SOLVENT: 0.05 M Potassium phosphate buffer, pH 7.3
COMMENTS: Spectra of reduced and oxidized forms
REFERENCE: K. T. Shanmugam, B. B. Buchanan, and D. I. Arnon, Biochim. Biophys. Acta, 256, 477 (1972)

922. Ferredoxin

SOURCE: Rhodospirillum rubrum, dark grown cells
SOLVENT: 20 mM Phosphate buffer, pH 7.4
REFERENCE: D. C. Yoch, D. I. Arnon, and W. V. Sweeney, J. Biol. Chem., 250, 8330 (1975)

923. Ferredoxin

SOURCE: Ruminococcus albus
REFERENCE: F. J. Tewes and R. K. Thauer, FEMS Microbiol. Lett., 6, 375 (1979)

924. Ferredoxin

SOURCE: Ruminococcus albus
SOLVENT: 0.05 M Potassium phosphate, pH 7.0
COMMENTS: Spectra of reduced and oxidized forms
REFERENCE: T. L. Glass, M. P. Bryant, and M. J. Wolin, J. Bacteriol., 131, 463 (1977)

925. Ferredoxin

SOURCE: Sambucus racemosa

SOLVENT: 0.1 M Tris-HCl, pH 7.5, containing 0.8 M NaCl and 1 mM mercaptoethanol
COMMENTS: Spectra of native and deteriorated protein
REFERENCE: I. Altosaar, B. A. Bohm, and I. E. P. Taylor, Can. J. Biochem., 55, 159 (1977)

926. Ferredoxin

SOURCE: Spirochaeta aurantia
SOLVENT: 0.15 M Tris-HCl, pH 7.3
REFERENCE: P. W. Johnson and E. Canale-Parola, Arch. Mikrobiol., 89, 341 (1973)

927. Ferredoxin

SOURCE: Spirulina maxima
COMMENTS: Spectra of reduced and oxidized forms
REFERENCE: P. J. Stephens, A. J. Thomson, J. B. R. Dunn, T. A. Keiderling, J. Rawlings, K. K. Rao, and D. O. Hall, Biochemistry, 17, 4770 (1978)

928. Ferredoxin

SOURCE: Veillonella alcalescens
COMMENTS: Spectrum of oxidized form
REFERENCE: H. Dalton and J. Zubieta, Biochim. Biophys. Acta, 322, 133 (1973)

929. Ferredoxin

SOURCE: Petunia (Petunia inflata Eries.)
COMMENTS: Spectra of major and minor components
REFERENCE: J. G. Huisman, S. Stapel, and A. O. Muijsers, FEBS Lett., 85, 198 (1978)

930. Ferredoxin Reductase

SOURCE: Rat liver mitochondria
COMMENTS: Spectrum of oxidized form
REFERENCE: Y. Atsuta and K. Okuda, J. Biol. Chem., 253, 4653 (1978)

931. Ferredoxin: CO_2 Oxidoreductase

SOURCE: Clostridium pasteurianum
SOLVENT: 50 mM Imidazole acetate buffer, pH 7.5, containing 10 mM sodium azide and 2 mM DTT
COMMENTS: Spectra of reduced and oxidized forms
REFERENCE: P. A. Scherer and R. K. Thauer, Eur. J. Biochem., 85, 125 (1978)

932. Ferredoxin-NADP$^+$ Reductase

SOURCE: Bacillus polymyxa
SOLVENT: 0.05 M Tris-HCl buffer, pH 7.4
REFERENCE: D. C. Yoch, J. Bacteriol., 116, 384 (1973)

933. Ferredoxin-NADP$^+$ Reductase

SOURCE: Spinach leaves
SOLVENT: 0.05 M Tris-HCl buffer, pH 7.5
COMMENTS: Spectra of enzyme and enzyme-NADP$^+$ complex
REFERENCE: M. Shin, Biochim. Biophys. Acta, 292, 13 (1973)

934. Ferredoxin-NADP$^+$ Reductase

SOURCE: Spinach leaves
SOLVENT: 10 mM Tris-HCl, pH 7.4
COMMENTS: Spectra of forms b, c, and d
REFERENCE: C. Gozzer, G. Zanetti, M. Galliano, G. A. Sacchi, L. Minchiotti, and B. Curts, Biochim. Biophys. Acta, 485, 278 (1977)

935. Ferredoxin-NADP$^+$ Reductase

SOURCE: Spinach
SOLVENT: 10 mM Tris-2-(N-morpholino) ethane sulfonate, pH 7.5
COMMENTS: Effect of polylysine on spectrum
REFERENCE: R. Schneeman and D. W. Krogmann, J. Biol. Chem., 250, 4965 (1975)

936. Ferredoxin-NADP$^+$ Reductase

SOURCE: Spinach chloroplasts
COMMENTS: Spectra of native and dimethylaminonaphthlysulfonyl modified
REFERENCE: G. Zanetti, Biochim. Biophys. Acta, 445, 14 (1976)

937. Ferredoxin Reductase, NADPH-

SOURCE: Rat liver
REFERENCE: J. I. Pedersen and H. K. Godager, Biochim. Biophys. Acta, 525, 28 (1978)

938. Ferredoxin-NADP$^+$ Reductases I and II

SOURCE: Blue-green alga
SOLVENT: 50 mM Tris-HCl buffer, pH 7.5, containing 0.35 M NaCl
REFERENCE: R. Masaki, K. Wada, and H. Matsubara, J. Biochem. Tokyo, 86, 951 (1979)

939. Ferredoxin-NADP$^+$ Reductase

SOURCE: Vicia faba
SOLVENT: 50 mM Tris-HCl, pH 8.0
REFERENCE: K. H. Suss, FEBS Lett., 101, 305 (1979)

940. Ferredoxin Nitrite Reductase

SOURCE: Spinach
SOLVENT: 50 mM Potassium phosphate buffer, pH 7.7
REFERENCE: M. J. Murphy, L. M. Siegel, S. R. Tove, and H. Kamin, Proc. Nat. Acad. Sci., 71, 612 (1974)

941. Ferridin

SOURCE: Azotobacter vinelandii
COMMENTS: Spectra of reduced and oxidized forms
REFERENCE: L. A. Syrtsova, V. A. Yakovlev, and V. F. Rachek, Biokhimiya (Biochemistry), 33, 615 (1968)

942. Ferritin

SOURCE: Octopus vulgaris LAM. hepatopancreas
REFERENCE: G. Nardi, E. O. Muzii and M. Puca, Comp. Biochem. Physiol., 40B, 199 (1971)

943. Ferritin, myco-

SOURCE: Mortierella alpina
REFERENCE: R. F. Bozarth and A. Goenaga, Can. J. Microbiol., 18, 619 (1972)

944. Ferritin, Apo-, Subunits

SOURCE: Horse spleen
SOLVENT: 0.01 M Glycine-HCl, pH 3.0
REFERENCE: B. S. Leach, M. E. May, and W. W. Fish, J. Biol. Chem., 251, 3856 (1976)

945. Ferritin and apo-Ferritin

SOURCE: Horse spleen
SOLVENT: 0.03 M Phosphate buffer, containing 0.1 M NaCl, pH 7.4
REFERENCE: M. M. May and W. W. Fish, Arch. Biochem. Biophys., 190, 720 (1978)

946. Ferroperoxidase

SOURCE: Horseradish
SOLVENT: 50 mM Sodium borate buffer, pH 9.1
COMMENTS: Spectra of protein and protein-ethyl isocyanide complex
REFERENCE: C. Phelps, E. Antonini, and M. Brunori, Biochem. J., 128, 377 (1972)

947. α-Fetoprotein

SOURCE: Beef
SOLVENT: Water and 0.1 N NaOH
REFERENCE: S. Aliau, J. Marti, and J. Moretti, Biochimie, 60, 663 (1978)

948. Fibrinogen

SOURCE: Beef blood
SOLVENT: a. 25 mM Borate buffer, pH 7.5
b. 0.1 M Phosphate buffer, pH 6, containing 7 M urea
COMMENTS: Spectra of protein and N-acetylimidazole treated protein before and after addition of hydroxylamine in solvent a; spectra of diethylpyrocarbonate treated and untreated protein in solvent b
REFERENCE: N. M. C. Kaye and P. Jolles, Molec. and Cell. Biochem., 20, 173 (1978)

949. Fibrinogen

SOURCE: Human blood
COMMENTS: Spectra of fibrinogen treated with different concentrations of hydrogen peroxide (0-10 mM) in the presence of Mn^{2+}
REFERENCE: Y. Ishida, H. Taiuchi, A. Matsushima, and Y. Inada, Biochim. Biophys. Acta, 536, 70 (1978)

950. Flagellin

SOURCE: Salmonella SJ25
SOLVENT: 0.15 M NaCl-0.01 M Tris-HCl buffer, pH 8.2
REFERENCE: H. Kagawa, K. Owaribe, S. Asakura, and N. Takahashi, J. Bacteriol., 125, 68 (1976)

951. Flavin Reductase, NADPH-

SOURCE: Human erythrocytes
SOLVENT: 50 mM Phosphate buffer, pH 7.5, containing 1 mM EDTA and 0.1 mM DTT

REFERENCE: T. Yubisui, T. Matsuki, M. Takeshita, and Y. Yoneyama, J. Biochem. Tokyo, 85, 719 (1979)

952. Flavodoxin

SOURCE: Red seaweed (Chondrus crispus)
SOLVENT: 0.3 M Tris-HCl, 0.55 M NaCl, pH 7.4
COMMENTS: Spectra of oxidized, semiquinone, and reduced (dithionite) forms
REFERENCE: M. P. Fitzgerald, A. Husain, and L. J. Rogers, Biochem. Biophys. Res. Comm., 81, 630 (1978)

953. Flavodoxin

SOURCE: Desulfovibrio gigas
COMMENTS: Spectra of native, dissociated, and reassociated proteins
REFERENCE: K. Irie, K. Kobayashi, M. Kobayashi, and M. Ishimoto, J. Biochem. Tokyo, 73, 353 (1973)

954. Flavodoxin

SOURCE: Desulfovibrio vulgaris
SOLVENT: 0.025 M Phosphate buffer, pH 8.1
COMMENTS: Spectra of oxidized, semiquinone, and reduced (dithionite) forms
REFERENCE: M. Dubourdieu, J. LeGall, and V. Favaudon, Biochim. Biophys. Acta, 376, 519 (1975)

955. Flavodoxin

SOURCE: Peptostreptococcus elsdenii
SOLVENT: 0.1 M Pyrophosphate buffer, pH 8.5
COMMENTS: Spectra of oxidized and reduced (dithionite) forms
REFERENCE: V. Massey, J. M. L. Hoste, and S. G. Mayhew, Biochemistry, 13, 589 (1974)

956. Flavoprotein, Fe-S

SOURCE: Beef heart mitochondria
SOLVENT: 0.1 M Tris-HCl, pH 7.5
COMMENTS: Spectra of oxidized, partly reduced, and fully reduced (dithionite) forms
REFERENCE: F. J. Ruzicka and H. Beinert, J. Biol. Chem., 252, 8440 (1977)

957. Flavoprotein

SOURCE: Egg yolk
SOLVENT: 0.05 M Phosphate buffer, pH 7.0
COMMENTS: Spectra of native material and pepsin-treated material
REFERENCE: Z. Zak, M. Weber, and W. Ostrowski, Bull. Soc. Chem. Biol., 50, 795 (1968)

958. Flavoprotein, T-

SOURCE: Hansenula anomala
COMMENTS: Spectra of oxidized and lactate-reduced forms of long proteolysis
REFERENCE: M. Gervais, O. Groudinsky, Y. Risler, and F. Labeyrie, Biochem. Biophys. Res. Comm., 77, 1543 (1977)

959. Flavoprotein

SOURCE: Rice embryo
REFERENCE: S. Ida and Y. Morita, J. Biochem. Tokyo, 34, 1470 (1970)

960. Flavoprotein

SOURCE: Alga (Synechoccus lividus)
SOLVENT: pH 7 in presence of EDTA
COMMENTS: Changes in the spectrum upon ultraviolet irradiation (anaerobic)
REFERENCE: H. L. Crespi, V. Smith, L. Gajda, T. Tisue, and R. M. Ammeraal, Biochim. Biophys. Acta, 256, 611 (1972)

961. Flavoprotein

SOURCE: Yeast
SOLVENT: 0.1 M Phosphate, pH 7
COMMENTS: Spectra of oxidized and reduced (L-lactate) forms
REFERENCE: M. Iwatsubo, M. Mevel-Ninio, and F. Labeyrie, Biochemistry, 16, 3558 (1977)

962. Formaldehyde Dehydrogenase

SOURCE: Yeast
SOLVENT: 0.01 M Potassium phosphate buffer, pH 7.5, containing 0.001 M DTT
REFERENCE: N. Kato, T. Tamaoki, Y. Tani, and K. Ogata, Agr. Biol. Chem., 36, 2411 (1972)

963. Formate Dehydrogenase

SOURCE: Pea seeds (Pisum sativum)
REFERENCE: T. Ohyama and I. Yamazaki, J. Biochem. Tokyo, 75, 1257 (1974)

964. Formate Dehydrogenase

SOURCE: Escherichia coli
SOLVENT: 10 mM Tris-HCl, pH 8, 0.2 M NaCl
COMMENTS: Spectra of reduced (formate) and oxidized forms
REFERENCE: H. G. Enoch and R. L. Lester, J. Biol. Chem., 250, 6693 (1975)

965. N-Formimino-L-glutamate Iminohydrolase

SOURCE: Pseudomonas sp.
REFERENCE: R. B. Wickner and H. Tabor, J. Biol. Chem., 247, 1605 (1972)

966. Formyltetrahydrofolate Synthetase

SOURCE: Spinach leaves
REFERENCE: K. Iwai, N. Suzuki, and S. Mizoguchi, Agr. Biol. Chem., 31, 267 (1967)

967. Fraction, Subcellular

SOURCE: Blue-green alga (Anacystis nidulans)
SOLVENT: Tris-HCl buffer, pH 8.0
REFERENCE: O. Hirayama, Agr. Biol. Chem., 31, 947 (1967)

968. Fractions

SOURCE: Pig brain, cerebellar material
REFERENCE: A. Flangas, J. Perrin, R. Gaison, and G. Scott, Prep. Biochem., 2, 167 (1972)

969. Fructose-1,6-diphosphatase

SOURCE: Pig kidney
SOLVENT: a. 0.05 M Tris-HCl, pH 7.5
b. 0.1 N NaOH
REFERENCE: J. Mendicino, N. Kratowich, and R. M. Oliver, J. Biol. Chem., 247, 6643 (1972)

970. Fructose-1,6-diphosphate Aldolase

SOURCE: Drosophila melanogaster and rabbit muscle
SOLVENT: Water, pH 6.5

REFERENCE: O. Brenner-Holzach, Arch. Biochem. Biophys., 194, 321 (1979)

971. Fumarase

SOURCE: Chicken heart muscle
SOLVENT: 0.01 M Sodium phosphate buffer, pH 7.3
REFERENCE: C. Reyns, J. Leonis, and J. Schlusselber, Biochimie, 57, 123 (1975)

972. D-Galactose Dehydrogenase

SOURCE: Pseudomonas fluorescens
SOLVENT: 66 mM Potassium sodium phosphate buffer, pH 7.6
REFERENCE: E. O. Blachnitzky, F. Wengenmayer, and G. Kurz, Eur. J. Biochem., 47, 235 (1974)

973. D - Galactose Dehydrogenase

SOURCE: Pseudomonas saccharophila
SOLVENT: 50 mM Potassium sodium phosphate buffer, pH 6.8
REFERENCE: F. Wengenmayer, K. H. Ueberschar, G. Kurz, and H. Sund, Eur. J. Biochem., 40, 49 (1973)

974. Galactose Oxidase

SOURCE: Dactylium dendroides
SOLVENT: 0.1 M Sodium phosphate buffer, pH 7.0
REFERENCE: L. D. Kwiatkowski, L. Siconolfi, R. E. Weiner, R. S. Giordano, R. D. Berema, M. J. Ettinger, and D. J. Kosman, Arch. Biochem. Biophys. 182, 172 (1977)

975. β-Galactosidase

SOURCE: Curvularia inaequalis
COMMENTS: Spectra of protein and bright orange KH dyed protein
REFERENCE: V. S. Zolotova, A. T. Khomirova, L. V. Kozlov, and V. K. Antonov, Biokhimiya (Biochemistry), 41, 1360 (1976)

976. Gelatin

COMMENTS: Changes in the spectrum when gelatin is treated with hydrogen peroxide for 0, 4, and 21 hours
REFERENCE: P. M. Gallop, S. Seifter, and E. Medman in Recent Advances in Gelatin and Glue Research (G. Stainsby, ed.), 82 (1958), Pergammon, New York

977. Globin

SOURCE: Human hemoglobin
SOLVENT: 0.05 M Potassium phosphate buffer, pH 7.0
COMMENTS: Spectral changes upon addition of porphyrin
REFERENCE: A. Treffy and S. Ainsworth, Biochem. J., 137, 319 (1974)

978. Globin

SOLVENT: 0.5% Formic acid
COMMENTS: Spectrum of pyridoxal-phosphate β-globin
REFERENCE: R. E. Benesch, R. Benesch, A. Bank, R. Renthal, and B. A. Bray, in Genetical, Functional, and Physical Studies of Hemoglobin (T. Arends, G. Bemski, and B. L. Nagel, eds.), 134 (1971), S. Karger, Basel

979. γ-Globulin

SOURCE: Beef
SOLVENT: 0.1 M Glycine buffer, pH 9.0
COMMENTS: Changes seen in spectrum upon addition of salicylic acid (0-104 moles)
REFERENCE: P. Nuhn and M. Strecker, Pharmazie, 29, 10 (1974)

980. Glucagon

SOURCE: Commercial
SOLVENT: pH 2.1, 10.0, and 11.5
REFERENCE: C. Conti and L. S. Forster, Biochem. Biophys. Res. Comm., 65, 1257 (1975)

981. Glucagon

SOURCE: Beef and pig
SOLVENT: 0.2 M Acetic acid
COMMENTS: Spectra of glucagon, $(glucagon)_2$-dinitrophenyldisulfenyl and glucagon-nitrophenylsulfenyl
REFERENCE: R. M. Epand and T. E. Cote, Biochim. Biophys. Acta, 453, 365 (1976)

982. 12-Homoarginine-Glucagon

SOLVENT: 0.1 M Potassium chloride
COMMENTS: Spectra of glucagon and modified protein
REFERENCE: J. B. A. Ross, K. W. Rousslang, C. Dehaen, V. R. Lavis, and D. R. Deranleau, Biochim. Biophys. Acta, 576, 372 (1979)

983. Glucanase

SOURCE: Rhizopus sp.
SOLVENT: a. 0.01 M Acetate buffer, pH 7.0
b. 0.1 N NaOH
REFERENCE: S. Yamamoto, R. Kobayashi, and S. Nagasaki, Agric. Biol. Chem., 38, 1563 (1974)

984. β-1,4-Glucan Cellobiohydrolase

SOURCE: Trichoderma viride
SOLVENT: 0.05 M Sodium citrate buffer, pH 5.0
REFERENCE: L. E. R. Berghem, L. G. Pettersson, and U. B. Axio-Fredriksson, Eur. J. Biochem., 53, 55 (1975)

985. 1,4-α-Glucan Phosphorylase

SOURCE: Klebsiella pneumoniae
SOLVENT: 50 mM Potassium sodium phosphate buffer, pH 7.4
REFERENCE: D. Linder, G. Kurz, H. Bender, and K. Wallenfels, Eur. J. Biochem., 70, 291 (1976)

986. Glucoamylase

SOURCE: Aspergillus oryzae
REFERENCE: M. Ohga, K. Shimizu, and Y. Morita, Agric. Biol. Chem., 30, 967 (1966)

987. Glucoamylase

SOURCE: Endomyces sp. IFO 0111
SOLVENT: a. 0.05 M Potassium phosphate buffer, pH 7.0
b. 0.1 N NaOH
REFERENCE: T. Fukui and Z. Nikuni, Agric. Biol. Chem., 33, 884 (1969)

988. D-Gluconate Dehydrogenase

SOURCE: Serratia marcescens
SOLVENT: 0.1 M Potassium phosphate buffer, pH 6.0
COMMENTS: Spectra of oxidized and reduced (sodium gluconate) forms
REFERENCE: E. Shinagawa, K. Matsushita, O. Adachi, and M. Ameyama, Agric. Biol. Chem., 42, 2355 (1978)

989. D-Gluconate Dehydrogenase

SOURCE: Pseudomonas aeruginosa
SOLVENT: 0.01 M Potassium phosphate buffer, pH 6.0

COMMENTS: Spectra of oxidized form and dithionite-reduced form
REFERENCE: K. Matsushita, E. Shinagawa, O. Adachi, and M. Ameyama, J. Biochem. Tokyo, 85, 1173 (1979)

990. D-Glucono-γ-lactone Dehydrogenase

SOURCE: Penicillium cyaneofulvum
SOLVENT: 0.01 M Phosphate buffer, pH 6.5
COMMENTS: Spectra of oxidized and reduced (D-glucono-γ-lactone)forms
REFERENCE: T. Takahashi, H. Yamashita, E. Kato, M. Mitsumoto, and S. Murakawa, Agr. Biol. Chem., 40, 121 (1976)

991. D-Glucose-isomerizing Enzyme

SOURCE: Bacillus coagulans strain HN-68
SOLVENT: a. 0.1 M Acetate buffer, pH 6.0
b. 0.1 M NaOH
REFERENCE: G. Danno, Agric. Biol. Chem., 34, 1795 (1970)

992. Glucose Oxidase

SOURCE: Aspergillus niger
COMMENTS: Spectra of enzyme and periodate-oxidized enzyme
REFERENCE: M. Gerber, O. Bodman, and G. V. Schulz, FEBS Lett., 80, 294 (1977)

993. Glucose Oxidase

SOURCE: Penicillium amagasakiense
SOLVENT: 0.1 M Sodium acetate buffer, pH 5.6
COMMENTS: Spectra of oxidized and reduced (glucose and dithionite) forms
REFERENCE: K. Kusai, I. Sekuzu, B. Hagihara, K. Okunuki, S. Yamauchi, and M. Nakai, Biochim. Biophys. Acta, 40, 555 (1960)

994. Glucose Oxidase

SOURCE: Penicillium vitale
REFERENCE: T. A. Abalikhina, A. D. Morozkin, V. P. Bogdanov, and E. D. Kaerzneva, Biokhimiya (Biochemistry), 36, 160 (1971)

995. Glucose Oxidase, deazaFAD-

SOURCE: Commercial
SOLVENT: 0.1 M Sodium phosphate, pH 7.0 and pH 7.6, containing 0.3 mM EDTA

COMMENTS: Spectral changes during the titration of the enzyme with each of the following: sulfite, cyanide, hydroxylamine, and glyoxylate
REFERENCE: M. S. Jorns and L. B. Hersh, J. Biol. Chem., 251, 4872 (1976)

996. Glucose-6-phosphate Dehydrogenase

SOURCE: Leuconostoc mesenteroides
SOLVENT: 0.03 M Phosphate buffer, pH 7.5
COMMENTS: Spectra of unmodified and protein modified with pyridoxal-phosphate and borohydride reduction
REFERENCE: M. Milhausen and H. R. Levy, Eur. J. Biochem., 50, 453 (1975)

997. α-Glucosidase

SOURCE: Mucor rouxii
SOLVENT: 0.05 M Phosphate buffer, pH 7.2
REFERENCE: A. Flores-Careon and J. Ruiz-Herrero, Biochim. Biophys. Acta, 258, 496 (1972)

998. α-Glucosidase

SOURCE: Mucor javanicus
SOLVENT: 0.05 M Acetate buffer, pH 5.3
REFERENCE: Y. Yamasai, T. Miyake, and Y. Suzuki, Agric. Biol. Chem., 37, 31 (1973)

999. α-Glucosidase

SOURCE: Human liver lysosomes
SOLVENT: a. pH 6.7
b. pH 12.4
REFERENCE: A. K. Murray, B. I. Brown, and D. H. Brown, Arch. Biochem. Biophys., 185, 511 (1978)

1000. α-Glucoside-synthesizing Enzyme

SOURCE: Pig liver
SOLVENT: 0.01 M Phosphate buffer, pH 6.0
REFERENCE: K. Uchida and Y. Suzuki, Agr. Biol. Chem., 38, 195 (1974)

1001. Glutamate Dehydrogenase

SOURCE: Escherichia coli
SOLVENT: a. pH 4.8
b. pH 6.5

COMMENTS: Spectra of native enzyme, enzyme plus hydroxylamine and, plus α-methyl-DL-glutamate
REFERENCE: B. S. Sukhareva and Y. M. Torchinsky, in Pyridoxal Catalysis: Enzyme and Model Systems (E. E. Snell, A. E. Braunitzer, E. S. Severin, and Y. M. Torchinsky, eds.), 470 (1968), Interscience, New York

1002. Glutamate Dehydrase

SOURCE: Commercial; beef liver
SOLVENT: 0.2 M Potassium phosphate buffer, pH 7.4
COMMENTS: Effect of diethylstilbesterol, ADP, GTP, ADP plus diethylstilbesterol, and GTP plus ADP on the spectrum of the protein
REFERENCE: H. Hillar, Arch. Intern. Physiol. Biochem., 86, 11 (1978)

1003. Glutamate Dehydrogenase

SOURCE: Beef liver
COMMENTS: Spectra of native and 5-fluoro-sulfonylbenzoyl adenosine-modified enzyme
REFERENCE: P. K. Pal, W. J. Wechter, and R. F. Colman, J. Biol. Chem., 250, 8140 (1975)

1004. Glutamate Dehydrogenase

SOURCE: Commercial
SOLVENT: 0.067 M Phosphate buffer, pH 7.6
COMMENTS: Change in spectrum due to photo (light)-irradiation in presence of pyridoxal phosphate
REFERENCE: F. Hucho, U. Markau, and H. Sund, Eur. J. Biochem., 32, 69 (1973)

1005. Glutamate Dehydrogenase

SOURCE: Beef liver
COMMENTS: Spectra of enzyme-NADPH-2-keto-glutarate complex and enzyme-NADPH-complex
REFERENCE: J. M. Jallon (1969) Thesis, 3rd cycle, Univ. of Paris-Sud, France, cited in A. DiFranco and M. Iwatsubo, Eur. J. Biochem., 30, 517 (1972)

1006. Glutamate Dehydrogenase, NADP

SOURCE: Yeast (Saccharomyces cerevisiae)
SOLVENT: 0.1 M Phosphate buffer, pH 7.6
REFERENCE: R. Venard and A. Fourcade, Biochimie, 54, 1381 (1972)

1007. Glutamate Synthase

SOURCE: Escherichia coli
SOLVENT: 20 mM Potassium N-2-hydroxyethyl-piperazine-N'-2-ethane sulfonate, pH 7.2, containing 2 mM α-ketoglutarate, and 1 mM EDTA
COMMENTS: Spectra of reduced (dithionite) and oxidized forms
REFERENCE: R. E. Miller and G. R. Stadtman, J. Biol. Chem., 247, 7407 (1972)

1008. Glutamate Synthase

SOURCE: Thiobacillus thioparus
REFERENCE: K. Adachi and L. Suzuki, J. Bacteriol., 129, 1173 (1977)

1009. Glutamate Synthase

SOURCE: Lupin nodules
SOLVENT: Phosphate buffer
COMMENTS: Spectra of reduced and oxidized enzyme
REFERENCE: M. J. Boland and A. G. Benny, Eur. J. Biochem., 79, 355 (1977)

1010. Glutamine Synthetase

SOURCE: Escherichia coli
SOLVENT: 20 mM Imidazole-HCl buffer, pH 7.1, containing 100 mM KCl and 1.0 mM manganese chloride
COMMENTS: Spectra of unmodified and pyridoxylated enzyme
REFERENCE: E. J. Whitley, Jr. and A. Ginsburg, J. Biol. Chem., 253, 7017 (1978)

1011. Glutamine Synthetase

SOURCE: Azotobacter vinelandii
SOLVENT: 10 mM Imidazole-HCl buffer, pH 7.0, containing 100 mM KCl and 1 mM manganese chloride
COMMENTS: Spectra of phosphodiesterase-treated and untreated enzyme
REFERENCE: J. Siedel and E. Shelton, Arch. Biochem. Biophys., 192, 214 (1979)

1012. Glutaredoxin

SOURCE: Escherichia coli
SOLVENT: a. 0.05 M Tris-Cl, pH 7.0, containing 0.10 M sodium acetate, and 1 mM EDTA
b. pH 13
REFERENCE: A. Holmeren, J. Biol. Chem., 254, 3664 (1979)

1013. Glutathione Peroxidase

SOURCE: Rat lung
REFERENCE: D. T. Y. Chiu, F. H. Stutts, and A. L. Tappel, Biochim. Biophys. Acta, 445, 558 (1976)

1014. Glutathione Reductase

SOURCE: Calf liver
SOLVENT: 0.1 M Potassium phosphate, containing 1 mM EDTA, pH 7.5
COMMENTS: Spectra of oxidized enzyme, sodium borohydride-reduced enzyme (BRE), and BRE after reaction with trinitrobenzenesulfonate
REFERENCE: I. Carlberg and B. Mannervik, FEBS Lett., 98, 263 (1979)

1015. Glutathione Reductase

SOURCE: Rice kernel
SOLVENT: 0.1 M Phosphate buffer, pH 7.0
COMMENTS: Changes in spectrum due to NADPH and dithionite
REFERENCE: S. Ida and Y. Morita, Agr. Biol. Chem., 35, 1542 (1971); 35, 1550 (1971)

1016. Glutathione Reductase

SOURCE: Sea urchin eggs (Hemicentrotus pulcherrimus)
SOLVENT: 20 mM Potassium phosphate buffer, pH 7.5
REFERENCE: I. Ii and H. Sakai, Biochim. Biophys. Acta, 350, 141 (1974)

1017. Glutathione Reductase

SOURCE: Human erythrocytes
SOLVENT: Phosphate buffer, pH 7.0, containing 1 mM EDTA
COMMENTS: Spectra of reduced (NADPH) and oxidized forms
REFERENCE: D. J. Worthington and M. A. Rosemeyer, Eur. J. Biochem., 48, 167 (1974)

1018. Glutathione Reductase

SOURCE: Rat liver
COMMENTS: Spectra of reduced and spontaneously reoxidized enzyme
REFERENCE: I. Carlberg and B. Mannewik, J. Biol. Chem., 250, 5475 (1975)

1019. Glutathione Reductase

SOURCE: Escherichia coli
SOLVENT: 50 mM Phosphate buffer, pH 7.6, containing 3 mM EDTA
REFERENCE: V. P. Pigiet and R. R. Conley, J. Biol. Chem., 252, 6367 (1977)

1020. Glutathione Reductase

SOURCE: Rabbit liver
SOLVENT: 0.1 M Potassium phosphate, pH 6.8, containing 1 mM EDTA
REFERENCE: G. Zanetti, Arch. Biochem. Biophys., 198, 241 (1979)

1021. Glutelin

SOURCE: Rice endosperm
SOLVENT: a. 0.05 N NaOH containing 8 M urea, pH 13.4
b. 0.05 N NaOH-0.05 M Potassium phosphate buffer, containing 8 M urea, pH 7.5
REFERENCE: H. Sawai and Y. Morita, Agr. Biol. Chem., 32, 76 (1968)

1022. Glyceraldehyde Phosphate Dehydrogenase

SOURCE: Escherichia coli
SOLVENT: 20 mM Tris-chloride buffer, pH 8, containing 2 mM EDTA
COMMENTS: Spectra of native enzyme, charcoal-treated enzyme, and NAD^+-loaded enzyme
REFERENCE: G. D'Alessio and J. Josse, J. Biol. Chem., 246, 4326 (1971)

1023. Glyceraldehyde 3-Phosphate Dehydrogenase

SOURCE: Chicken breast muscle
SOLVENT: 0.1 M Acetate buffer, pH 5.0
COMMENTS: Spectra of apoenzyme-2,6-dichlorophenol indophenol complex and apoenzyme-2,6-dibromophenol indophenol complex
REFERENCE: L. V. Benitez and W. S. Allison, Arch. Biochem. Biophys., 159, 89 (1973)

1024. Glyceraldehyde 3-Phosphate Dehydrogenase

SOURCE: Bacillus stearothermophilus
SOLVENT: a. 50 mM HCl-6 M guanidine-HCl
b. 50 mM NaOH-6 M guanidine-HCl

COMMENTS: Spectra of apoenzyme iodinated to different extents
REFERENCE: G. Allen and J. I. Harris, Biochem. J., 155, 523 (1976)

1025. Glyceraldehyde 3-Phosphate Dehydrogenase

SOURCE: Thermus thermophilus Str. HB8
COMMENTS: Spectra of holo- and apoenzymes
REFERENCE: S. C. Fujita, T. Oshima, and K. Imahori, Eur. J. Biochem., 64, 57 (1976)

1026. Glyceraldehyde 3-Phosphate Dehydrogenase

SOURCE: Palinurus vulgaris
SOLVENT: 10 mM Ethylenediamine-1 mM EDTA, pH 7
COMMENTS: Spectrum of β-(2-furyl) acryloyl-enzyme
REFERENCE: R. Berni, A. Mozzarelli, L. Pellacani, and G. L. Rossi, J. Mol. Biol., 110, 405 (1977)

1027. Glycerol-3-phosphate Dehydrogenase

SOURCE: Rabbit liver and muscle
COMMENTS: Spectra of nucleotide-containing and nucleotide-free enzyme
REFERENCE: J. Otto, A. Raggi, W. Machleidt, and T. Bucher, Hoppe-Seyler's Z. Physiol. Chem., 353, 332 (1972)

1028. L-Glycerol-3-phosphate Dehydrogenase

SOURCE: Rabbit adipose tissue and muscle
SOLVENT: 0.02 M Phosphate, pH 7.6
REFERENCE: D. L. Warkentin and T. P. Fondy, Eur. J. Biochem., 30, 97 (1973)

1029. Glycerol-3-phosphate Dehydrogenase

SOURCE: Rabbit muscle mitochondria
SOLVENT: 10 mM Sodium phosphate, pH 7.5, containing 0.5% Triton X-100, 0.02% sodium azide, and 10 μM 2-mercaptoethanol
COMMENTS: Spectra of untreated, dithionite-treated, and mersalyl-dithionite-treated enzyme
REFERENCE: E. S. Cole, C. A. Lepp, P. D. Holohan, and T. P. Fondy, J. Biol. Chem., 253, 7952 (1978)

1030. L-Glycerol-3-phosphate Dehydrogenase.

SOURCE: Ceratitis capitata
REFERENCE: J. M. Fernandez-Sousz, J. G. Gavilanes, A. M. Municio, and A. Perez-Aranda, Biochim. Biophys. Acta, 481, 6 (1977)

1031. Glycine Reductase

SOURCE: Clostridium sticklandii
SOLVENT: 0.2 M Tris-HCl, pH 8
COMMENTS: Spectra of native, potassium borohydride-reduced, and air-oxidized enzyme
REFERENCE: J. E. Cone, R. Martin del Rio, J. N. Davis, and T. C. Stadtman, Proc. Nat. Acad. Sci., 73, 2659 (1976)

1032. Glycine Reductase

SOURCE: Clostridium sticklandii
SOLVENT: a. 50 mM Potassium phosphate, pH 7.2
b. 50 mM Tris-HCl, pH 8.2, containing 1 mM DTT and 1 mM MgK_2 EDTA
c. 20 mM Potassium phosphate, pH 7.2
d. 0.1 N HCl
e. 0.1 N KOH
COMMENTS: Spectra of enzyme and carboxamidomethyl-enzyme before and after addition of potassium borohydride; spectra of enzyme in presence of 4.8 M guanidine in solvents c, d, and e
REFERENCE: J. E. Cone, R. Martin del Rio, and T. C. Stadtman, J. Biol. Chem., 252, 5337 (1977)

1033. Glycogen Phosphorylase B

SOURCE: Rabbit liver
SOLVENT: 4 mM Tris buffer, pH 7.0, containing 1 mM EDTA and 1 mM 2-mercaptoethanol
REFERENCE: T. B. Eronina, G. V. Silonova, and N. B. Livanova, Biokhimiya (Biochemistry), 42, 197 (1977)

1034. Glycogen Synthetase B

SOURCE: Rat liver
SOLVENT: 50 mM glycyl-glycine, pH 7.4, containing 0.25 M sucrose, 3 mM DTT, and 4 mM glucose 6-phosphate
REFERENCE: D. C. Lin and H. L. Segal, J. Biol. Chem., 248, 7007 (1973)

1035. Glycolate Oxidase

SOURCE: Human liver
SOLVENT: 0.1 M Sodium phosphate, pH 7
COMMENTS: Spectra of enzyme and enzyme-sulfite complex
REFERENCE: H. Schwam, S. Michelson, W. C. Randall, J. M. Sondey, and R. Hirschmann, Biochemistry, 18, 2828 (1979)

1036. Glycolic Acid Oxidase

SOURCE: Pig liver
SOLVENT: 0.1 M Sodium phosphate buffer, pH 7.0, containing 0.3 mM Na_2 EDTA
COMMENTS: Changes in the spectra of the oxidized, deaza FMN-enzyme upon reaction with glycolate, glyoxylate, sulfite, oxalate, cyanide, hydroxylamine, and oxalate on the enzyme-hydroxylamine complex
REFERENCE: M. S. Jorns and L. B. Hersh, J. Biol. Chem., 251, 4872 (1976)

1037. Glycoprotein

SOURCE: Human ovarian cyst
SOLVENT: Water
REFERENCE: K. Ramakrishinan Bhaskar, and J. M. Creeth, Biochem. J., 143, 669 (1974)

1038. Glycoprotein

SOURCE: Rat intestinal mucosa
REFERENCE: J. Frot-Coutaz, P. Louisot, and R. Got, Biochim. Biophys. Acta, 264, 362 (1972)

1039. Glycoprotein Immunogen

SOURCE: Horseradish peroxidase (commercial)
REFERENCE: L. A. Moroz, J. R. Joubert, and J. C. Hogg, J. Immunol., 112, 1094 (1974)

1040. Glyoxalate Carboligase

SOURCE: Escherichia coli
COMMENTS: Spectra of apoenzyme and apoenzyme plus 5-deaza FAD
REFERENCE: T. H. Cromartie and C. T. Walsh, J. Biol. Chem., 251, 329 (1976)

1041. Gramicidin C

SOURCE: Commercial
COMMENTS: Spectrum of O-dimethylaminonaphthylsulfonyl derivative
REFERENCE: W. R. Veatch and E. R. Blout, Biochemistry, 15, 3026 (1976)

1042. L-Gulono-γ-lactone Oxidase

SOURCE: Rat liver/goat liver
REFERENCE: M. Nishikimi, B. M. Tolbert, and S. Udenfriend, Arch. Biochem. Biophys., 175, 427 (1976)

1043. Gulonolactone Oxidase

SOURCE: Rat liver microsomes
SOLVENT: 20 mM Potassium phosphate buffer, pH 7.0, containing 0.4% Emulgen 109P and 20 mM KCl
REFERENCE: H. Nakagawa, A. Asano, and R. Sato, J. Biochem. Tokyo, 77, 22 (1975)

1044. Haptoglobin

SOURCE: Eel (Anguila japonica)
SOLVENT: Water
REFERENCE: M. Kadama, K. Hashimoto, and F. Matsuura, Bull. Jap. Soc. Sci. Fish., 41, 1021 (1975)

1045. Haptoglobin

SOURCE: Human
SOLVENT: 0.01 M Tris-HCl buffer, pH 7.0
COMMENTS: Spectra of native haptoglobin and α-chain, β-chain, and N-benzoyl-valine haptoglobin
REFERENCE: J. Osada and W. Dobryszycka, Biochim. Biophys. Acta, 412, 306 (1975)

1046. Hemagglutinin A

SOURCE: Castor bean (Ricinus communis)
REFERENCE: S. Ueno, G. Funatsu, and M. Funatsu, Agr. Biol. Chem., 41, 1069 (1977)

1047. Heme Oxygenase

SOURCE: Pig spleen microsomes
SOLVENT: 10 mM Potassium phosphate buffer, pH 7.4
COMMENTS: Spectra of oxidized and reduced (NADPH and cytochrome c reductase) forms
REFERENCE: T. Yoshida and G. Kikuchi, J. Biochem. Tokyo, 81, 265 (1977)

1048. Heme Oxygenase

SOURCE: Pig spleen microsomes
SOLVENT: 110 mM Potassium phosphate buffer, pH 7.4, containing 1 mM EDTA, 0.1% Triton X-100, and 0.1% sodium cholate

COMMENTS: Spectra of oxidized, reduced (dithionite) CO-reduced forms, and ferric heme-heme oxygenase complex, complex plus cyanide, and complex plus azide
REFERENCE: T. Yoshida and G. Kikuchi, J. Biol. Chem., 253, 4224 (1978)

1049. Heme Oxygenase

SOURCE: Rat liver microsomes
COMMENTS: Spectra of enzyme and of the oxidized, reduced, and CO-reduced forms of the heme-heme oxygenase complex
REFERENCE: T. Yoshida and G. Kikuchi, J. Biol. Chem., 254, 4487 (1979)

1050. Hemerythrin

SOURCE: Worm (Dendrostonum pyroides)
COMMENTS: Spectra of coelomic-, vascular-, and myohemerythrin
REFERENCE: G. L. Klippenstein, D. A. Van Riper, and E. A. Oosterom, J. Biol. Chem., 247, 5959 (1972)

1051. Hemerythrin

SOURCE: Marine worm (Golfingia gouldii)
SOLVENT:
a. 0.1 M Sodium bicarbonate and different concentrations of sodium perchlorate, pH 8.62-pH 6.36
b. 0.1 M Tris-cacodylate buffer, 0.1 sodium perchlorate, pH 6.55-pH 9.32

COMMENTS:
1. Spectra changes upon conversion of methydroxohemerythrin to metaquohemerythrin
2. Spectra of succinylated methydroxohemerythrin and metaquohemerythrin
3. Spectra of methydroxohemerythrin and metaquohemerythrin
4. Spectrum of metazidohemerythrin

REFERENCE: K. Garbett, D. W. Darnall and I. M. Klotz, Arch. Biochem. Biophys., 142, 455 (1971)

1052. Hemerythrin

SOURCE: Marine worm (Golfingia gouldii)
SOLVENT: 0.1 M Sodium carbonate
COMMENTS: Spectra of iron-free dinitrohemerythrin and extensively nitrated protein
REFERENCE: R. L. Rill and I. M. Klotz, Arch. Biochem. Biophys., 147, 226 (1971)

1053. Hemerythrin A and Hemerythrin B

SOURCE: Marine worm (Golfingia gouldii)
COMMENTS: Spectra of metazide form of proteins
REFERENCE: G. L. Klippenstein, Biochemistry, 11, 372 (1972)

1054. Hemerythrin

SOURCE: Lingula unguis
SOLVENT: 0.05 Borate-0.5 M NaCL, pH 8.0
COMMENTS: Spectra of oxyhemerythrin, deoxyhemerythrin, and methemerythrin
REFERENCE: J. G. Joshi and B. Sullivan, Comp. Biochem. Physiol., 44B, 857 (1973)

1055. Hemerythrin

SOURCE: Marine worm (Phascolopsis gouldii)
COMMENTS: Spectra of metazidohemerythrin and oxyhemerythrin
REFERENCE: J. B. R. Dunn, A. W. Addison, R. E. Bruce, J. S. Loehr, and T. M. Loehr, Biochemistry, 16, 1743 (1977)

1056. Hemerythrin

SOURCE: Marine worm (Golfingia gouldii)
SOLVENT: a. 0.1 M Tris-chloride, pH 8.3
b. 0.05 M Tris-borate, pH 7.5
COMMENTS: Spectra of oxy-, O-acetylated amidinated-, and tetranitromethane-treated O-acetylaminidinated forms
REFERENCE: C. C. Fan and J. L. York, Biochem. Biophys. Res. Comm., 47, 472 (1972)

1057. Hemerythrin, Met-

SOURCE: Phascolopsis (syn. Golfingia) gouldii
SOLVENT: 0.5 M Tris-Acetate, pH 8.0
COMMENTS: Spectral changes upon formation of sulfidomethemerythrin
REFERENCE: S. M. Freier, L. L. Duff, R. P. Van Duyne, and I. M. Klotz, Biochemistry, 18, 5372 (1979)

1058. Hemerythrin

SOURCE: Phascolosoma lurco
SOLVENT: 0.1 M Tris-acetate buffer, pH 8.2
COMMENTS: Spectra of oxy-, metazido forms and metselenocyanato- and metdicyanamido-complexes
REFERENCE: A. W. Addison and R. E. Bruce, Arch. Biochem. Biophys., 183, 328 (1977)

1059. Hemerythrin

SOURCE: Marine worm (Themiste pyroides)
SOLVENT: 0.05 M Tris pH 7.0, 0.05 M sodium sulfate
COMMENTS: Changes in spectrum upon addition of methemerythrin to deoxymyoglobin (time = 1 minute-100 minutes)
REFERENCE: Z. Bradic, P. C. Harrington, and R. G. Wilkins, Biochemistry, 18, 889 (1979)

1060. Hemiglobin

SOURCE: Beef erythrocytes
SOLVENT: pH 7.2-11.8
COMMENTS: Spectrum of hemiglobin cyanide at pH 7.2-pH 11.8
REFERENCE: L. H. J. Van Kempen and F. Kreuzer in Oxygen Affinity of Hemoglobin and Red Cell Acid Base Status (M. Rorth and P. Astrup, eds.), 219 (1972), Academic Press, New York

1061. Hemiglobin

SOURCE: Beef erythrocytes
SOLVENT: a. Disodium phosphate-potassium acid phosphate buffer, pH 6.41-7.72
b. NaOH-glycine buffer, pH 8.70-9.43
REFERENCE: H. Sternberg and A. I. Virtanen, Acta Chem. Scand., 6, 1342 (1952)

1062. Hemocyanin

SOURCE: Busycon canaliculatum
COMMENTS: Spectrum at different oxygen pressures (0-619 mm)
REFERENCE: H. A. Dephillips cited by K. E. Van Holde and E. F. J. Van Bruggen in Subunits in Biological Systems (J. N. Timasheff and G. D. Fasman, eds.), 1 (1971), Marcel Dekker, Inc., New York

1063. Hemocyanin

SOURCE: Snail (Achatina fulica) hemolymph
REFERENCE: A. A. Aboderin and H. A. Kareem, Comp. Biochem. Physiol., 40A, 113 (1971)

1064. Hemocyanin, 25.7 S

SOURCE: Blue crab (Callinectes sapidus)
SOLVENT: 0.1 Tris buffer, pH 8.8, containing 0.01 $MgCl_2$
REFERENCE: L. M. Hamlin and W. W. Fish, Biochim. Biophys. Acta, 491, 46 (1977)

1065. Hemocyanin

SOURCE: Crayfish (Cherax destructor)
SOLVENT: 6 M Urea
COMMENTS: Changes in spectrum with time after dilution into 6 M urea
REFERENCE: D. I. Marlborough, Arch. Biochem. Biophys., 196, 248 (1979)

1066. Hemocyanin

SOURCE: Snail (Helix pomatia)
COMMENTS: Spectra of oxyhemocyanin, oxy- plus KF,and oxy- plus sodium azide; spectra of these solutions after treatment with hydrogen peroxide
REFERENCE: R. Witters and R. Lontie, FEBS Lett., 60, 400 (1975)

1067. Hemocyanin

SOURCE: Snail (Helix pomatia)
SOLVENT: 0.1 M Borate-HCl buffer, pH 8.2
REFERENCE: R. Lontie, M. DeLey, H. Robberecht, and R. Witters, Nature, New Biology, 242, 180 (1972)

1068. Hemocyanin

SOURCE: Snail (Helix pomatia)
SOLVENT: pH 6.2
COMMENTS: Spectra of unirradiated and irradiated with U.V. light; spectra of undissociated oxidized form, dissociated into eighths oxidized form, undissociated reduced form, dissociated into eighths, and copper free form
REFERENCE: I. M. Claesson, Arkiv Kemi, 10, 4 (1956)

1069. Hemocyanin

SOURCE: Crayfish (Jasus edwardsii)
SOLVENT: 0.1 ionic strength Tris buffer, pH 7.0
COMMENTS:
1. Spectrum of oxy form in presence of $CaCl_2$
2. Spectra of oxy form and deoxy form in presence and absence of $MgCl_2$
3. Spectrum of apo-hemocyanin

REFERENCE: H. D. Ellerton, L. B. Collins, J. S. Gale, and A. Y. P. Yung, Biophys. Chem., 6, 47 (1977)

1070. Hemocyanin

SOURCE: Horseshoe crab (Limulus polyphemus)
SOLVENT: 0.052 Tris-Glycine, pH 8.9
COMMENTS: Spectra of oxy-, deoxy-, and CO-forms
REFERENCE: C. Bonaventura, B. Sullivan, J. Bonaventura, and S. Bourne, Biochemistry, 13, 4784 (1974)

1071. Hemocyanin

SOURCE: Lymnaea stagnalis hemolymph
SOLVENT: 0.05 M Borate buffer, pH 9.2
COMMENTS: Spectrum of oxy form
REFERENCE: R. L. Hall, J. S. Pearson, and E. J. Wood, Comp. Biochem. Physiol., 52B, 211 (1975)

1072. Hemocyanin

SOURCE: Octopus vulgaris
COMMENTS: Spectrum of oxyhemocyanin
REFERENCE: L. Tallandini and B. Salvato, FEBS Lett., 54, 283 (1975)

1073. Hemocyanin

SOURCE: a. Octopus vulgaris
b. Eriphia spinifrons
SOLVENT: 1 M Tris, pH 8.0
COMMENTS: Spectra of native and reconstituted forms
REFERENCE: A. Ghiretti-Magaldi, G. Nardi, F. Ghiretti, and R. Zito, Bull. Soc. Ital. Biol. Sper., 38, 1839 (1962)

1074. Hemocyanin

SOURCE: Seploteuthis lessoniana
COMMENTS: Spectra made at 77°K, 198°K, and 293°K
REFERENCE: W. Mori, O. Yamauchi, Y. Nakao, and A. Nakahara, Biochem. Biophys. Res. Comm., 66, 725 (1975)

1075. Hemoglobin

SOLVENT: pH 6.5
COMMENTS: Spectra of methemoglobin over the temperature range 165°K-250°K
REFERENCE: K. Gray and E. F. Slade, Biochem. Biophys. Res. Comm., 48, 1019 (1972)

1076. Hemoglobin

SOLVENT: 0.01 M Potassium phosphate buffer, pH 7.0
COMMENTS: Spectra of hemoglobin and CO-hemoglobin before and after photo-irradiation at 4.2°K
REFERENCE: T. Yonetani, Y. Itzuka, H. Yamamoto, and B. Chance, in International Symposium on Oxidases and Related Redox Systems (T. E. King, H. S. Mason, and M. Morrison, eds.), Vol. 1, 401 (1972), University Park Press, Baltimore

1077. Hemoglobin

SOURCE: a. Aphrodite nerve
b. Aplysia nerve
c. Halosydna nerve
d. Tubifex nerve
SOLVENT: pH 7.35-9.22
COMMENTS: Spectra of ferrous deoxyhemoglobins; spectra of ferric hemoglobins (Aphrodite and Aplysia); spectra of oxyhemoglobin, CO-hemoglobin, ferrous-hemoglobin, cyanide-,and deoxygenated hemoglobin (Aphrodite, Aplysia, and Halosydna)
REFERENCE: B. A. Wittenberg, R. W. Briehl, and J. B. Wittenberg, Biochem. J., 96, 363 (1965)

1078. Hemoglobin

SOURCE: Artemia salina
COMMENTS: Spectra of adult hemoglobins I, II, and III
REFERENCE: L. Moens and M. Kondo, Eur. J. Biochem., 67, 397 (1976)

1079. Hemoglobin

SOURCE: Artemia salina
SOLVENT: 5 mM Tris-glycine, pH 8.3
COMMENTS: Spectra of adult, nauplus, and 0.5 hour, and 10 hours encysted embryos
REFERENCE: L. Moens and M. Kondo, Biochem. J., 165, 111 (1977)

1080. Hemoglobin

SOURCE: Ascaris lumbricoides
SOLVENT: 0.05 M Potassium phosphate buffer, pH 6.2-7.0
COMMENTS: 1. Spectra of Hb IV, HbO_2 at 0°C and 77°K
2. Spectrum of product of reaction of Hb with H_2O_2 at 0°C

REFERENCE: B. A. Wittenberg, J. B. Wittenberg, and R. W. Noble, J. Biol. Chem., 247, 4008 (1972)

1081. Hemoglobin

SOURCE: Snail (Australorbis glabratus)
COMMENTS: Spectra of oxyhemoglobin, carboxy-hemoglobin, and cyanmethemoglobin
REFERENCE: G. A. T. Targett, J. Helminth, 36, 201 (1962)

1082. Hemoglobin

SOURCE: Beef erythrocytes
SOLVENT: 0.1 M Potassium phosphate buffer, pH 7.0, containing 60 mM EDTA
COMMENTS: Change in spectrum with oxygen bubbled through solution in the presence of riboflavin and illumination (448 mm)
REFERENCE: R. E. Lynch, G. R. Lee, and G. E. Cartwright, J. Biol. Chem., 251, 1015 (1976)

1083. Hemoglobin

SOURCE: Beef erythrocytes
SOLVENT: 0.1 M Tris-HCl buffer, pH 7.0
COMMENTS: Spectral change with time of nitrite-treated hemoglobin treated with NADH and NADH dehydrogenase with and without methylene blue
REFERENCE: K. Adachi, Biochim. Biophys. Acta, 289, 262 (1972)

1084. Hemoglobin

SOURCE: Beef erythrocytes
COMMENTS:
1. Changes in spectrum of methemoglobin upon addition of benzoate (0-1 M)
2. Changes in spectrum of reduced hemoglobin upon addition of benzoate (0-2.0 M)
3. Effect of glycerol on spectra

REFERENCE: K. Tsushima, J. Biochem. Tokyo, 41, 215 (1954)

1085. Hemoglobin

SOURCE: Rainbow trout erythrocytes
SOLVENT: pH 6.0-pH 8.0
COMMENTS: Spectra of oxy- and deoxy- forms
REFERENCE: I. Iuchi, Comp. Biochem. Physiol., 44B, 1087 (1973)

1086. Hemoglobin

SOURCE: Midge (Chironomus thummi thummi)
SOLVENT: pH 7.0
COMMENTS: Spectrum of globin
REFERENCE: G. Amiconi, E. Antonini, M. Brunori, H. Formaneck, and R. Huber, Eur. J. Biochem., 31, 52 (1972)

1087. Hemoglobin

SOURCE: Midge (Chironomus thummi thummi)
SOLVENT: pH 5.75-pH 10.44
COMMENTS: Spectrum of oxidized form I
REFERENCE: A. Wollmer, G. Buse, H. Sick, and K. Gersonde, Eur. J. Biochem., 24, 547 (1972)

1088. Hemoglobin

SOURCE: Dog/human erythrocytes
COMMENTS: Spectra of $\alpha_2^{Dog}\gamma_2^F$ and $\alpha_2^F\beta_2^{Dog}$
REFERENCE: E. Kleihauer and N. Miltenyi, Acta Haematol., 32, 280 (1964)

1089. Hemoglobin

SOURCE: Dog/human erythrocytes
COMMENTS: Spectra of $\alpha_2^{Dog}\beta_2^{A(Mk)}$, $\alpha_2^{Mk}\beta_2^{Dog}$, and $\alpha_2^{Dog}\beta_2^{Dog}$
REFERENCE: P. Heller, H. G. Wrinstein, V. J. Yakulis, and I. M. Rosenthal, Blood, 20, 287 (1962)

1090. Hemoglobin

SOURCE: Blue-white dolphin (Stenella caeraleo-alba)
COMMENTS: Spectrum of ferrihemoglobin
REFERENCE: T. Matsui, C. Shimizu, and F. Matsuura, Bull. Jap. Soc. Sci. Fish., 41, 771 (1975)

1091. Hemoglobin

SOURCE: Guinea pig blood in male tick
SOLVENT: 3 M Ammonium sulphate, 0.1 M potassium phosphate buffer, pH 6.7 saturated with CO
COMMENTS: Spectra of oxy, met- and metcyanide forms
REFERENCE: J. D. G. Smit, O. Grandjean, R. Guggenheim, and K. H. Winterhalter, Nature, 266, 536 (1977)

1092. Hemoglobin

SOURCE: Horseleech (Hagmopsis sanguisuga)
COMMENTS: Spectra of hemoglobin, oxyhemoglobin and carbon monoxyhemoglobin
REFERENCE: E. J. Wood, L. J. Mozby, and M. S. Robinson, Biochem. J., 153, 589 (1976)

1093. Hemoglobin

SOURCE: Horse erythrocytes
SOLVENT: a. Water
b. Cyanide solution
c. pH 4.68 and pH 10.8
COMMENTS: Spectra of hemoglobin (solvent a and b), methemoglobin and cyanmethemoglobin (solvent c), green hemoglobin (solvent c), green hemoglobin HCl, and coal gas reduced green hemoglobin
REFERENCE: H. F. Holden, Aus. J. Exp. Biol. Med. Sci., 21, 158 (1944)

1094. Hemoglobin, Manganese-

SOLVENT: 0.05 Tris HCl, pH 7.0, in presence of toluidine blue-O
COMMENTS: Spectra of fully oxidized, partially reduced, fully reduced, and after addition of DPG
REFERENCE: C. Bull, R. G. Fisher, and B. M. Hoffman, Biochem. Biophys. Res. Comm., 59, 140 (1974)

1095. Hemoglobin, Cobalt-

SOURCE: Synthetic
SOLVENT: 0.025 M Borate buffer, pH 9.2
COMMENTS: Spectra of cobaltiglobin and cyanide complex; spectrum of change with time after cyanide addition
REFERENCE: S. Ridsdale, J. C. Cassatt, and J. Steinhardt, J. Biol. Chem., 248, 771 (1973)

1096. Hemoglobin

SOURCE: Mermis nigrescens
SOLVENT: 0.1 M Phosphate buffer, pH 7.0
COMMENTS: Spectra of oxy-, deoxy-, and CO-forms
REFERENCE: A. H. Burr, R. Schiefke, and G. Bollerup, Biochim. Biophys. Acta, 405, 404 (1975)

1097. Hemoglobin

SOURCE: Hagfish (Myxine glutinosa L)
COMMENTS: Spectrum of oxy-hemoglobin I
REFERENCE: S. Paleus and G. Liljeqvist, Comp. Biochem. Physiol., 42B, 611 (1972)

1098. Hemoglobin

SOURCE: Opossum (Didelphius marsupialis)
SOLVENT: 0.1 M Phosphate buffer, pH 7.0
COMMENTS: Spectra of deoxy-, oxy-, met-, and cyanmet-forms
REFERENCE: M. R. Waterman and P. Stenzel, Biochem. Biophys. Acta, 359, 401 (1974)

1099. Hemoglobin

SOURCE: Opossum (Didelphius marsupialis)
SOLVENT: pH 7.4
COMMENTS: Spectra of reduced, partially saturated, and fully oxygen saturated
REFERENCE: C. Petty, N. C. Bethlefalvay, and T. Bageant, Comp. Biochem. Physiol., 50A, 273 (1975)

1100. Hemoglobin

SOURCE: Opossum (Didelphius marsupialis)
SOLVENT: 0.2 M Bis (2-hydroxyethyl) imino-tris (hydroxymethyl) methane, pH 6.7
COMMENTS: Spectral change upon addition of inositol phosphate to nitrosyl hemoglobin
REFERENCE: M. E. John and M. R. Waterman, J. Biol. Chem., 254, 11953 (1979)

1101. Hemoglobin

SOURCE: Pig blood
COMMENTS: Spectra of adult and fetal hemoglobins in oxy- and meta-forms
REFERENCE: E. Kleihauer and C. Tautz, Res. Exp. Med., 188, 219 (1972)

1102. Hemoglobin

SOURCE: Pig blood
SOLVENT: a. 60 mM EDTA, pH 6.8
b. 0.1 M Phosphate buffer, pH 7.0, 10 mM EDTA
COMMENTS: 1. Changes in spectrum due to photochemically generated superoxide anions (solvent a)

2. Changes in spectrum of methemoglobin upon illumination in presence of riboflavin (solvent b)

REFERENCE: R. E. Lynch, J. E. Thomas, and G. R. Lee, Biochemistry, 16, 4563 (1977)

1103. Hemoglobin

SOURCE: Pig erythrocytes
SOLVENT: 0.2 M Sodium phosphate buffer, pH 6.8, containing equal volume of glycerol
COMMENTS: Spectra of deoxyhemoglobin at room temperature and liquid nitrogen temperature
REFERENCE: Y. A. Sharonov, A. P. Mineyev, M. A. Livshitz, N. A. Sharonova, V. B. Zhurkin, and Y. P. Lysov, Biophys. Str. Mech., 4, 139 (1978)

1104. Hemoglobin

SOURCE: Rabbit erythrocytes
COMMENTS: Spectra of reconstituted cyanmet form with and without dithionite; spectrum of aeration of deoxyhemoglobin
REFERENCE: C. Baglioni, T. Campana, and B. Colombo, Arch. Biochem. Biophys., 117, 515 (1976)

1105. Hemoglobin

SOURCE: Rabbit erythrocytes
COMMENTS: Spectra of CO-hemoglobin and after separate additions of cyanide plus ferricyanide, and hydroxylamine plus cyanide
REFERENCE: J. R. Shaeffer and L. C. Alternburg, J. Biol. Chem., 249, 2242 (1974)

1106. Hemoglobin

SOURCE: Yeast (Candida mycoderma)
SOLVENT: 0.1 M Potassium phosphate buffer, pH 7.0
COMMENTS: Effect of ethyl hydrogen peroxide on spectrum
REFERENCE: R. Oshino, N. Oshino, B. Chance, and B. Hagihara, Eur. J. Biochem., 35, 23 (1973)

1107. Hemoglobin

SOURCE: Yeast (Candida mycoderma)
SOLVENT: 0.1 M Phosphate buffer, pH 6.0
COMMENTS: Spectra of hemoglobin, dithionite reduced, and dithionite reduced-CO forms

REFERENCE: R. Oshino, T. Asakura, K. Takio, N. Oshino, B. Chance, and B. Hagihara, Eur. J. Biochem., 39, 581 (1973)

1108. Hemoglobin

SOURCE: Human erythrocytes, adult and fetal
COMMENTS: Spectra of oxyhemoglobin
REFERENCE: G. H. Beaven, H. Hoch, and E. R. Holiday, Biochem. J., 49, 374 (1951)

1109. Hemoglobin

SOURCE: Human erythrocytes, fetal
REFERENCE: W. Kunzer and H. Drescher, Klin. Wochen., 918 (1954)

1110. Hemoglobin

SOURCE: Human erythrocytes
SOLVENT: Mercaptoacetic acid
COMMENTS: Spectral changes with time
REFERENCE: A. Shoden and P. Sturgeon, Acta Haemat., 23, 376 (1960)

1111. Hemoglobin

SOURCE: Human erythrocytes
COMMENTS: Spectra of Hb γ_4 and Hb-X
REFERENCE: D. Todd, M. C. S. Lai, G. H. Beaven, and E. R. Huehns, Brit. J. Hematol., 19, 27 (1970)

1112. Hemoglobin

SOURCE: Human erythrocytes, fetal
SOLVENT: pH 7.0
COMMENTS: Spectrum of oxyhemoglobin
REFERENCE: G. Melki, Biochim. Biophys. Acta, 263, 226 (1972)

1113. Hemoglobin

SOURCE: Human erythrocytes, fetal
SOLVENT: a. Phosphate buffer, pH 7.0
b. Phosphate buffer, pH 5.89
c. Borate buffer, pH 9.06
COMMENTS: Spectra of ferrihemoglobin F (solvents b and c) after addition of potassium ferricyanide
REFERENCE: E. C. Abraham and J. F. Taylor, J. Biol. Chem., 250, 3929 (1975)

1114. Hemoglobin A, Met-

SOURCE: Human erythrocytes
SOLVENT: 0.05 M Bis (2-hydroxyethyl) imino-tris-(hydroxymethyl)-methane, pH 6.0-pH 6.5
COMMENTS: Spectra of methemoglobin A with sodium thiocyanate or sodium cyanate in the presence and absence of inositol hexaphosphate at 0°C; spectra with imidazole in the presence and absence of inositol hexaphosphate
REFERENCE: M. F. Perutz, J. K. M. Sanders, D. H. Chenery, R. W. Noble, R. R. Pennelly, L. W. M. Fung, C. Ho, I. Giannini, D. Porschke, and H. Winkler, Biochemistry, 17, 3640 (1978)

1115. Hemoglobin

SOURCE: Human erythrocytes, glucose-deleted
SOLVENT: Phosphate buffer, pH 7.5
COMMENTS: Spectra of intermediate hemoglobin under aerobic and anaerobic conditions
REFERENCE: A. Tomoda, M. Kateshita, and Y. Yoneyama, FEBS Lett., 88, 247 (1978)

1116. Hemoglobin A

SOURCE: Human erythrocytes
COMMENTS: Spectrum of hemoglobin A in various oxygenation states
REFERENCE: J. D. Mahoney, S. C. Ross, and S. J. Gill, Analyt. Biochem., 78, 535 (1977)

1117. Hemoglobin, met-

SOURCE: Human erythrocytes
COMMENTS: Spectra of films at pH 6 of methemoglobin at 0-99.7% relative humidity
REFERENCE: S. Bohm and L. V. Abaturov, FEBS Lett., 77, 21 (1977)

1118. Hemoglobin

SOURCE: Human erythrocytes
COMMENTS: Spectra of oxy and deoxy forms at 4°C
REFERENCE: P. L. San Biagio, E. Vitrano, A. Cupane, F. Madonia, and M. U. Palma, Biochem. Biophys. Res. Comm., 77, 1158 (1977)

1119. Hemoglobin

SOURCE: Human erythrocytes
COMMENTS: Spectra of hemoglobin from erythrocytes before and after 24 hour treatment with potassium ferricyanide
REFERENCE: P. D. Morse, Biochem. Biophys. Res. Comm., 77, 1486 (1977)

1120. Hemoglobin

SOURCE: Human erythrocytes
COMMENTS: Spectral changes associated with haptoglobin binding to oxy- and deoxy-hemoglobin
REFERENCE: S. H. C. Ip, M. L. Johnson, and G. K. Ackers, Biochemistry, 15, 654 (1976)

1121. Hemoglobin

SOURCE: Human erythrocytes
SOLVENT: Tris-HCl, pH 7.0-8.0 and glycine-NaOH, pH 9.0
COMMENTS: Spectra of ferric hemoglobin at 82°K at different pH values
REFERENCE: S. Yoshida, T. Iizuka, T. Nozawa, and M. Hatano, Biochim. Biophys. Acta, 405, 122 (1975)

1122. Hemoglobin

SOURCE: Human erythrocytes
SOLVENT: 50 mM Tris-Cl, pH 7.6, containing 0.1 M KCl and 1 mM EDTA
COMMENTS: Spectra of oxyhemoglobin in the presence of liposomes at 25°C and 37°C
REFERENCE: L. Bossi, S. Alema, P. Colissaino, and E. Marra, Biochim. Biophys. Acta, 375, 477 (1975)

1123. Hemoglobin, met-

SOURCE: Human erythrocytes
COMMENTS: Spectra at different stages of potential reduction
REFERENCE: F. Scheller, M. Janchen, J. Lampe, H. J. Prumke, J. Blanck, and E. Palecek, Biochim. Biophys. Acta, 452, 157 (1975)

1124. Hemoglobin

SOURCE: Human erythrocytes
SOLVENT: 0.5 M Acetate buffer, pH 5.55
COMMENTS: Spectral changes with time (0-45 min.) upon addition of sodium azide
REFERENCE: W. J. Wallace, J. C. Maxwell, and W. S. Caughey, Biochem. Biophys. Res. Comm., 57, 1104 (1974)

1125. Hemoglobin

SOURCE: Human erythrocytes
SOLVENT: Citrate-phosphate buffer, pH 4.0
COMMENTS: Spectra in presence and absence of haptoglobin 2-1
REFERENCE: T. Sasazuki, H. Tsunoo, H. Nakajima, and K. Imai, J. Biol. Chem., 249, 2441 (1974)

1126. Hemoglobin

SOURCE: Human erythrocytes
SOLVENT: Citrate-phosphate buffer, pH 4.5 and pH 5.3
COMMENTS:
1. Spectral changes associated with regeneration of acid-denatured hemoglobin in presence of haptoglobin
2. Spectra of hemoglobin-haptoglobin complexes in the presence and absence of hydrogen peroxide, and upon addition of ferrocyanide

REFERENCE: T. Sasazuki, H. Tsunoo, H. Nakajima, and K. Imai, J. Biol. Chem., 249, 2441 (1974)

1127. Hemoglobin

SOURCE: Human erythrocytes
SOLVENT: pH 7.4
COMMENTS: Spectra of oxyhemoglobin, plus ATP, and plus AMP, and CO-hemoglobin plus ATP and changes in spectra with time (0-26 hours)
REFERENCE: M. Loeffler, F. K. Schweighovelt, S. S. Chong, S. C. Chiang, and N. C. Li, J. Carbohydrates Nucl. Nucl., 1, 121 (1974)

1128. Hemoglobin

SOURCE: Human erythrocytes
COMMENTS: Spectral changes seen upon addition of methylene blue to oxyhemoglobin and CO-hemoglobin
REFERENCE: C. Bernasconi, G. C. Gerli, R. Stabilini and A. Agostini, Experentia, 30, 601 (1974)

1129. Hemoglobin

SOURCE: Human erythrocytes
SOLVENT: 0.1 M Potassium phosphate buffer, pH 7.0
COMMENTS: Spectra of ferrous hemoglobin and CO-hemoglobin
REFERENCE: T. Iizuka, H. Yamamoto, M. Kotani, and T. Yonetani, Biochim. Biophys. Acta, 371, 126 (1974)

1130. Hemoglobin A

SOURCE: Human erythrocytes
SOLVENT: 0.1 M Phosphate buffer, pH 7.0
COMMENTS: Spectra of hemoglobin-haptoglobin complexes in deoxy forms
REFERENCE: K. Kawamura, S. Kagiyama, A. Ogawa, and T. Yanase, Biochim. Biophys. Acta, 285, 12 (1972)

1131. Hemoglobin A, Cobalt

SOURCE: Human erythrocytes
SOLVENT: 0.1 M Phosphate buffer, pH 7.0-pH 7.2
COMMENTS:
1. Spectra in presence and absence of carbon dioxide
2. Spectrum of cyanide treated form
3. Spectra of deoxy- and oxygenated forms
4. Spectrum of cobaltihemoglobin

REFERENCE: L. C. Dickinson and J. C. W. Chien, J. Biol. Chem., 248, 5005 (1973)

1132. Hemoglobin A_o, A_2, A_2'

SOURCE: Human erythrocytes
REFERENCE: I. Tyuma and Y. Ueda, in Oxygen Affinity of Hemoglobin and Red Cell Acid Base Status (M. Rorth and P. Astrup, eds.), 157 (1972), Academic Press, New York

1133. Hemoglobin A, met-

SOURCE: Human erythrocytes
SOLVENT: Phosphate buffer
COMMENTS: Spectra of non-deionized and deionized forms
REFERENCE: O. Ristau and H. Rein, VIth International Symposium on Structure and Function of Erythrocytes, 129 (1972), Akad. Verlag, Berlin

1134. Hemoglobin

SOURCE: Human erythrocytes
COMMENTS: Spectral changes due to potassium linoleate and time
REFERENCE: Y. Nakamura and T. Nishido, J. Lipid Res., 12, 149 (1971)

1135. Hemoglobin

SOURCE: Human erythrocytes
SOLVENT: pH 7

COMMENTS: Spectra of oxy-, deoxy-, and carbon monoxide forms
REFERENCE: E. Antonini and M. Brunori, in Hemoglobin and Myoglobin in Their Reactions with Ligands (Frontiers of Biology #21), 18 (1971), American Elsevier Publishing Co., New York

1136. Hemoglobin

SOURCE: Human erythrocytes
SOLVENT: 0.025 M Tris-acetate buffer prepared in 0.1 M NaCL, pH 7
COMMENTS: Spectra of "stripped" oxyhemoglobin in presence and absence of ATP or diphosphoglycerate or N-ethylmaleimide
REFERENCE: S. A. Morell, V. E. Ayers, and S. Patkar, Physiol. Chem. and Phys., 2, 467 (1970)

1137. Hemoglobin

SOURCE: Human erythrocytes
COMMENTS: Spectral changes in oxyhemoglobin on oxidation in air (0 to 10 days) and upon addition of catalase at tenth day
REFERENCE: Y. V. Shtamm, A. P. Purmal, and Y. I. Shurlatov, Biophysics, 15, 1164 (1970)

1138. Hemoglobin

SOURCE: Human erythrocytes
SOLVENT: 0.1 M Potassium phosphate buffer, pH 7.0
COMMENTS:
1. Spectra of deoxygenated form and carbon monoxide form
2. Spectral changes upon photo-dissociation of carbon monoxide form

REFERENCE: T. Iizuka, H. Yamamoto, M. Kotani, and T. Yonetani, Biochim. Biophys. Acta, 371, 126 (1974)

1139. Hemoglobin

SOURCE: Human erythrocytes
SOLVENT:
a. 0.02 M Acetate buffer, pH 5.5
b. 0.067 M Acetate buffer, pH 4.0 and pH 4.5
c. 0.1 N HCl
d. Glycine buffer, pH 7.7

COMMENTS:
1. Spectra of oxyhemoglobin oxidation (air) product and of its reduction (dithionite) product in the presence and absence of mercurous ion
2. Spectra of oxyhemoglobin (solvent b and solvent c)

3. Spectrum of methemoglobin (solvent a)
REFERENCE: K. Murawski and K. Zakrzewski, Acta Biochim. Polon., 11, 379 (1964)

1140. Hemoglobin

SOURCE: Human erythrocytes
SOLVENT: a. 8 M Urea-0.1 M sodium acetate, pH 4.0
b. 0.05 M Sodium phosphate buffer, pH 7.0
COMMENTS: 1. Spectral changes due to reaction with N-bromosuccinimide
2. Spectral changes associated with the addition of urea (0-8 M)
3. Spectral changes associated with the addition of sucrose (50% and 70%)
REFERENCE: G. J. S. Rao and H. R. Cama, Biochim. Biophys. Acta, 71, 139 (1963)

1141. Hemoglobin

SOURCE: Human erythrocytes
COMMENTS: Spectral changes associated with the addition of sodium dodecylsulfate (0.008% and 0.4%) to oxy-hemoglobin
REFERENCE: L. Konieczny and J. Domanski, Acta Biochim. Polon., 10, 325 (1963)

1142. Hemoglobin

SOURCE: Human erythrocytes
SOLVENT: Acetate buffer, pH 4.8, pH 4.1, and pH 3.8
COMMENTS: Changes in the absorption spectrum of oxyhemoglobin with time at pH 4.8, pH 4.1, and pH 3.8
REFERENCE: E. Bucci and C. Fronticelli, Ital. J. Biochem., 10, 313 (1961)

1143. Hemoglobin

SOURCE: Human erythrocytes
COMMENTS: Changes in the spectrum of oxyhemoglobin due to ultrasonic treatment in the presence and and absence of ether
REFERENCE: A. Weissler, J. Acoustical Soc., 32, 1208 (1960)

1144. Hemoglobin

SOURCE: Human erythrocytes
COMMENTS: Spectrum of carbon monoxide hemoglobin
REFERENCE: H. Kromphardt and D. Lubbers, Biochem. Z., 330, 342 (1958)

1145. Hemoglobin

SOURCE: Human erythrocytes
SOLVENT: 0.2 M Phosphate, pH 6.5
REFERENCE: M. Kiese, H. Kurz, and C. Schneider, Klin. Woch., 34, 957 (1956)

1146. Hemoglobin

SOURCE: Human erythrocytes
SOLVENT: M/15 Phosphate buffer, pH 7.38
COMMENTS: 1. Spectral changes seen during the following reactions: oxyhemoglobin plus cyanide and hydrogen peroxide, then dithionite, then sodium hydroxide
2. Spectral changes seen in the system oxyhemoglobin-cyanide-ascorbic acid-oxygen
REFERENCE: G. Kikuchi, M. Makita, and T. Saito, J. Biochem. Tokyo, 41, 227 (1954)

1147. Hemoglobin

SOURCE: Human erythrocytes
SOLVENT: M/15 Phosphate buffer, pH 7.38
COMMENTS: Spectral changes seen in the system oxyhemoglobin-cyanide-dithionite-ascorbic acid-oxygen
REFERENCE: G. Kikuchi, J. Biochem. Tokyo, 41, 247 (1954)

1148. Hemoglobin, met-

SOURCE: Human erythrocytes
SOLVENT: 1.2 M Phosphate buffer, pH 7.2
COMMENTS: Spectra in presence of sodium cyanate, potassium thiocyanate and sodium chloride
REFERENCE: R. Bader, P. Dirnhuber, and F. Schutz, Biochim. Biophys. Acta, 5, 105 (1950)

1149. Hemoglobin

SOURCE: Human erythrocytes
COMMENTS: Spectra of oxyhemoglobin, sulfhemoglobin, methemoglobin, and cyanmethemoglobin
REFERENCE: K. A. Evelyn and H. T. Malloy, J. Biol. Chem., 126, 655 (1938)

1150. Hemoglobin A, sulf-

SOURCE: Human hemoglobin
SOLVENT: 0.1 M Potassium phosphate, pH 8.2 and pH 5.4

COMMENTS: Spectra of oxy- and deoxysulfhemoglobins, and ferric sulfhemoglobin
REFERENCE: R. J. Carrico, J. Peisach, and J. O. Allen, J. Biol. Chem., 253, 2386 (1978); R. J. Carrico, W. E. Blumberg, and J. Peisach, J. Biol. Chem., 253, 7212 (1978)

1151. Hemoglobin (intermediate)

SOURCE: Human erythrocytes
COMMENTS: Intermediates made by partial oxidation of hemoglobin with ferricyanide and identified as $\alpha^{2+}\ \beta^{3+}$ and $\alpha^{3+}\ \beta^{2+}$ hybrids
REFERENCE: A. Tomoda and Y. Yoneyama, Biochim. Biophys. Acta, 581, 128 (1979)

1152. Hemoglobin

SOURCE: Human erythrocytes
SOLVENT: Phosphate buffer, pH 7.4, containing EDTA and propan-2-ol
COMMENTS: Spectral changes associated with menadione oxidation of oxyhemoglobin
REFERENCE: C. C. Winterbourn, J. K. French, and R. F. C. Claridge, Biochem. J., 179, 665 (1979)

1153. Hemoglobin (intermediate)

SOURCE: Human hemoglobin
SOLVENT: Phosphate buffer, pH 7.55-7.65
COMMENTS: Intermediate made by oxidation of methmoglobin with ascorbic acid and identified as $\alpha^{3+}\ \beta^{2+}$ hybrid (oxy and deoxy forms)
REFERENCE: A. Tomoda, M. Takeshita, and Y. Yoneyama, J. Biol. Chem., 253, 7415 (1979)

1154. Hemoglobin

SOURCE: Human hemoglobin
COMMENTS: Spectra of hemoglobin before and after treatment with ascorbic acid
REFERENCE: T. Murakami and T. Murachi, J. Biochem. Tokyo, 84, 83 (1978)

1155. Hemoglobin, Dimethylsuberimidate Cross-linked

SOURCE: Hemoglobin
SOLVENT: pH 8.5
COMMENTS: Spectrum of cross-linked oxyhemoglobin

REFERENCE: B. Horowitz and A. Mazur, in Blood Substitutes and Plasma Expanders (G. A. Jamieson and T. J. Greenwalt eds.), 149 (1978), A. R. Liss, Inc., New York

1156. Hemoglobin

SOURCE: Human erythrocytes
COMMENTS: Spectral changes upon addition of phenylhydrazine in the presence and absence of cyanide
REFERENCE: C. E. Castro, R. S. Wade, and N. O. Belser, Biochemistry, 17, 225 (1978)

1157. Hemoglobin, dextran-

SOLVENT: 0.05 M Sodium phosphate, pH 7.0
REFERENCE: S. C. Tam, J. Blumenstein, and J. T. F. Wong, Proc. Nat. Acad. Sci., 73, 2128 (1976)

1158. Hemoglobin

SOURCE: Human erythrocytes
SOLVENT: Oxygen-saturated phosphate buffer, pH 7.0, containing 6.7 mM EDTA
COMMENTS: Spectra of methemoglobin and oxyhemoglobin before and after exposure to superoxide for different times
REFERENCE: C. C. Winterbourn, B. M. McGrath, and R. W. Carrell, Biochem. J., 155, 493 (1976)

1159. Hemoglobin

SOURCE: Human erythrocytes
COMMENTS: Spectral changes of ferrihemoglobin upon reaction with phenyldiazene
REFERENCE: H. A. Itano and S. Mannen, Biochim. Biophys. Acta, 421, 87 (1976)

1160. Hemoglobin, NO-

SOURCE: Human hemoglobin
SOLVENT: 0.05 M Potassium phosphate, pH 7.0
COMMENTS: Spectral changes during dissociation of nitric oxide from hemoglobin (0-21 hours)
REFERENCE: E. G. Moore and Q. H. Gibson, J. Biol. Chem., 251, 2788 (1976)

1161. Hemoglobin, met-

SOURCE: Human hemoglobin
SOLVENT: a. 0.1 M Bis (2-hydroxyethyl) imino-tris (hydroxymethyl) methane, 0.1 M NaCl, pH 6.0 and pH 7.1
b. 0.1 M Tris, 0.1 M NaCl, pH 8.0
c. 0.05 M Borate, pH 9.1
COMMENTS: Spectra of methemoglobin and p-mercuribenzoate treated methemoglobin
REFERENCE: J. S. Olson, J. Biol. Chem., 251, 441 (1976)

1162. Hemoglobin

SOURCE: Human erythrocytes
COMMENTS: Spectra of oxygenated, deoxygenated, and reoxygenated hemoglobin reacted with dimethyl-dithiobis propionimidate
REFERENCE: A. Ruoko, P. A. Bartlett, A. Dutton, and S. J. Singer, Biochem. Biophys. Res. Comm., 63, 417 (1975)

1163. Hemoglobin, ferri-

SOLVENT: 0.05 M Sodium phosphate buffer, pH 6.84
COMMENTS: Spectral changes upon reaction with the following oxidized phenyhydrazines: 2,3-dichloro-, 3,4-dichloro-, 3,5-dichloro-, and 2,6-dichloro-
REFERENCE: H. A. Itano, K. Hirota, and K. Hosokawa, Nature, 256, 665 (1975)

1164. Hemoglobin

SOURCE: Human erythrocytes
COMMENTS: Spectral changes due to interaction of oxy- and deoxyhemoglobin with p-aminophenol
REFERENCE: J. Peisach, W. E. Blumberg, and E. Rachmilewitz, Biochim. Biophys. Acta, 393, 404 (1975)

1165. Hemoglobin

SOURCE: Human erythrocytes
SOLVENT: pH 8.3
COMMENTS: Spectral changes due to reaction of oxyhemoglobin with phenol
REFERENCE: W. J. Wallace and W. S. Caughey, Biochem. Biophys. Res. Comm., 62, 561 (1975)

1166. Hemoglobin, Modified

SOURCE: Human hemoglobin
SOLVENT: 0.1 M Potassium phosphate buffer, pH 7.0
COMMENTS: Spectra of oxygenated, reduced (dithionite) and CO-reduced forms of 2-formyl 4-vinyl hemin hemoglobin (spirographis), 2-vinyl-4-formyl hemin hemoglobin (isospirographis), and 2,4-diformyl hemin hemoglobin
REFERENCE: T. Asakura and M. Sono, J. Biol. Chem., 249, 7087 (1974)

1167. Hemoglobin, Cobalt, Modified

SOURCE: Human hemoglobin
SOLVENT: 0.1 M Phosphate buffer, pH 7.0
COMMENTS:
1. Spectra of oxyhemoglobin and deoxyhemoglobin (dithionite) having the following cobaltous containing porphyrins: proto-, meso-, deutero-porphyrin IX
2. The effect of inositol hexaphosphate on protoporphyrin IX-containing spectra

REFERENCE: T. Yonetani, H. Yamamoto, and G. V. Woodrow, J. Biol. Chem., 249, 682 (1974)

1168. Hemoglobin Hybrids

SOLVENT: 0.1 M Phosphate buffer, pH 7.0
COMMENTS:
1. Spectral changes during oxygenation of α(proto) β(meso) and α(meso) β(proto) porphyrin hemoglobin
2. Spectra of oxy-, deoxy-, carbon monoxy-, and m ethemoglobins containing the following porphyrins: α(meso) β(proto), α(proto) β(meso), and a mixture of proto- and meso- hemoglobins

REFERENCE: H. Yamamoto and T. Yonetani, J. Biol. Chem., 249, 7964 (1974)

1169. Hemoglobin

SOURCE: Human hemoglobin
SOLVENT:
a. 0.1 M Phosphate buffer, pH 7.0
b. 0.05 M Sodium phosphate
c. 0.05 M Bis (2-hydroxyethyl) imino-tris-(hydroxymethyl) methane

COMMENTS:
1. Spectra of N-ethylsuccinimide-des Arg hemoglobin, hemoglobin Kempsey, and des-Arg-des-Tyr hemoglobin in the presence and absence of inositol hexaphosphate (solvent a)

2. Spectra of oxyhemoglobin A in the presence and absence of inositol hexaphosphate (solvent b)
3. Spectrum of N-ethylsuccinimide des-Arg-oxyhemoglobin (solvent b)
4. Spectra of des-Arg(114α)-des Tyr (140α) hemoglobin in the presence and absence of inositol hexaphosphate (solvent c)

REFERENCE: M. F. Perutz, J. E. Ladner, S. R. Simon, and C. Ho, Biochemistry, 13, 2163 (1974)

1170. Hemoglobin, met-

SOURCE: Human hemoglobin

SOLVENT:
a. 0.01 M Bis (2-hydroxyethyl) imino-tris (hydroxymethyl) methane, pH 6.5
b. 0.1 Acetate buffer, 0.1 NaCl, pH 5
c. 0.05 M Bis (2-hydroxyethyl) imino-tris (hydroxymethyl) methane, 0.1 M NaCl, pH 7.0 or 0.2 M Bis (2-hydroxyethyl) imino-tris (hydroxymethyl) methane, pH 7.0

COMMENTS:
1. Spectra of fluoromethemoglobin in the presence and absence of inositol hexaphosphate (solvent a)
2. Spectra of methemoglobin in the presence and absence of inositol hexaphosphate (solvent b and solvent c)
3. Spectra of azidomethemoglobin in the presence and absence of inositol hexaphosphate (solvent c)
4. Spectral changes due to increasing inositol hexaphosphate: methemoglobin ratio at pH 5.0, 6.0, 7.2, and 7.55

REFERENCE: M. F. Perutz, E. J. Heidner, J. E. Ladner, J. G. Beetlestone, C. Ho, and E. F. Slade, Biochemistry, 13, 2187 (1974)

1171. Hemoglobin, 2,2,5,5-tetramethyl-3-amino-pyrrolidine-1-oxyl-(SL)

SOURCE: Human erythrocytes

SOLVENT:
a. 0.1 M Potassium phosphate buffer, pH 6.0
b. 1 mM Potassium phosphate buffer, pH 6.2
c. 0.1 M Potassium phosphate buffer, pH 7.0

COMMENTS:
1. Spectra of 2,2,5,5-tetramethyl-3-amino-pyrrolidine-1-oxyl (SL) labelled protein in the presence of cyanide, azide, and fluoride (solvent a)
2. Spectral changes due to changes in pH of a solution of SL-hemoglobin (solvent b plus NaOH)

3. Spectra of SL-hemoglobin at different stages of purification
4. Spectra of SL-hemoglobin in the oxy,deoxy (dithionite), and CO-deoxy states (solvent c)

REFERENCE: T. Asakura, J. Biol. Chem., 249, 4495 (1974); T. Asakura and M. Tamura, J. Biol. Chem., 249, 4504 (1974)

1172. Hemoglobin, Sephadex Bound

SOURCE: Human erythrocytes
COMMENTS: Spectra of oxy and deoxy forms
REFERENCE: A. Colosimo, S. Stefanini, M. Brunori, and E. Antonini, Biochim. Biophys. Acta, 328, 74 (1973)

1173. Hemoglobin, meso-, proto-, and deutero-

SOURCE: Mammalian erythrocytes
COMMENTS: Spectra of mesohemoglobin, protohemoglobin and deuterohemoglobin in the oxy and deoxy states
REFERENCE: Y. Sugita, S. Bannai, Y. Yoneyama, and T. Nakamura, J. Biol. Chem., 247, 6092 (1972)

1174. Hemoglobin, Manganese-protoporphyrin IX-

SOURCE: Human hemoglobin
COMMENTS: Spectra of Mn^{III}, Mn^{II}, and Mn^{II}-NO derivatives
REFERENCE: T. Yonetani, H. Yamamoto, J. E. Erman, J. S. Leigh, and G. H. Reed, J. Biol. Chem., 247, 2447 (1972)

1175. Hemoglobin, 2,2,5,5-tetramethyl-3-amino pyrrolidine-oxyl (SL-Hb)

SOURCE: Human globin
SOLVENT: 0.1 M Potassium phosphate buffer, pH 6.3
COMMENTS: Spectral changes seen upon reduction of ferric SL-Hb by ferredoxin reductase system
REFERENCE: T. Asakura, M. Tamura, and M. Shin, J. Biol. Chem., 247, 3693 (1972)

1176. Hemoglobin

SOURCE: Human erythrocytes
SOLVENT: 0.1 M Tris buffer, pH 7.4
COMMENTS: Spectral changes associated with binding of nitrosobenzene to oxyhemoglobin
REFERENCE: M. Murayama, J. Biol. Chem., 235, 1024 (1960)

1177. Hemoglobin

SOURCE: Human erythrocytes
SOLVENT: Phosphate, pH 7.0-pH 7.10
COMMENTS: Spectral changes associated with binding of nitrosobenzene to oxyhemoglobin and reduced hemoglobin
REFERENCE: 1. W. Scheler, Acta Biol. Med. Germ., 5, 382 (1960)
2. T. W. Szezepkowski, Acta Biochim. Polon., 2, 385 (1955)

1178. Hemoglobin

SOLVENT: 0.05 M Phosphate, pH 7.0
COMMENTS: 1. Spectral changes upon removal of CO from α_2 spin-labeled hemoglobin-CO
2. Spectral changes upon removal of oxygen from oxygenated α_2-spin labeled hemoglobin
3. Spin label is 2,2,5,5-tetramethyl-3-aminopyrrolidine-1-oxyl group
REFERENCE: P. W. Lau and T. Asakura, J. Biol. Chem., 255, 1617 (1980)

1179. Hemoglobin (hybrid)

SOURCE: Human hemoglobin
SOLVENT: pH 6
COMMENTS: Spectrum of α^{3+} β^{2+} CO-hemoglobin
REFERENCE: R. Banerjee and R. Cassoly, J. Mol. Biol., 42, 351 (1969)

1180. Hemoglobin Bart's, Hemoglobin A, and Hemoglobin F

SOURCE: Human erythrocytes
SOLVENT: pH 6.5
COMMENTS: Spectra of methemoglobin A and methemoglobin Bart's
REFERENCE: P. Ronisch and E. Kleihauer, Klin. Woch., 45, 1193 (1967)

1181. Hemoglobin Bart's, Hemoglobin A, and Hemoglobin H

SOURCE: Human erythrocytes
SOLVENT: pH 7.1
COMMENTS: Spectra after oxidation with potassium ferricyanide
REFERENCE: E. A. Rachmilewitz and E. Harari, Brit. J. Hematol., 22, 357 (1972)

1182. Hemoglobin M Boston, Hemoglobin A, and Hemoglobin M Saskatoon

SOURCE: Human erythrocytes
SOLVENT: Sodium phosphate, pH 7.0

COMMENTS: Spectra of met forms and spectrum of methemoglobin A fluoride complex
REFERENCE: P. S. Gerald and P. George, Science, 129, 393 (1959)

1183. Hemoglobin M Boston, met-

SOURCE: Human erythrocytes
SOLVENT: 0.1 M Potassium phosphate buffer, pH 7.0
COMMENTS: Spectral changes associated with increase in cyanide binding to metHb M Boston
REFERENCE: A. Hayashi, A, Shimizu, T. Suzuki, and Y. Yamamura, Biochim. Biophys. Acta, 140, 251 (1967)

1184. Hemoglobin M Boston, met- and Hemoglobin A, met-

SOURCE: Human erythrocytes
COMMENTS: Spectra taken before and after enzyme (glucose-6-phosphate dehydrogenase) reduction
REFERENCE: A. Hayashi, T. Suzuki, and M. Shin, Biochim. Biophys. Acta, 310, 309 (1973)

1185. Hemoglobin Freiburg and Hemoglobin A

SOURCE: Human erythrocytes
SOLVENT: 0.05 M Phosphate buffer, pH 7.0
COMMENTS:
1. Spectra of HbA and Hb Freiburg after oxidation with potassium ferricyanide
2. Spectrum of oxidized whole hemolysate of Hb Freiburg
3. Spectral changes upon reduction with sodium ascorbate of oxidized Hb Freiburg

REFERENCE: E. A. Rachmilewitz and E. Harari, Hematologie and Blut Transfusion, 10, 241 (1972)

1186. Hemoglobin Hammersmith and Hemoglobin A

SOURCE: Human erythrocytes
COMMENTS:
1. Spectra of oxidized forms
2. Spectrum of total hemolysate

REFERENCE: D. Labie and H. Wajcman, Biochimie, 54, 625 (1972)

1187. Hemoglobin M Hyde Park and Hemoglobin A

SOURCE: Human erythrocytes
SOLVENT: Phosphate buffer, pH 6.6
COMMENTS: Spectra of met- and cyanmet- forms
REFERENCE: P. Heller, R. Coleman, and V. J. Yakulis, Proc. XI Cong. Internat. Soc. Hematology, 427 (1966)

1188. Hemoglobin M Hyde Park, Hemoglobin M. Novi-Sad (Saskatoon), and Hemoglobin A (All Met-)

SOURCE: Human erythrocytes
REFERENCE: G. D. Efremov, T. H. J. Huisman, M. Stanulovic, M. Zurovec, H. Duma, J. B. Wilson, and J. Jeremic, Scand. J. Haematol., 13, 48 (1974)

1189. Hemoglobin M Iwate and Hemoglobin A

SOURCE: Human erythrocytes
SOLVENT: 0.01 M Phosphate buffer, pH 7.0
COMMENTS: Spectra of oxidized, cyanide, azide, and fluoride derivatives of oxidized, reduced (dithionite), and CO-reduced forms
REFERENCE: Y. Motokawa, N. Hayashi, and G. Kikuchi, Arch. Biochem. Biophys., 105, 612 (1964)

1190. Hemoglobin M Iwate and Hemoglobin A

SOURCE: Human erythrocytes
SOLVENT: 0.1 M Phosphate buffer, pH 6.5
COMMENTS: Spectra of met Hb A, met Hb M Iwate, cyanmet Hb A, and cyanmet Hb M Iwate
REFERENCE: E. Jaffe and P. Heller, Proc. Hematol., 4, 48 (1964)

1191. Hemoglobin M Iwate

SOURCE: Human erythrocytes
SOLVENT: pH 7.35-pH 7.45
COMMENTS:
1. Spectrum of α_2 (M-met) β(met) $\beta(O_2)$ treated with diphosphoglycerate and oxygenated
2. Spectral changes with time of stripped α_2 (M-met), β_2(CO)

REFERENCE: K. Gersonde, M. Overkamp, H. Sick, E. Trittelvitz, and W. Junge, Eur. J. Biochem., 39, 403 (1973)

1192. Hemoglobin M Iwate

SOURCE: Human erythrocytes
SOLVENT: pH 7.3
COMMENTS: Spectra of nitrosyl Hb M Iwate with and without 2,3-biphosphoglycerate
REFERENCE: E. Trittelvitz, K. Gersonde, and K. Winterhalter, Eur. J. Biochem., 51, 33 (1975)

1193. Hemoglobin M Iwate, met-

SOURCE: Human erythrocytes
SOLVENT: 0.2 M Sodium phosphate buffer, pH 7.0
COMMENTS: Spectra of sodium ascorbate and sodium dithionite treated material
REFERENCE: K. Wishikura, Y. Sugita, M. Nagai, and Y. Yoneyama, J. Biol. Chem., 250, 6679 (1975)

1194. Hemoglobin M Iwate

SOURCE: Human erythrocytes
SOLVENT: 0.2 M Bis (2-hydroxyethyl) imino-tris-(hydroxymethyl)-methane, pH 7.4
COMMENTS: Spectra of reduced, CO-reduced, met, and CO-met forms
REFERENCE: H. T. Gaud, S. J. Gill, B. G. Barisas, and K. Gersonde, Biochemistry, 14, 4584 (1975)

1195. Hemoglobin M Iwate

SOURCE: Human erythrocytes
SOLVENT: 10 mM or 50 mM Potassium phosphate buffer, pH 6.5
COMMENTS:
1. Spectra of deoxy- and carbonmonoxy hemoglobin
2. Spectral changes during titration of Hb M Iwate and of Hb CO M Iwate with dithionite in presence of inositol hexaphosphate

REFERENCE: J. Peisach and K. Gersonde, Biochemistry, 16, 2539 (1977)

1196. Hemoglobin M Kankakee

SOURCE: Human erythrocytes
COMMENTS: Spectra of untreated, oxidized and cyanferrihemoglobins
REFERENCE: P. Heller, H. G. Weinstein, V. J. Yakulis and I. M. Rosenthal, Blood, 20, 287 (1962)

1197. Hemoglobin Kansas

SOURCE: Human erythrocytes
SOLVENT:
a. 0.1 M Bis (2-hydroxyethyl) imino-tris (hydroxymethyl) methane buffer, pH 6.5
b. 0.1 M N-2-Hydroxyethyl piperazine-N'-2-ethane sulfonic acid buffer, pH 7.0

COMMENTS:
1. Spectra of HbO_2 Kansas in presence and absence of inositol hexaphosphate
2. Spectra of HbCO Kansas in presence and absence of inositol hexaphosphate

REFERENCE: M. F. Perutz, J. V. Kilmartin, K. Nagai, A. Szabo, S. R. Simon, Biochemistry, 15, 378 (1976)

1198. Hemoglobin Kempsey and Hemoglobin A

SOURCE: Human erythrocytes
SOLVENT: 0.05 M Bis (2-hydroxyethyl) imino-tris (hydroxymethyl) methane buffer, pH 7.0
COMMENTS: Spectra of oxy- and deoxy- forms
REFERENCE: S. Matsukawa, M. Nishibu, M. Nagai, K. Mawatari, and Y. Yoneyama, J. Biol. Chem., 254, 2358 (1979)

1199. Hemoglobin Koln

SOURCE: Human erythrocytes
SOLVENT: 0.5 M Phosphate buffer, pH 7.0
COMMENTS: Changes in spectrum after oxidation with potassium ferricyanide
REFERENCE: E. A. Rachmilewitz and E. Harari, Hematol. and Blood Trans., 10, 241 (1972)

1200. Hemoglobin Koln and Hemoglobin A

SOURCE: Human erythrocytes
SOLVENT: pH 5.5
COMMENTS: Spectra of Hb Koln, Hb A and Met Hb A
REFERENCE: R. W. Carrel, H. Lehmann, and W. Pribilla, Klin. Woch., 45, 1189 (1967)

1201. Hemoglobin M

SOURCE: Human erythrocytes
SOLVENT: M/15 Sodium phosphate buffer, pH 6.5
COMMENTS: Spectra of HbO_2 M before and after treatment with potassium ferricyanide, of met Hb M, and after addition of cyanide
REFERENCE: P. S. Gerald, Blood, 13, 936 (1958)

1202. Hemoglobin McKees Rocks and Hemoglobin A

SOURCE: Human erythrocytes
SOLVENT: 0.15 M Bis (2-hydroxyethyl) imino tris (hydroxymethyl) methane buffer, pH 7, 0.1 M NaCl
COMMENTS: Spectra of oxy- and deoxy- (dithionite) forms in presence and absence of inositol hexaphosphate
REFERENCE: R. M. Winslow, M. L. Swenberg, E. Gross, P. A. Chervenick, R. R. Buchman, and W. F. Anderson, J. Clin. Inves., 57, 772 (1976)

1203. Hemoglobin M Milwaukee, et- and Hemoglobin A, et-

SOURCE: Human erythrocytes
SOLVENT: a. 0.1 M Citrate buffer, pH 5.0
b. 0.1 M Citrate buffer, pH 6.1
c. 0.1 M Bis (2-hydroxyethyl) imino tris (hydroxymethyl) methane buffer, pH 7.3
d. 0.1 M Glycine buffer, pH 8.1
COMMENTS: Effect of inositol hexaphosphate on spectra of metHb Milwaukee and metHb A
REFERENCE: M. E. John and M. R. Waterman, Biochim. Biophys. Acta, 578, 269 (1979)

1204. Hemoglobin M Milwaukee

SOURCE: Human erythrocytes
SOLVENT: a. 0.1 M Sodium phosphate buffer, pH 6.5
b. 0.05 M Bis (2-hydroxyethyl) imino tris (hydroxymethyl) methane buffer, pH 6.5, 0.1 M NaCl
COMMENTS: 1. Spectra of metHb Milwaukee, metHb CO Milwaukee
2. Spectra of metHb Milwaukee in presence and absence of inositol hexaphosphate
REFERENCE: M. F. Perutz, P. D. Pulsinelli, and H. M. Ranney, Adv. Exp. Med. Biol., 28, 3 (1972)

1205. Hemoglobin Osler and Hemoglobin A

SOURCE: Human erythrocytes
SOLVENT: 0.05 M Bis (2-hydroxyethyl) imino tris (hydroxymethyl) methane buffer, pH 7.8
COMMENTS: Spectra of Hb A, Hb Osler, and Hb Osler plus inositol hexaphosphate
REFERENCE: E. Bucci, C. Fronticelli, J. Nicklas, and S. Charache, J. Biol. Chem., 254, 10811 (1978)

1206. Hemoglobin M Radom, met- and Hemoglobin A, met-

SOURCE: Human erythrocytes
SOLVENT: 0.0167 M Phosphate buffer, pH 6.6
COMMENTS: Spectra of metHb A, metHb Radom and cyanmet forms
REFERENCE: K. Murawski, Z. Szymanowska, and J. Kozlowska, Biochim. Biophys. Acta, 69, 442 (1963)

1207. Hemoglobin Riverdale-Bronx

SOURCE: Human erythrocytes
SOLVENT: Phosphate buffer, pH 8.4
COMMENTS: Spectral changes with time after oxidation with potassium ferricyanide

REFERENCE: E. A. Rachmilewitz and E. Harari, Hematol. and Blut Trans., 10, 241 (1972)

1208. Hemoglobin S

SOURCE: Human erythrocytes
COMMENTS: Spectrum of HbO_2 S and change upon addition of NaOH
REFERENCE: R. Asakura, K. Minakata, K. Adachi, M. Russel, and E. Schwartz, J. Clin. Inves., 59, 641 (1977)

1209. Hemoglobin S

SOURCE: Human erythrocytes
SOLVENT: 2.48 M Potassium phosphate buffer, pH 7.0
COMMENTS: Spectra of oxy- and deoxy- forms
REFERENCE: K. Adachi and T. Asakura, J. Biol. Chem., 254, 4079 (1979)

1210. Hemoglobin S

SOURCE: Human erythrocytes
SOLVENT: 0.1 M Bis (2-hydroxyethyl) imino tris (hydroxymethyl) methane buffer, pH 7.4
COMMENTS: Effect of serum albumin on spectrum of HbO_2 S after shaking
REFERENCE: K. Adachi and T. Asakura, Biochim. Biophys. Acta, 427, 536 (1976)

1211. Hemoglobin S and Hemoglobin A

SOURCE: Human erythrocytes
COMMENTS: Effect of prostaglandin E_2 on Hb S, Hb A, HbO_2 A, and HbO_2 S spectra
REFERENCE: M. Johnson, I. Rabinowitz, A. L. Willis, and P. L. Wolff, Clin. Chem., 19, 23 (1973)

1212. Hemoglobin Saint Etienne, met- and Hemoglobin A, met-

SOURCE: Human erythrocytes
REFERENCE: Y. Beuzard, J. C. Couvolin, M. C. Solal, M. C. Garel, J. Rosa, C. P. Brizard, and A. Gibaud, FEBS Lett., 27, 76 (1972)

1213. Hemoglobin St. Louis, met- and Hemoglobin A, met-

SOURCE: Human erythrocytes
SOLVENT: 0.15 M Phosphate buffer, pH 7.0
REFERENCE: M. Cohen-Solal, M. Seligman, J. Thillet, and J. Rosa, FEBS Lett., 33, 37 (1973)

1214. Hemoglobin M Saskatoon, met-

SOURCE: Human erythrocytes
SOLVENT: 0.1 M Potassium phosphate buffer, pH 7.0
COMMENTS: Spectral changes due to added fluoride
REFERENCE: A. Hayashi, A. Shimizu, T. Suzuki, and Y. Yamamura, Biochim. Biophys. Acta, 140, 251 (1967)

1215. Hemoglobin Yakima

SOURCE: Human erythrocytes
SOLVENT: 0.05 M Bis (2-hydroxyethyl) imino tris (hydroxymethyl) methane buffer, pH 7.0
COMMENTS: Spectra of hemoglobin in presence and absence of inositol hexaphosphate and spectra of oxy- and deoxy- forms
REFERENCE: S. Matsukawa, M. Nishibu, M. Nagai, K. Mawatari, and Y. Yoneyama, J. Biol. Chem., 254, 2358 (1979)

1216. Hemoglobin Zurich and Hemoglobin A

SOURCE: Human erythrocytes
SOLVENT: 0.2 M Phosphate buffer, pH 7.0
COMMENTS: Spectra of oxy- and deoxy- forms
REFERENCE: K. H. Winterhalter, E. E. Dilorio, J. G. Beetlestone J. B. Kushimo, J. Eubelhack, H. Eicher, and A. Mayer, J. Mol. Biol., 70, 665 (1972)

1217. Hemoglobin Zurich

SOURCE: Human erythrocytes
SOLVENT: pH 6.7 and pH 5.7
COMMENTS: Spectral changes with time due to addition of azide
REFERENCE: W. J. Wallace, J. A. Volpe, J. C. Maxwell, W. S. Caughey, and S. Charache, Biochem. Biophys. Res. Comm., 68, 1379 (1976)

1218. Hemoglobin, semi-

SOURCE: Heme-containing gamma subunits plus apo-alpha subunits
SOLVENT: 0.1 M Phosphate buffer, pH 7.0
COMMENTS: Spectra of met-, deoxy-, and carbonmonoxy forms
REFERENCE: R. Gondko, M. J. Obreska, and M. R. Waterman, Biochem. Biophys. Res. Comm., 56, 444 (1974)

1219. $\alpha^- \beta^+$-Semi Hemoglobin

SOURCE: Human hemoglobin
SOLVENT: 0.1 M Phosphate buffer, pH 7.0
COMMENTS: Spectra of $\alpha^- \beta^+$-semi hemoglobin in met-, deoxy-, oxy-, and carbonmonoxy states
REFERENCE: M. R. Waterman, R. Gendko, and T. Yonetani, Arch. Biochem. Biophys., 145, 448 (1971)

1220. Hemoglobin-cysteine Complexes

SOURCE: Human hemoglobin
SOLVENT: pH 6.78; NH_3, pH 9.84; sodium carbonate, pH 9.90; pH 9.91
COMMENTS: Spectra of ferro-hemoglobin-cysteine complex and ferri-hemoglobin-cysteine complex
REFERENCE: W. R. Holmquist and J. R. Vinograd, Biochim. Biophys. Acta, 69, 337 (1963)

1221. Hemoglobin Subunits

SOURCE: Human hemoglobin
SOLVENT:
a. Phosphate buffer, pH 7.0
b. Borate buffer, pH 9.06
c. Phosphate buffer, pH 5.89

COMMENTS:
1. Spectra of ferri Hb A, ferric α-p-mercuribenzoate chain, and ferric β-p-mercuribenzoate chain (solvent a)
2. Spectra of ferri Hb F, ferric γ-p-mercuribenzoate chain (solvents b and c)

REFERENCE: E. C. Abraham and J. F. Taylor, J. Biol. Chem., 250, 3929 (1975)

1222. Hemoglobin A, Hemoglobin H, and α- and β- subunits

SOURCE: Human erythrocytes
SOLVENT: 0.1 M Phosphate buffer, pH 6.8-7.2
COMMENTS:
1. Spectral changes with time of HbO_2 A and HbO_2 H due to oxidation with ferricyanide
2. Spectral change in oxidized Hb H due to ascorbate
3. Spectral change in ferri Hb A due to added sodium salicylate
4. Spectra of oxidized Hb A and oxidized α-subunits
5. Spectral changes with time after oxidation of oxy-α-p-mercuribenzoate and oxy-β-p-mercuribenzoate subunits by ferricyanide

REFERENCE: E. A. Rachmilewitz, J. Reisach, and W. E. Blumberg, J. Biol. Chem., 246, 3356 (1971)

1223. Globin, α- and β-Chains

SOURCE: Human hemoglobin
SOLVENT: a. 0.02 M Phosphate buffer, pH 6.7
b. 0.1 M Tris-HCl buffer, pH 8
REFERENCE: Y. K. Yip, M. Waks, and S. Beychock, J. Biol. Chem., 247, 7237 (1972)

1224. Hemoglobin: α- and β-cyanmet and α-met Subunits

SOURCE: Human hemoglobin
SOLVENT: 0.1 M Potassium phosphate, pH 7.5
COMMENTS: Spectra of native and reconstituted α-cyanmet, β-cyanmet and reconstituted α-met subunits
REFERENCE: M. Waks, Y. K. Yip, and S. Beychock, J. Biol. Chem., 248, 6462 (1973)

1225. Hemoglobin, α-chain-p-mercuribenzoate and β-chain-p-mercuribenzoate

SOURCE: Human hemoglobin
SOLVENT: a. 0.1 M Phosphate buffer, pH 6
b. 0.4 M Acetate buffer, pH 5 and pH 5.5
COMMENTS: 1. Spectral changes with time
2. Spectral changes with α-p-mercuribenzoate chain upon addition of epinepherine
REFERENCE: M. Brunori, G. Falcioni, E. Fioretti, B. Giardina, and G. Rotillo, Eur. J. Biochem., 53, 99 (1975)

1226. Hemoglobin αNOβ

SOURCE: Human hemoglobin
SOLVENT: 0.1 M Sodium phosphate buffer, pH 7.0
COMMENTS: Spectra of αNOβ in presence and absence of haptoglobin 2-2 with dithionite present
REFERENCE: S. Bannai and Y. Sugita, J. Biol. Chem., 248, 7527 (1973)

1227. Hemoglobin: α-p-mercuribenzoate and β-p-mercuribenzoate Subunits

SOURCE: Human hemoglobin
SOLVENT: 0.20 Phosphate buffer, pH 7
COMMENTS: Spectra of CO and deoxy forms
REFERENCE: E. Antonini, E. Bucci, C. Fronticelli, J. Wyman, and A. Rossi-Fannelli, J. Mol. Biol., 12, 375 (1965)

1228. Hemoglobin: ferri α-chains, ferri β-chains and met-

SOURCE: Human hemoglobin
SOLVENT: 1 M Glycine, pH 5.9 and pH 9.5
REFERENCE: R. Banerjee, Y. Alpert, F. Leterrier, and R. J. P. Williams, Biochemistry, 8, 2862 (1969)

1229. Hemoglobin: α- and β-p-mercuribenzoate chains and γ-chains

SOURCE: Human hemoglobin
SOLVENT: a. pH 7.2
b. pH 6
COMMENTS: 1. Spectral changes due to addition of sodium nitrate, pH 7.2, to α- and β-p-mercuribenzoate chains
2. Change in spectrum of oxy-γ-chain upon oxidation with ferricyanide
3. Spectra of α-p-mercuribenzoate and β-p-mercuribenzoate chains before and after addition of potassium ferricyanide plus cyanide
REFERENCE: E. A. Rachmilewitz in Red Cells Structure and Metabolism (B. Ramat, ed.), 94 (1969), Academic Press, New York

1230. Hemoglobin: α- and β-chains

SOURCE: Human hemoglobin
SOLVENT: 50 mM Sodium phosphate, pH 7.0
COMMENTS: Spectra of oxy-α-, oxy-β-, CO-α-, CO-β-, met-α-, met-β-, deoxy-α-, deoxy-β chains, and recombined α-, and β-chains of all forms
REFERENCE: G. Geraci and T. K. Li, Biochemistry, 8, 184x (1969)

1231. Hemoglobin: Apo-α- and Apo-β-chains

SOURCE: Human globin
SOLVENT: 0.1 N NaOH
REFERENCE: B. Alpert, Compte Rend. Acad. Sci., 275, 1179 (1972)

1232. Hemoglobin and α- and β-chains

SOURCE: Human hemoglobin
SOLVENT: 0.1 M Potassium phosphate, pH 7.0
COMMENTS: Spectra of Hb, isolated chains, and n-isobutylisocyanide Hb
REFERENCE: J. S. Olson, Proc. Nat. Acad. Sci., 73, 1140 (1976)

1233. Hemoglobin: α- and β-subunits

SOURCE: Human hemoglobin
COMMENTS: Spectra of hydrogen peroxide oxidized α- and β-subunits
REFERENCE: A. Tomoda, K. Sugimoto, M. Suhara, M. Takeshita, and Y. Yoneyama, Biochem. J., 171, 329 (1978)

1234. Hemoglobin: α- and β- subunits

SOURCE: Methemoglobin
COMMENTS: Spectra after reduction with NADH methemoglobin reductase of methemoglobin
REFERENCE: A. Mansouri, Biochim. Biophys. Acta, 579, 191 (1979)

1235. Hemoglobin: α- and β-globin chains

SOURCE: Toad (Bufo bufo) hemoglobin
REFERENCE: A. A. Christomanos, Enzymologia, 42, 385 (1972)

1236. Hemoglobin

SOURCE: Phoronopsis viridis
COMMENTS: Spectra of carboxy- and deoxy-states
REFERENCE: R. L. Garlick, B. J. Williams, and A. F. Riggs, Arch. Biochem. Siophys., 194, 13 (1979)

1237. Hemolymph Fractions

SOURCE: Achatina fulica
SOLVENT: 0.01 N NaOH
COMMENTS: Spectra of fractions A, B, and C
REFERENCE: H. A. Kareem, A. M. Gilles, T. Nguyen-Thanh, and B. Keil, Comp. Biochem. Physiol., 44B, 963 (1973)

1238. Hemolysate

SOURCE: Human erythrocytes
SOLVENT: 0.05 M Phosphate buffer, pH 7.0
COMMENTS: Spectrum of hemolysate after incubation with acetylphenylhydrazine and changes in spectrum with time
REFERENCE: J. Peisach, W. E. Blumberg, and E. A. Rachmilewitz, Biochim. Biophys. Acta, 393, 404 (1975)

1239. Hemolysate and Ghosts (Hb Koln)

SOURCE: Human erythrocytes
COMMENTS: Spectral changes with time after oxidation with potassium ferricyanide (hemolysate)

REFERENCE: E. A. Rachmilewitz and J. M. White, Nature, New Biology, 241, 115 (1973)

1240. Hemolysate, Suspension, and Dried Blood Stains

SOURCE: Human erythrocytes
SOLVENT: 1% Universal buffer, pH 5.15-9.05 and other solutions
COMMENTS:
1. Spectrum of methemoglobin in hemolysate at pH 5.15-9.05
2. Spectra of erythrocyte suspension and dried blood stains
3. Spectrum of hemolysate in alkaline sol (ammoniacal)
4. Spectrum of oxidized (ferricyanide) hemolysate

REFERENCE: S. S. Kind, D. Patterson, and G. W. Owen, Forensic Sci., 1, 27 (1972)

1241. Hemolysate: Hemoglobin A and Hemoglobin M Kanakee

SOURCE: Human erythrocytes
COMMENTS: Spectra of untreated, oxidized, and cyanferri-hemoglobins
REFERENCE: P. Heller, H. G. Weinstein, V. J. Yakulis, and I. M. Rosenthal, Blood, 20, 287 (1962)

1242. Hemolysate: Hemoglobin M Chicago

SOURCE: Human erythrocytes
SOLVENT: Phosphate buffer, pH 6.5 and pH 8.0
COMMENTS: Spectra of Hb M Chicago in following forms: oxy, ferri, and cyanferri
REFERENCE: A. J. Josephson, H. G. Weinstein, V. J. Yakulis, L. Singer, and P. Heller, J. Lab. Clin. Med., 59, 918 (1980)

1243. Hemolysate: Hemoglobin M and Hemoglobin A

SOURCE: Human erythrocytes
SOLVENT: Phosphate buffer, pH 6.8
COMMENTS: Spectra of destromatized hemolysate in metHb and cyanmetHb forms
REFERENCE: K. Betke, E. Groschen, and K. Bock, Nature, 188, 864 (1960)

1244. Hemolysate: Hemoglobin M

SOURCE: Human erythrocytes
SOLVENT: pH 5.3

COMMENTS: Spectra of potassium ferricyanide treated normal hemolysate and hemolysates from two patients' erythrocytes
REFERENCE: A. V. Pisciotta, S. N. Ebbe, and J. E. Hinz, J. Lab. Clin. Med., 54, 73 (1959)

1245. Hemolysate: Hemoglobin M

SOURCE: Human erythrocytes
SOLVENT: M/15 Sodium phosphate buffer, pH 6.5
COMMENTS: Spectrum after treatment with potassium ferricyanide
REFERENCE: P. S. Gerald, Blood, 13, 936 (1958)

1246. Hemolysate

SOURCE: Human erythrocytes
SOLVENT: pH 6.29-9.20
COMMENTS: Spectra of hemolysate under various conditions and compared with spectrum of calf hemoglobin
REFERENCE: B. L. Horecker, J. Biol. Chem., 148, 173 (1943)

1247. Hemolysate: Dried Blood Stain

SOURCE: Human blood
COMMENTS: Spectra of dried blood stains made from aerated blood and nonaerated blood
REFERENCE: G. Sensbaugh, A. Wilson, and P. Kirk, Int. J. Biochem., 2, 545 (1971)

1248. Hemolysin

SOURCE: Listeria monocytogenes
REFERENCE: H. Siddique, I. F. Lin, r. A. Chung, Am. J. Vet. Res., 35, 289 (1974)

1249. Hemopexin

SOURCE: Rabbit blood
SOLVENT: 0.1 M 2-(N-morpholino) ethane sulfonic acid
COMMENTS: Spectra of protein, bromoacetate, and deuteroheme at zero time and 120 hours
REFERENCE: W. T. Morgan and U. Muller-Eberhard, Arch. Biochem. Biophys., 176, 431 (1976)

1250. Hemoprotein

SOURCE: Rhizobium japonicum
SOLVENT: 0.05 M Phosphate, pH 7.0

COMMENTS: Spectra of preparations before and after affinity chromatography purification
REFERENCE: K. Dus, R. Goewert, C. C. Weaver, D. Carey, and C. A. Appleby, Biochem. Biophys. Res. Comm., 69, 437 (1971)

1251. Hemoprotein

SOURCE: Human erythrocytes
SOLVENT: 0.1 M Tris buffer, pH 7.2
COMMENTS: Spectra of protein, protein plus dithionite, and protein plus cyanide
REFERENCE: H. K. Prins, J. A. Loos, and C. Zurcher, in Glutathione Deficiency (E. Beutler, ed.), 165 (1978), Grune and Stratton, New York

1252. Hemoprotein, Green

SOURCE: Beef erythrocytes
SOLVENT: 0.05 M Potassium phosphate buffer, pH 7.2
COMMENTS: 1. Spectral changes due to bisulfite, dithionite in the presence and absence of superoxide dismutase, and dithionite plus catalase
2. Spectra of ferriheme protein in presence of CO and O_2, ferrohemeprotein plus CO and effect of light irradiation
REFERENCE: L. J. DeFillippi, D. P. Ballou, and D. E. Hultquist, J. Biol. Chem., 254, 6917 (1979)

1253. Hemoprotein (H-450)

SOURCE: Pig liver
SOLVENT: a. 0.2 M Potassium phosphate buffer, pH 7.4, 1 mM EDTA
b. 0.05 M Tris-HCl buffer, pH 7.4, 1 mM EDTA, mixed 1:1 with glycerol
COMMENTS: 1. Spectra of oxidized, reduced (dithionite) and reaerated forms at room temperature (solvent a)
2. Spectra of oxidized and reduced (dithionite) H-450 at 77°K (solvent b)
REFERENCE: I. C. Kim and W. C. Deal, Biochemistry, 15, 4925 (1976)

1254. Hemoprotein

SOURCE: Azotobacter vinelandii OP ATCC13705
COMMENTS: Spectrum of oxidized form
REFERENCE: W. A. Bulen, J. R. Lecomte, and S. Lough, Biochem. Biophys. Res. Comm., 54, 1274 (1973)

1255. Hemosiderin and Aposiderin

SOURCE: Rat liver
SOLVENT: 80% Mercaptoacetic acid
COMMENTS: Spectra of hemosiderin and aposiderin
REFERENCE: A. Vidnes and L. Helgeland, Biochim. Biophys. Acta, 328, 365 (1973)

1256. Hemosiderin

SOURCE: Human liver
SOLVENT: Mercaptoacetic acid
REFERENCE: A. Shoden and P. Sturgeon, Acta Haemat., 23, 376 (1960)

1257. Hepatocuprein

SOURCE: Horse liver
SOLVENT: Water
REFERENCE: A. Finazzi Agro, V. Albergoni, and A. Cassini, FEBS Lett., 39, 164 (1974)

1258. Hepatoredoxin Reductase

SOURCE: Rabbit
SOLVENT: 0.1 M Potassium phosphate buffer, pH 7.4
COMMENTS: Spectra of oxidized and reduced (dithionite) forms
REFERENCE: Y. Ichikawa and A. Hiwatashi, Biochem. Biophys. Res. Comm., 79, 443 (1977)

1259. Hexokinase

SOURCE: Beef brain
SOLVENT: 0.05 M Potassium phosphate, 0.002 M EDTA, pH 7.8
REFERENCE: G. P. Schwartz and R. E. Basford, Biochemistry, 6, 1070 (1967)

1260. Hexokinase

SOURCE: Beef brain mitochondria
SOLVENT: 50 mM Tris-Chloride buffer, 0.5 mM EDTA, pH 8.0
REFERENCE: V. Redkar and U. Kenkare, J. Biol. Chem., 247, 7576 (1972)

1261. Hexokinase

SOURCE: Beef heart
SOLVENT: 1 mM Phosphate buffer, pH 7.5
REFERENCE: S. V. Paranjpe and V. Jagannathan, Ind. J. Biochem. Biophys., 8, 227 (1971)

1262. Hexokinase

SOURCE: Pig heart
SOLVENT: Phosphate (0.1 ionic strength), pH 7
REFERENCE: J. S. Easterby and M. J. O'Brien, Eur. J. Biochem., 38, 201 (1973)

1263. Histidine Ammonia Lyase

SOURCE: Pseudomonas ATCC 11,299b
SOLVENT: 0.05 M Potassium phosphate buffer, pH 7.5
COMMENTS: Spectra of L-cysteine-inactivated enzyme in the presence and absence of guanidine hydrochloride
REFERENCE: C. B. Klee, Biochemistry, 13, 4501 (1974)

1264. Histidyl-tRNA Synthetase

SOURCE: Salmonella typhimurium
SOLVENT: 50 mM Tris-HCl, 50 mM NaCl, 0.1 mM 2-mercaptoethanol, pH 7.0
REFERENCE: F. DeLorenzo, P. Dinatale, and A. N. Schechter, J. Biol. Chem., 249, 908 (1974)

1265. Histone, H3, Arginine Rich

SOURCE: Rat thymus
COMMENTS: Spectral changes due to gamma-irradiation
REFERENCE: M. S. Path, S. Narasimhan, and A. Sreenivasan, Ind. J. Biochem. Biophys., 14, 44 (1977)

1266. Histone, f-1 DNA Complexes and Histone IV-DNA Complexes

SOURCE: Calf thymus
SOLVENT: 0.14 M Sodium fluoride-1 mM Tris, pH 7.0
REFERENCE: T. Y. Shih and G. D. Fasman, Biochemistry, 11, 398 (1972)

1267. Histone F2a

SOURCE: Calf hepatocytes
SOLVENT: Water, pH 6.5
REFERENCE: M. J. Pieri and M. F. Kergueris, Compte. Rend. Acad. Sci. Paris, 274D, 2366 (1972)

1268. Histone

SOURCE:
a. Calf Thymus
b. Pea (Pisum sativum)
c. Vicia faba
d. Allium cepa

REFERENCE: N. S. Cohen and R. S. Quatrano, Experientia, 21, 237 (1965)

1269. Histone

SOURCE: a. Calf thymus
b. Volvox carteri
REFERENCE: D. M. Bradley, H. H. Goldin, and J. R. Claybrook, FEBS Lett., 41, 219 (1974)

1270. Homarine

SOURCE: Limulus polyphemus
SOLVENT: 0.2 M Acetic acid
REFERENCE: N. J. Larsen, Comp. Biochem. Physiol., 51B, 323 (1975)

1271. Hormone: Adrenocorticotropic ($ACTH_{1-24}$)

COMMENTS: Modified on lysine-24 with a dimethylaminonaphthyl sulfonyl group
REFERENCE: P. W. Schiller, Proc. Nat. Acad. Sci., 69, 975 (1972)

1272. Hormone: Choriomammotropin

SOURCE: Human
SOLVENT: 5% Acetic acid
COMMENTS: Spectra of unmodified and carboxymethyl-cysteine modified fragments
REFERENCE: C. H. Li and R. Houghten, Int. J. Pep. Prot. Res., 12, 114 (1978)

1273. Hormone: Follicle Stimulating

SOURCE: Human
SOLVENT: 0.01 M Potassium phosphate buffer, pH 7.0
REFERENCE: P. Roos, Acta Endocrinol., Supplement, 131, 9 (1968)

1274. Hormone: Interstitial Cell Stimulating

SOURCE: Sheep
SOLVENT: 0.1 M Sodium phosphate, pH 7.0
REFERENCE: M. R. Sairam and C. H. Li, Arch. Biochem. Biophys., 165, 709 (1974)

1275. Hormone: Lutenizing and Subunits

SOURCE: Human
SOLVENT: 0.20 M NaCl-0.01 M phosphate, pH 7.5
COMMENTS: Spectra of hormone, α, and β subunits
REFERENCE: W. H. Bishop and R. J. Ryan, Biochemistry, 72, 3076 (1973)

1276. Hormone: Lutenizing

SOURCE: Human pituitary
SOLVENT: a. 50 mM KCl, 10 mM sodium phosphate, pH 7.5
b. 4 M Guanidine HCl, 10 mM sodium phosphate, pH 5.8
REFERENCE: D. Puett, A. Nureddin, and L. A. Holladay, Int. J. Pep. Prot. Res., 8, 183 (1976)

1277. Hormone: Lutropin Subunits, nitrated

SOURCE: Lutropin
SOLVENT: 0.1 M Ammonium hydroxide, pH 10.7
COMMENTS: Spectra of fully nitrated α-subunits, after performic acid oxidation, and fully nitrated β-subunit
REFERENCE: W. K. Liu and D. N. Ward, J. Biol. Chem., 251, 316 (1976)

1278. Hydrogenase

SOURCE: Alcaligenes eutrophus H16
SOLVENT: 50 mM Potassium phosphate, pH 7.0
COMMENTS: Spectra of native enzyme and apo-enzyme
REFERENCE: K. Schneider and H. G. Schlegel, Biochim. Biophys. Acta, 452, 66 (1976); B. Schink and H. G. Schlegel, Biochim. Biophys. Acta, 567, 315 (1979)

1279. Hydrogenase

SOURCE: Chromatium
REFERENCE: P. H. Gitlitz and A. I. Krasna, Biochemistry, 14, 256 (1975)

1280. Hydrogenase

SOURCE: Clostridium pasteurianum W5
SOLVENT: 20 mM Tris-HCl, pH 8
COMMENTS: Spectra of oxidized (air) and reduced (hydrogen) forms
REFERENCE: D. L. Erbes, R. H. Burris, and W. H. Orme-Johnson, Proc. Nat. Acad. Sci., 72, 4795 (1975)

1281. Hydrogenase

SOURCE: Clostridium pasteurianum W5
SOLVENT: 0.1 M Phosphate buffer, pH 7.2
COMMENTS: Spectrum of reduced (hydrogen) form
REFERENCE: J. S. Chen and L. E. Mortenson, Biochim. Biophys. Acta, 371, 283 (1974)

1282. Hydrogenase

SOURCE: Desulfovibrio gigas
SOLVENT: 20 mM Tris-HCl, pH 7.6
COMMENTS: 1. Spectra of oxidized and reduced (dithionite) forms
2. Spectral changes due to benzenethiol, and benzenethiol plus potassium ferricyanide
REFERENCE: E. C. Hatchikian, M. Bruschi, and J. LeGall, Biochem. Biophys. Res. Comm., 82, 451 (1972)

1283. Hydrogenase

SOURCE: Desulfovibrio vulgaris
SOLVENT: 100 mM Potassium phosphate buffer, pH 7.6
COMMENTS: Spectra of oxidized and reduced (dithionite) forms
REFERENCE: J. LeGall, D. V. Dervartanian, E. Spilker, J. P. Lee, and H. D. Peck, Jr., Biochim. Biophys. Acta, 234, 525 (1971)

1284. Hydrogenase

SOURCE: Desulfovibrio vulgaris
SOLVENT: 20 mM Phosphate buffer, pH 7.0
COMMENTS: Spectra of enzyme and enzyme plus Cytochrome C_3 under hydrogen
REFERENCE: T. Yagi, K. Kimura, H. Daidoji, F. Sakai, S. Tamura, and K. Inokuchi, J. Biochem. Tokyo, 79, 661 (1976)

1285. Hydrogenase

SOURCE: Proteus mirabilis
SOLVENT: 25 mM Tris-HCl, buffer, pH 7.2
COMMENTS: Spectra of enzyme and dithionite-treated enzyme
REFERENCE: G. S. Schoenmaker, L. F. Oltmann, and A. H. Stouthamer, Biochim. Biophys. Acta, 567, 511 (1979)

1286. Hydrogenase

SOURCE: Thiocapsa roseopersicina
COMMENTS: Spectra of oxidized and reduced (dithionite) forms

REFERENCE: I. N. Gogotov, N. A. Zorin, L. T. Serebriakova, and E. N. Kondratieva, Biochim. Biophys. Acta, 523, 335 (1978); I. N. Gogotov, N. A. Zorim, and E. N. Kondratieva, Biokhimiya (Biochemistry), 41, 685 (1976)

1287. Hydroperoxidase I

SOURCE: Escherichia coli B
SOLVENT: 0.01 mM Potassium phosphate buffer, pH 6.5
REFERENCE: A. Claiborne and I. Fridovich, J. Biol. Chem., 254, 4245 (1979)

1288. Hydroperoxidase II

SOURCE: Escherichia coli B
SOLVENT: 50 mM Potassium phosphate buffer, pH 7.0
REFERENCE: A. Clairborne, D. P. Malinowski, and I. Fridovich, J. Biol. Chem., 254, 11664 (1979)

1289. L-α-Hydroxyacid Oxidase, isozyme B

SOURCE: Rat kidney
COMMENTS: Spectra of enzyme and apoenzyme
REFERENCE: J. A. Duley and R. S. Holmes, Eur. J. Biochem., 63, 163 (1976)

1290. L-α-Hydroxyacid Oxidase

SOURCE: Rat kidney
COMMENTS: Spectral changes due to anaerobic reaction with D L-2-hydroxy-3-butynoate, and with vinyl glycolate
REFERENCE: T. H. Cromartie and C. Walsh, Biochemistry, 14, 3482 (1975)

1291. L-α-Hydroxyacid Oxidase

SOURCE: Rat kidney
COMMENTS:
1. Spectra of soluble enzyme, after anaerobic reaction with D L-α-hydroxybutyrate, after anaerobic reaction with L-leucine, and after reaction with dithionite
2. Spectrum of mitochondrial enzyme

REFERENCE: T. H. Cromatrie and C. T. Walsh, Biochemistry, 14, 2588 (1975)

1292. p-Hydroxybenzoate Hydroxylase

SOURCE: Pseudomonas fluorescens
SOLVENT: 0.033 M Tris-HCl, 0.2 mM EDTA, pH 8.0

COMMENTS: 1. Spectra of enzyme in presence and absence of 2,4-dihydroxybenzoate and 3,4-dihydroxybenzoate complex
2. Spectral changes during EDTA-light reduction
3. Spectral changes during TPNH anaerobic reduction of enzyme-p-hydroxybenzoate complex
4. Spectral changes during dithionite anaerobic reduction of enzyme-2,4-dihydroxybenzoate complex

REFERENCE: T. Spector, and V. Massey, J. Biol. Chem., 247, 4679 (1972)

1293. p-Hydroxybenzoate Hydroxylase

SOURCE: Pseudomonas fluorescens
SOLVENT: 0.033 M Tris-HCl, 0.3 mM EDTA, pH 8.0
COMMENTS: 1. Spectral changes during anaerobic TPNH reduction of enzyme-2,4-dihydroxybenzoate complex
2. Spectral changes due to addition of oxygen after reduction

REFERENCE: T. Spector and V. Massey, J. Biol. Chem., 247, 5632 (1972)

1294. p-Hydroxybenzoate Hydroxylase

SOURCE: Pseudomonas fluorescens
SOLVENT: 0.02 M Tris-phosphate, 0.1 mM EDTA, pH 7.6
COMMENTS: Spectral changes during anaerobic reduction of enzyme-p-hydroxybenzoate complex
REFERENCE: L. G. Howell, T. Spector, and V. Massey, J. Biol. Chem., 247, 4340 (1972)

1295. p-Hydroxybenzoate Hydroxylase

SOURCE: Pseudomonas fluorescens
COMMENTS: Spectral changes upon addition of oxygen to reduced enzyme-p-hydroxybenzoate complex
REFERENCE: T. Spector and V. Massey, J. Biol. Chem., 247, 7123 (1972)

1296. p-Hydroxybenzoate Hydroxylase

SOURCE: Pseudomonas fluorescens
COMMENTS: Spectrum of reduced enzyme-2,4-dihydroxybenzoate complex and changes upon addition of oxygen
REFERENCE: B. Entsch, V. Massey, and D. P. Ballou, Biochem. Biophys. Res. Comm., 57, 1018 (1974)

1297. p-Hydroxybenzoate Hydroxylase

SOURCE: Pseudomonas fluorescens
SOLVENT: 50 mM Potassium phosphate, 20 mM EDTA, pH 6.5
COMMENTS: 1. Spectral changes due to binding of p-hydroxybenzoate by reduced (dithionite) enzyme
2. Spectra of reduced enzyme-complexes upon addition of oxygen; substances in complex are p-hydroxybenzoate and azide, 2,4-dihydroxybenzoate, p-amino-benzoate, and 6-hydroxynicotinate
3. Spectra of intermediates I and II formed during reaction of reduced enzyme-complexes with oxygen
REFERENCE: B. Entsch, D. P. Ballou, and V. Massey, J. Biol. Chem., 251, 2550 (1976); M. Husain and V. Massey, J. Biol. Chem., 254, 6657 (1979)

1298. p-Hydroxybenzoate Hydroxylase

SOURCE: Pseudomonas fluorescens
SOLVENT: a. 50 mM Potassium phosphate, 0.5 mM EDTA buffer, pH 6.5
b. pH 8.6
COMMENTS: 1. Spectral changes upon addition of p-hydroxybenzoate to 1-deazaFAD-enzyme in solvent a
2. Spectral changes upon reaction of 1-deazaFAD-enzyme with oxygen in presence of each of the following: p-hydroxybenzoate, p-aminobenzoate, and 6-hydroxynicotinate
REFERENCE: B. Entsch, M. Husan, D. P. Ballou, V. Massey, and C. Walsh, J. Biol. Chem., 255, 1420 (1980)

1299. p-Hydroxybenzoate Hydroxylase

SOURCE: Pseudomonas desmolytica IAM1123
SOLVENT: a. pH 6.0
b. pH 8.0
COMMENTS: 1. Spectra in presence and absence of NADPH (solvent a)
2. Spectral changes during anaerobic reduction by NADPH (solvent b) in presence of excess p-hydroxybenzoate
REFERENCE: H. Shoun, N. Higashi, T. Beppu, S. Nakamura, K. Hiromi, and K. Arima, J. Biol. Chem., 254, 10944 (1979)

1300. m-Hydroxybenzoate 4-Hydroxylase

SOURCE: Aspergillus niger
SOLVENT: 0.125 M Sodium phosphate buffer, pH 7.0

REFERENCE: R. P. Kumar, P. V. S. Rao, and C. S. Vaidyanathan, Ind. J. Biochem. Biophys., 10, 184 (1973)

1301. α-Hydroxy-γ-carboxymuconic ε-Semialdehyde Dehydrogenase

SOURCE: Pseudomonas ochraceae
REFERENCE: K. Maruyama, N. Ariga, M. Tsuda, and K. Deguchi, J. Biochem. Tokyo, 83, 1125 (1978)

1302. Hydroxyindole Oxidase

SOURCE: Mussel (Mytilus edulis)
COMMENTS: Spectra of oxidized and hydrosulfite-reduced-CO forms
REFERENCE: L. Kampa and J. Peisach, J. Biol. Chem., 255, 595 (1980)

1303. Hydroxylamine Oxidoreductase

SOURCE: Nitrosomonas europaea
SOLVENT: 0.1 M Phosphate buffer, pH 7.0
COMMENTS: Spectra of oxidized and dithionite-reduced forms
REFERENCE: T. Yamanaka, M. Shinra, K. Takahashi, and M. Shibasaka, J. Biochem. Tokyo, 86, 1101 (1979)

1304. Hydroxylamine Oxidoreductase

SOURCE: Nitrosomonas europaea
SOLVENT: 0.05 M Phosphate, pH 7.5
COMMENTS: Spectra of enzyme, enzyme plus dithionite, and addition of carbon monoxide
REFERENCE: A. B. Hooper, P. C. Maxwell, and K. R. Terry, Biochemistry, 17, 2984 (1978)

1305. ω-Hydroxylase

SOURCE: Pseudomonas oleovorans
REFERENCE: R. T. Ruettinger, S. T. Olson, R. F. Boyer, and M. J. Coon, Biochem. Biophys. Res. Comm., 57, 1011 (1974)

1306. L-4-Hydroxymandelate Oxidase

SOURCE: Pseudomonas convexa
SOLVENT: 0.025 M Sodium phosphate buffer, pH 7.0
REFERENCE: S. G. Bhat and C. S. Vaidyanathan, Eur. J. Biochem., 68, 323 (1976)

1307. D-6-Hydroxynicotine Oxidase

SOURCE: Arthrobacter oxidans
SOLVENT: 0.1 M Glycine, NaOH buffer, pH 9.2
COMMENTS: 1. Spectral changes during anaerobic reaction with D-6-hydroxynicotine and reoxidation by oxygen
2. Changes in spectrum due to anaerobic EDTA-light reduction
REFERENCE: M. Bruhmuller, H. Mohler, and K. Decker, Eur. J. Biochem., 29, 143 (1972)

1308. Hydroxynitril Lyase

SOURCE: Prunus amygdalus and Malus communis
SOLVENT: a. 0.1 M Pyrophosphate buffer, pH 7.5
b. Water, pH 1.5
REFERENCE: E. Gerstner and E. Pfeil, Hoppe-Seyler's Z. Physiol. Chem., 353, 271 (1972)

1309. 3β -Hydroxysteroid Oxidase

SOURCE: Streptomyces violascens
SOLVENT: Water
REFERENCE: H. Tomioka, M. Kagawa, and S. Nakamura, J. Biochem. Tokyo, 79, 903 (1976)

1310. 3β -Hydroxysteroid Oxidase

SOURCE: Streptomyces violascens
SOLVENT: 0.1 M Phosphate-citrate buffer, pH 6.0
COMMENTS: Spectra of enzyme and reduced (anaerobic, cholesterol) enzyme
REFERENCE: T. Kamei, Y. Takiguchi, H. Suzuki, M. Matsuzaki, and S. Nakamura, Chem. Pharm. Bull., 26, 2799 (1978)

1311. 3β -Hydroxysteroid Oxidase

SOURCE: Brevibacterium sterolicum
SOLVENT: 0.1 M Phosphate buffer, pH 7.0, pH 7.5
COMMENTS: Spectra of enzyme and reduced enzyme (anaerobic and dithionite or dehydro-epi-androsterone)
REFERENCE: T. Uwajima, H. Yagi, and O. Terada, Agric. Biol. Chem., 38, 1149 (1974)

1312. I-Band

SOURCE: Crab muscle
COMMENTS: Optical density parallel and perpendicular to I-Band

REFERENCE: Y. Tanagida, M. Taniguchi, and F. Oosawa, J. Mol. Biol., 90, 509 (1974)

1313. Immunoglobulin, Ig G-Cu Complex

SOURCE: Human myeloma protein
SOLVENT: 0.01 M Sodium phosphate, 0.04 M NaCl, pH 7.2
REFERENCE: B. L. Baker and D. E. Hultquist, J. Biol. Chem., 253, 1195 (1978)

1314. Immunoglobulin, γG

SOURCE: Goat
SOLVENT: 0.1 M Borate-buffered saline, pH 9.5
COMMENTS: Spectra of protein and 2-methoxy-2,4-diphenyl-3(2H) furanone derivative
REFERENCE: U. E. Handschin and W. J. Ritschard, Anal. Biochem., 71, 143 (1976)

1315. Immunoglobulin: Protein 315 Fraction Fv

SOURCE: Mouse Ig A melanoma
SOLVENT: 0.01 M Sodium phosphate, 0.15 M NaCl, pH 7.4
REFERENCE: J. Schlessinger, D. Z. Steinberg, D. Givol, and J. Hochman, FEBS Lett., 52, 231 (1975)

1316. Immunoglobulin: Antiovalbumin

SOURCE: Rabbit
SOLVENT: pH 7.0, pH 8.0, pH 9.6, pH 11.0, pH 12.5, and pH 13.0
COMMENTS: Spectral changes with time at pH 12.5 and pH 13.0
REFERENCE: A. A. Ansari and A. Salahuddin, Eur. J. Biochem., 35, 290 (1973)

1317. Immunoglobulin: Ig G

SOURCE: Normal and myeloma (human)
SOLVENT: Sodium phosphate, pH 7.7, 0.1 ionic strength
COMMENTS: Spectra of citraconylated and carbamylated proteins
REFERENCE: Y. Nakagawa, S. Capetillo, and B. Jirgensons, J. Biol. Chem., 247, 5703 (1972)

1318. Immunoglobulin, Tetramethylrhodamine Thiocarbamyl Derivative

SOURCE: Sheep
REFERENCE: L. Amante, A. Ancona, and L. Forni, J. Immunol. Methods, 1, 289 (1972)

1319. Immunoglobulin, fluoresceinthiocarbamyl derivative

SOURCE: Rabbit and goat
SOLVENT: 0.01 M Phosphate buffer, pH 7.4
REFERENCE: H. J. Breustedt, in Immunological Methods in Endocrinology (K. Federlin, C. N. Hales, and J. Kracht, eds.), 18 (1971), Academic Press, New York

1320. Immunoglobulin: γG and Macroglobulin (Waldenstrom)

SOURCE: Human
REFERENCE: H. D. Grumer, E. Grunze, and H. Horner, Klin. Wochen., 33, 588 (1955)

1321. Indoleamine 2,3-Dioxygenase

SOURCE: Rabbit small intestine
SOLVENT: 50 mM Potassium phosphate buffer, pH 8
COMMENTS:
1. Spectral change during reaction between ferrous enzyme and oxygen
2. Spectral change upon conversion of ternary complex (tryptophan-ferrous-oxygen) into ferric form

REFERENCE: T. Taniguchi, M. Sono, F. Hirata, O. Hayaishi, M. Tamura, K. Hayashi, T. Iizuka, and Y. Ishimura, J. Biol. Chem., 254, 3288 (1979)

1322. Indoleamine 2,3-Dioxygenase

SOURCE: Rabbit small intestine
SOLVENT: 0.1 M Potassium phosphate buffer, pH 7.0 and pH 8.0
COMMENTS: Spectral changes upon binding of L-tryptophan by the ferric, ferrous, and ferrous (CO) forms of the enzyme
REFERENCE: M. Sono, T. Taniguchi, Y. Watanabe, and O. Hayaishi, J. Biol. Chem., 255, 1339 (1980)

1323. Indoleamine 2,3-Dioxygenase

SOURCE: Rabbit intestine
SOLVENT: 0.05 M Potassium phosphate buffer, pH 8.0
COMMENTS:
1. Spectral changes upon reaction with O_2^- and hydrogen peroxide
2. Spectral changes of enzyme-O_2^- complex plus cyanide and carbon monoxide
3. Spectra of enzyme before and after oxygenation

REFERENCE: F. Hirata, T. Ohnishi, and O. Hayaishi, J. Biol. Chem., 252, 4637 (1977)

1324. Indoleamine 2,3-Dioxygenase

SOURCE: Rabbit intestine
SOLVENT: 0.1 M Phosphate buffer, pH 7.0
COMMENTS: Spectra of oxidized and reduced (dithionite) forms, and enzyme plus catalase, then sodium cyanide added
REFERENCE: F. O. Brady and A. Udom, Adv. Exp. Med. Biol., 74, 343 (1978)

1325. Indolyl-3-alkane α-Hydroxylase

SOURCE: Pseudomonas XA
SOLVENT: pH 5
REFERENCE: J. Roberts and H. J. Rosenfeld, J. Biol. Chem., 252, 2640 (1977)

1326. Inhibitor, Trypsin

SOURCE: Mung bean (Phaseolus aureus Roxb)
SOLVENT: a. 0.05 M Phosphate, pH 7.8
b. 0.1 N NaOH
COMMENTS: Spectra in solvents a and b; spectrum after reduction with mercaptoacetic acid and dialysis
REFERENCE: K. Wang, C. Chi, and T. Tsao, Acta Biochim. Biophys. Sinica, 5, 510 (1965)

1327. Inhibitor, FF1

SOURCE: Yeast
SOLVENT: Water
REFERENCE: E. Ebner and K. L. Maier, J. Biol. Chem., 252, 671 (1977)

1328. Inhibitor of Activation of cAMP Phosphodiesterase

SOURCE: Beef
SOLVENT: 0.04 M Tris-HCl buffer, pH 8, containing 1 mM $MgCl_2$, 0.1 M NaCl, 0.1 mM dithioerithritol, and 10% glycerol
REFERENCE: C. B. Klee and M. H. Krinks, Biochemistry, 17, 1204 (1978)

1329. Inhibitor of Influenza Virus Hemagglutination

SOURCE: Egg white
REFERENCE: F. Lanni, D. G. Sharp, E. A. Eckert, E. S. Dillon, D. Beard, and J. W. Beard, J. Biol. Chem., 179, 1275 (1949)

1330. Inhibitor of Protein Kinase

SOURCE: Rabbit skeletal muscle
SOLVENT: 50 mM Ammonium bicarbonate, pH 7.8
REFERENCE: J. G. Demaille, K. A. Peters, and E. H. Fischer, Biochemistry, 16, 3080 (1977)

1331. Inhibitor of Nuclease

SOURCE: Aspergillus oryzae mycelia
REFERENCE: T. Uozumi, G. Tamura, and K. Arima, Agric. Biol. Chem., 33, 636 (1969)

1332. Inhibitor of Tyrosinase

SOURCE: Agaricus hortensis Ckc.
REFERENCE: C. Madhosingh and L. Sundberg, FEBS Lett., 49, 156 (1974)

1333. Inhibitor of Chymotrypsin

SOURCE: Silkworm hemolymph (Bombyx mori)
SOLVENT: 10 mM Sodium phosphate buffer, pH 6.5
REFERENCE: T. Sasaki, J. Biochem. Tokyo, 84, 267 (1978)

1334. Inhibitor of Chymotrypsin

SOURCE: Potato (Solanum tuberosum)
SOLVENT: pH 7.0
REFERENCE: C. A. Ryan and A. K. Balls, Proc. Nat. Acad. Sci., 48, 1839 (1962)

1335. Inhibitor of Trypsin and Chymotrypsin

SOURCE: Black-eyed pea (Vigna sinensis)
SOLVENT: Phosphate buffer, pH 7.0
COMMENTS: Spectra of protein and of 2-methoxy-5-nitrobenzyl derivative
REFERENCE: M. M. Ventura and J. X. Filho, An. Acad. Brasil. Cienc., 38, 553 (1966); K. Mizuta and M. M. Ventura, An. Acad. Brasil. Cienc., 51, 349 (1979)

1336. Inhibitor of Pepsin

SOURCE: Ascaris
SOLVENT: a. 10 mM HCl, pH 2.0
b. 0.1 mM Tris-HCl, pH 7.0
COMMENTS: Spectra of inhibitors I, II or III, and IV
REFERENCE: G. M. Abu-Errgish and R. J. Peanasky, J. Biol. Chem., 249, 1558 (1974)

1337. Inhibitor of Proteinase

SOURCE: Lima bean
SOLVENT: 50 mM Sodium phosphate, pH 7.0
COMMENTS: 1. Spectra of protein and acetylated protein
2. Spectrum of acetylated protein in 6 M guanidine HCl
3. Spectra of protein irradiated in nitrous oxide saturated solutions of KBr and KCNS, and air-saturated solutions
REFERENCE: E. Kay and B. Gumbiner, J. Biol. Chem., 254, 7643 (1979)

1338. Inhibitor of Protease

SOURCE: Potato
REFERENCE: V. V. Moslov, E. L. Malova, T. A. Valueva, and A. I. Shulman, Bioorg. Chem., 1, 1449 (1975)

1339. Inhibitors of Protease

SOURCE: Potato
SOLVENT: pH 6 and pH 13
COMMENTS: Spectra of fragments of IIA and IIB
REFERENCE: T. Iwasaki, T. Kiyohara, and M. Yoshikawa, J. Biochem. Tokyo, 78, 1267 (1975)

1340. Inhibitors of Proteinase

SOURCE: Japanese radish seed (Raphanus sativus)
COMMENTS: Spectra of inhibitors I and II
REFERENCE: T. Ogawa, T. Higasa, and T. Hata, Agric. Biol. Chem., 32, 484 (1968)

1341. Inhibitors of Protease

SOURCE: Black-eyed peas
SOLVENT: 0.001 M HCl, pH 3 and 0.1 M Tris-HCl, pH 8.0
COMMENTS: Spectra of two inhibitors
REFERENCE: L. S. Gennis and C. R. Cantor, J. Biol. Chem., 251, 741 (1976)

1342. Inhibitors of Protease

SOURCE: Adjuki beans (Phaseolus angularis)
SOLVENT: a. 0.05 M Phosphate buffer, pH 6.0
b. 0.1 M NaOH
COMMENTS: Spectra of inhibitors I and II
REFERENCE: C. Yoshida and M. Yoshikawa, J. Biochem. Tokyo, 78, 935 (1975)

1343. Inhibitors of Proteinase B

SOURCE: Yeast
SOLVENT: 0.05 M Tris-HCl, pH 4.5-pH 9.0
COMMENTS: Spectra of inhibitor, 3-nitrotyr (41)-inhibitor, and 3-aminotyr (41)-inhibitor
REFERENCE: P. Bunning and H. Holzer, J. Biol. Chem., 252, 531 (1977)

1344. Inhibitors of Protease, modified

SOURCE: Beef inhibitor
SOLVENT: pH 5.3
COMMENTS: Spectra of tryptophan (15) and arginine (15) inhibitors
REFERENCE: H. Jering and H. Tschesche, Eur. J. Biochem., 61, 453 (1976)

1345. Inhibitor of Subtilisin

SOURCE: Streptomyces albogriseolus S-3253
SOLVENT: a. 0.01 M Borate buffer, pH 7.2
b. 0.025 Phosphate buffer, pH 7.0
c. 1.0 M NaOH, pH 13.1
REFERENCE: S. Sato and S. Murao, Agric. Biol. Chem., 37, 1067 (1973); K. Inouye, B. Tonomura, K. Hiromi, S. Sato, and S. Murao, J. Biochem. Tokyo, 82, 961 (1977); K. Inouye, B. Tonomura, K. Hiromi, S. Sato, and S. Murao, J. Biochem. Tokyo, 82, 1207 (1977)

1346. Inhibitor of Subtilisin, nitrated

SOURCE: Streptomyces albogriseolus
SOLVENT: 25 mM Pyrophosphate buffer, pH 8.0
COMMENTS: Spectra of one- and two-nitrated tyrosyl residues inhibitors
REFERENCE: K. Inouye, B. Tonomura, and K. Hiromi, J. Biochem. Tokyo, 85, 1115 (1979)

1347. Inhibitor of Subtilisin

SOURCE: Barley
SOLVENT: a. 0.1 M Sodium acetate buffer, pH 6.0
b. 0.1 M NaOH
REFERENCE: M. Yoshikawa, T. Iwasaki, M. Fujii, and M. Oogaki, J. Biochem. Tokyo, 79, 765 (1976)

1348. Inhibitor of Trypsin

SOURCE: Hakuhenzu bean (Dolichos lablab)
SOLVENT: a. Water
b. 0.1 N NaOH
REFERENCE: Y. Furusawa, Y. Kurosawa, and I. Chuman, Agric. Biol. Chem., 38, 1157 (1974)

1349. Inhibitor of Trypsin

SOURCE: Soybeans
COMMENTS: Spectra of inhibitor, chlorine-modified inhibitor, and adventitiously (water) modified inhibitor
REFERENCE: S. Odani, T. Koide, and T. Ikenaka, J. Biochem. Tokyo, 70, 925 (1971)

1350. Inhibitor of Trypsin

SOURCE: Soybeans
COMMENTS: Spectra of dinitrophenyl derivative and of its complex with trypsin
REFERENCE: M. Wilchek and M. Gorecki, FEBS Lett., 31, 149 (1973)

1351. Inhibitor of Trypsin (Bowman-Birk)

SOURCE: Soybeans
SOLVENT: a. 50 mM Sodium phosphate, pH 7.0
b. 0.1 M NaOH, pH 12.8
COMMENTS: Spectra of native, acetylated, and deacetylated protein
REFERENCE: E. Kay, J. Biol. Chem., 251, 3411 (1976)

1352. Inhibitor I of Trypsin

SOURCE: Pig pancreatic juice
SOLVENT: 0.1 N NaOH
COMMENTS: Spectral changes with time
REFERENCE: L. J. Greene, J. J. Dicarlo, A. J. Sussman, D. C. Bartlett, and D. E. Roark, J. Biol. Chem., 243, 1804 (1968)

1353. Inhibitor of Trypsin, fragments

SOURCE: Inhibitor
COMMENTS: Spectra of various fragments from hydrogen peroxide oxidation
REFERENCE: T. Koide, T. Ikenaka, K. Ikeda, and K. Homaguchi, J. Biochem. Tokyo, 75, 805 (1974)

1354. Inhibitor of Trypsin

SOURCE: Beef pancreas
SOLVENT: 0.1 M Triethanolamine-HCl buffer, pH 7.8
COMMENTS: Spectra of inhibitor and inhibitor modified with acetamidomethanol in anhydrous hydrogen fluoride
REFERENCE: R. Rocchi, C. A. Benassi, R. Tomatis, R. Ferroni, and E. Menegatti, Int. J. Pep. Prot. Res., 8, 167 (1976)

1355. Inhibitor of Trypsin, Modified

SOURCE: Beef pancreas inhibitor
SOLVENT: 0.01 M Ammonium bicarbonate
COMMENTS: Spectrum of dimethylaminonaphthyl sulfonyl tetramaleyl derivative
REFERENCE: M. Rigbi, in Proceedings International Conference on Proteinase Inhibitors (H. Fritz and H. Tschesche, eds.), 75 (1970), Walter de Gruyter, Berlin

1356. Inhibitor of Trypsin

SOURCE: Beef pancreatic juice
SOLVENT: a. 0.1 N NaOH
b. 8M Urea
COMMENTS: 1. Spectral change with time in 0.1 N NaOH
2. Spectrum of protein in 8 M urea for 6 minutes
REFERENCE: L. J. Greene, M. Rigbi, and D. S. Fackre, J. Biol. Chem., 241, 5160 (1966)

1357. Insulin

SOURCE: Beef pancreas
SOLVENT: 0.1 M Phosphate buffer, pH 8.0
REFERENCE: T. Ichikawa and H. Terada, Biochim. Biophys. Acta, 580, 120 (1979)

1358. Insulin

SOURCE: Beef
SOLVENT: 0.01 M Ammonium bicarbonate, pH 8.0
COMMENTS: Spectra of insulin and glycine (A-1) azidonitrophenyl-insulin
REFERENCE: D. Levy, Biochim. Biophys. Acta, 322, 329 (1973)

1359. Insulin and Insulin-Hg

SOURCE: Beef
SOLVENT: 0.5 M Acetic acid
COMMENTS: Spectra of insulin and insulin-Hg

REFERENCE: R. Sperling and I. Z. Steinberg, Biochemistry, 13, 2007 (1974)

1360. Insulin

SOURCE: Commercial
SOLVENT: 0.05 M Phosphate buffer, pH 7.4
COMMENTS: Spectral changes associated with the oxidation of amorphous insulin and Zn-insulin by mushroom tyrosinase
REFERENCE: J. G. Cory, C. C. Bigelow, and E. Frieden, Biochemistry, 1, 419 (1962)

1361. Insulin

SOURCE: Beef
COMMENTS: Spectra of insulin and O-acylated insulin
REFERENCE: A. Previero, L. G. Barry, and M. A. Coletti-Previero, Biochim. Biophys. Acta, 263, 7 (1972)

1362. Insulin

SOLVENT: M/20 KCl buffer, pH 2.0
COMMENTS: Spectra of insulin before and after addition of potassium permanganate at 77°K
REFERENCE: W. J. Haas, I. W. Sizer, and J. R. Zoofbourow, Biochim. Biophys. Acta, 6, 601 (1951)

1363. Insulin

SOURCE: Beef
SOLVENT: 0.05 M Phosphate buffer, pH 7.8
COMMENTS: Spectra of insulins, amorphous and crystalline, N-ε B29-acetyl insulin, N-α Al-acetyl-insulin, and N-α1 Al, N-ε B29-diacetyl-insulin, crystalline and amorphous
REFERENCE: D. Brandenberg, H. G. Gattner, and A. Wollmer, Hoppe-Seyler's Z. Physiol. Chem., 353, 599 (1972)

1364. Insulin

SOURCE: Beef
SOLVENT: 0.01 M Ammonium bicarbonate, pH 8.5
COMMENTS: Spectra of insulin and glycyl (Al-p-mercuribenzoate)-insulin
REFERENCE: D. Levy, Biochim. Biophys. Acta, 317, 473 (1973)

1365. Insulin Derivatives

SOURCE: Bovine

SOLVENT: Phosphate buffer, pH 7.4
COMMENTS: Spectral changes seen when ε-N-B29-azidobenzoyl-insulin and α-N-A1, ε-N-B29-di(azidobenzoyl)-insulin are photolysed
REFERENCE: C. C. Yip, C. W. T. Yeung, and M. L. Moule, Biochemistry, 19, 70 (1980)

1366. Invertase

SOURCE: Candida utilis
SOLVENT: a. 0.01 M Phosphate buffer, pH 7.0
b. 0.1 N NaOH
REFERENCE: M. Iizuka, H. Chiura, and T. Yamamoto, Agric. Biol. Chem., 42, 1207 (1978)

1367. Invertase

SOURCE: Yeast
SOLVENT: 0.05 M Phosphate buffer, pH 6.0
COMMENTS: Spectral change upon incubation with tyrosinase
REFERENCE: I. W. Sizer and J. F. Fennessey, J. Biol. Chem., 188, 351 (1951)

1368. Iodotyrosine Deiodinase

SOURCE: Beef thyroid
SOLVENT: a. 0.01 M Potassium phosphate buffer, pH 7.4
b. 0.05 M Potassium phosphate buffer, pH 7.4
COMMENTS: Spectra of enzyme, apoenzyme, and reconstituted enzyme
REFERENCE: I. N. Rosenberg and A. Goswami, J. Biol. Chem., 254, 12318 (1979); A. Goswami and I. N. Rosenberg, J. Biol. Chem., 254, 12326 (1979)

1369. Isocitrate Dehydrogenase

SOURCE: Bacillus stearothermophilus
SOLVENT: 0.08 M Phosphate buffer, pH 8.0
COMMENTS: Spectra of enzyme, enzyme after modification with 5,5'-dithio-bis-(2-nitrobenzoic acid), and after treating modified enzyme with dithiothreitol
REFERENCE: T. Nagaoka, A. Hachimori, A. Takeda, and T. Samejima, J. Biochem. Tokyo, 81, 71 (1977)

1370. Isocitrate Dehydrogenase

SOURCE: Beef heart
SOLVENT: a. 0.2 M Glycine-NaOH, pH 8.6-pH 11.3
b. KOH, pH 12 and pH 13
REFERENCE: C. C. Fan, L. A. Tomcho, and G. W. E. Plaut, J. Biol. Chem., 250, 6197 (1975)

1371. Isomeroreductase

SOURCE: Salmonella typhimurium
SOLVENT: 0.05 M Potassium phosphate, 0.05 mM dithiothreitol, and 0.5 mM dimagnesium EDTA, pH 7.5
REFERENCE: J. G. Hofler, C. J. Decedue, G. H. Luginbuhl, J. A. Reynolds, and R. O. Burns, J. Biol. Chem., 250, 877 (1975)

1372. C-55-Isoprenoid Alcohol Phosphokinase

SOURCE: Staphylococcus aureus
SOLVENT: Methanol:1-butanol (1:1) 1 M in ammonium acetate
REFERENCE: H. Sandermann, Jr., and J. L. Strominger, J. Biol. Chem., 247, 5123 (1972)

1373. Kallikrein B

SOURCE: Pig pancreas
SOLVENT: Phosphate buffer, pH 7.0
REFERENCE: C. Kutzbach and G. Schmidt-Kastner, Hoppe-Seyler's Z. Physiol. Chem., 353, 1099 (1972)

1374. Kallikrein

SOURCE: Pig pancreas
SOLVENT: a. Water
b. 0.125 N NaOH
c. 0.0125 M Sodium acetate buffer, pH 5.5
REFERENCE: H. Fritz, I. Eckert, and E. Werle, Hoppe-Seyler's Z. Physiol. Chem., 348, 1120 (1967)

1375. Keratin, Solubilized

SOURCE: Chicken feathers
SOLVENT: Water
REFERENCE: Y. Nagi and Y. Nishikawa, Agric. Biol. Chem., 34, 16 (1970)

1376. Keratin

SOURCE: Chicken feathers
COMMENTS: Spectra of S-carboxymethyl rachis and rachis keratose
REFERENCE: B. S. Harrap and E. F. Woods, Biochem. J., 92, 8 (1964)

1377. Keratohyalin

SOURCE: Beef hoof epidermis
SOLVENT: 0.02 M Sodium decylsulfate
REFERENCE: A. R. Ugel, J. Inves. Derm., 65, 118 (1975)

1378. Kininogen, High-molecular Weight

SOURCE: Human plasma
SOLVENT: a. 0.1 M Acetate buffer, pH 6.5
b. 0.1 M NaOH
REFERENCE: T. Nakayasu and S. Nagasawa, J. Biochem. Tokyo, 85, 249 (1979)

1379. Kininogen, High-molecular Weight

SOURCE: Beef plasma
SOLVENT: a. 0.1 M Acetate buffer, pH 6.5
b. 0.1 M NaOH
REFERENCE: M. Komiya, H. Kato, and T. Suzuki, J. Biochem. Tokyo, 76, 811 (1974)

1380. Kojic Acid Oxidase

SOURCE: Arthrobacter ureafaciens (K-1)
COMMENTS: Spectra of oxidized and reduced (dithionite or enzymically produced hydrogen)
REFERENCE: J. Imose, S. Nonomura, and C. Tatsumi, Agric. Biol. Chem., 34, 1443 (1970); S. Nonomura, J. Imose, and C. Tatsumi, Agric. Biol. Chem., 33, 1223 (1969)

1381. Kynureninase

SOURCE: Pseudomonas marginalis
SOLVENT: 0.01 M Potassium phosphate buffer, pH 7.2
REFERENCE: M. Moriguchi, T. Yamamoto, and K. Soda, Biochem. Biophys. Res. Comm., 44, 725 (1971)

1382. Kynureninase

SOURCE: Pseudomonas marginalis
SOLVENT: 0.01 M Potassium phosphate buffer, pH 7.2
COMMENTS: Spectra of enzyme, reduced (sodium borohydride) enzyme, and apoenzyme
REFERENCE: M. Moriguchi, T. Yamamoto, and K. Soda, Biochemistry, 12, 2969 (1973)

1383. Kynureninase

SOURCE: Pseudomonas marginalis
SOLVENT: 0.03 M Tris-HCl buffer, pH 8.0
COMMENTS: Spectra of apoenzyme, plus pyridoxamine-5'-phosphate, and then addition of pyruvate
REFERENCE: M. Moriguchi and K. Soda, Biochemistry, 12, 2974 (1973)

1384. L-Kynurenine 3-Hydroxylase

SOURCE: Rat liver
SOLVENT: 0.05 M Tris-acetate buffer, pH 7.8
COMMENTS: Spectra of enzyme and reduced (dithionite) enzyme
REFERENCE: Y. Nisimoto, F. Takeuchi, and Y. Shibata, J. Biochem. Tokyo, 81, 1413 (1977)

1385. Laccase

SOURCE: Rhus vernicifera
SOLVENT: Potassium phosphate buffer, 0.1 ionic strength, pH 7.4
COMMENTS: 1. Spectra of native and copper-depleted enzyme
2. Spectra of enzyme, ascorbate and ferrocyanide added, then reoxidation with air
REFERENCE: B. Reinhammar and Y. Oda, J. Inorg. Biochem., 11, 115 (1979)

1386. Laccase

SOURCE: Rhus vernicifera
SOLVENT: 0.01 M Phosphate buffer, pH 6.0
COMMENTS: Spectral changes due to photoirradiation (358 nm) under aerobic and anaerobic conditions
REFERENCE: Y. Henry and J. Peisach, J. Biol. Chem., 253, 7751 (1978)

1387. Laccase

SOURCE: Chinese lacquer tree
COMMENTS: Spectra of apo- and reconstituted laccase
REFERENCE: B. L. Hauenstein, Jr. and D. R. McMillin, Biochem. Biophys. Res. Comm., 85, 505 (1978)

1388. Laccase

SOURCE: Lacquer tree (Rhus vernicifera)
SOLVENT: Phosphate buffer, pH 7.0
COMMENTS: Spectra of oxidized and reduced (ascorbate) forms
REFERENCE: T. Nakamura, Adv. Exp. Med. Biol., 74, 408 (1976)

1389. Laccase

SOURCE: Rhus vernicifera
SOLVENT: 0.1 M Phosphate buffer, pH 7.6
COMMENTS: Spectra of protein under argon, equilibrated with nitric oxide, then argon, and then air
REFERENCE: G. Rotilio, L. Morpurgo, M. T. Graziani, and M. Brunori, FEBS Lett., 54, 163 (1975)

1390. Laccase

SOURCE: Rhus vernicifera
SOLVENT: 0.05 M Phosphate buffer, pH 7.0
COMMENTS: Spectrum before and after treatment with dimethylglyoxime
REFERENCE: M. T. Graziani, L. Morpurgo, G. Rotilio, and B. Mondovi, FEBS Lett., 70, 87 (1976)

1391. Laccase

SOURCE: Fungus (Polyporus versicolor) and lacquer tree (Rhus vernicifera)
SOLVENT: 0.1 M Sodium phosphate buffer, pH 6.0 and pH 7.5
REFERENCE: K. E. Falk and B. Reinhammar, Biochim. Biophys. Acta, 285, 84 (1972)

1392. Laccase

SOURCE: Vietnamese lacquer tree (Rhus succedanea)
SOLVENT: 0.02 M Sodium phosphate buffer, pH 5.5
REFERENCE: R. Branden and J. Deinum, Biochim. Biophys. Acta, 524, 297 (1978)

1393. Laccase A and Laccase B

SOURCE: Fungus (Polyporous versicolor)
SOLVENT: 0.05 M Sodium phosphate buffer, pH 6.0
COMMENTS: Spectra of laccase A, laccase B, and reduced (ascorbate) laccase A
REFERENCE: R. Mosback, Biochim. Biophys. Acta, 73, 204 (1963)

1394. Laccase

SOURCE: Neurospora crassa, str. 69-1113a
REFERENCE: S. C. Froehner and K. E. Eriksson, J. Bacteriol., 120, 458 (1974)

1395. Lac Repressor and 3-Fluorotyrosine Derivative

SOURCE: Escherichia coli
SOLVENT: a. 0.05 M Citrate buffer, 8 M guanidine HCl, pH 6.9
b. 0.1 M Ammonium bicarbonate buffer, 8 M guanidine HCl, pH 9.0
REFERENCE: P. Lu, M. Jarema, K. Mosser, and W. E. Daniel, Proc. Nat. Acad. Sci., 73, 3471 (1976)

1396. Lac Repressor

SOURCE: Escherichia coli, wild type
REFERENCE: H. Sommer, P. Lu, and J. H. Miller, J. Biol. Chem., 251, 3774 (1974)

1397. α-Lactalbumin

SOURCE: Beef milk
SOLVENT: pH 2, pH 6, and pH 11
COMMENTS: Spectra of protein and protein-N-methyl nicotinamide chloride complex
REFERENCE: F. M. Robbins and L. G. Holmes, J. Biol. Chem., 247, 2062 (1972)

1398. L-Lactate Dehydrogenase

SOURCE: Baker's yeast
SOLVENT: 0.1 M Phosphate buffer, 1 mM EDTA, pH 7
COMMENTS: Spectra of protein and protein-reacted with 2-hydroxy-3-butynoic acid at 0°C
REFERENCE: F. Lederer, Eur. J. Biochem., 46, 393 (1974)

1399. Lactate Dehydrogenase

SOURCE: Beef heart
COMMENTS: Spectra of protein, protein-ε-aminocaproic acid complex, and protein-N^6-[(6-amino-hexyl) carbamoylmethyl] nicotinamide adenine dinucleotide complex
REFERENCE: R. F. Venn, P. O. Larsson, and K. Mossback, Acta Chem. Scand., 31B, 141 (1977)

1400. D-Lactate Dehydrogenase

SOURCE: Escherichia coli membranes
SOLVENT: 0.05 M Sodium phosphate buffer, 0.01% β-mercaptoethanol, and 20% ethylene glycol, pH 7.1
COMMENTS: Spectra in presence and absence of D-lactate
REFERENCE: M. Futai, Biochemistry, 12, 2468 (1973)

1401. D-Lactate Dehydrogenase

SOURCE: Escherichia coli
SOLVENT: 0.1 M Potassium phosphate, pH 7.5
COMMENTS: Spectra of protein, protein plus D-lactate, and protein plus L-lactate
REFERENCE: L. D. Kohn, and H. R. Kaback, J. Biol. Chem., 248, 7012 (1973)

1402. D-Lactate Dehydrogenase

SOURCE: Peptostreptococcus elsdenii
SOLVENT: 3 M Potassium phosphate buffer, pH 6.0
COMMENTS: Spectra in presence and absence of D-lactate
REFERENCE: H. L. Brockman and W. A. Wood, J. Bacteriol., 124, 1454 (1975)

1403. D-Lactate Dehydrogenase

SOURCE: Megasphaera elsdenii (Peptostreptococcus elsdenii)
SOLVENT: 0.1 M Potassium phosphate, pH 7.0
COMMENTS:
1. Spectra of zinc-depleted enzyme and after addition of $ZnCl_2$
2. Spectral changes due to anaerobic photochemical reduction by oxalate and 5-deazaflavin at 3°C
3. Spectral changes due to anaerobic (dithiothreitol) reduction by D-lactate and then addition of hydroxylamine
4. Spectral changes due to anaerobic (dithiothreitol) reaction of 2-hydroxyl-3-butynoate with enzyme

REFERENCE: S. T. Olson and V. Massey, Biochemistry, 18, 4714 (1979); S. T. Olson, J. Massey, S. Ghisla, and D. Whitfield, Biochemistry, 18, 4724 (1979)

1404. L-Lactate Dehydrogenase

SOURCE: Escherichia coli membranes
SOLVENT: 0.05 M Tris-HCl, 20% glycerol, pH 8.0
COMMENTS: Spectra of enzyme, enzyme plus lactate, and enzyme plus hydrosulfite
REFERENCE: M. Futai and H. Kimura, J. Biol. Chem., 252, 5820 (1977)

1405. L-Lactate Dehydrogenase

SOURCE: Polyporus circinatus
SOLVENT: 0.1 M Tris-HCl buffer, pH 6.7
COMMENTS: Spectra of oxidized and reduced (dithionite or L-lactate) forms

REFERENCE: S. Funayma and G. Zancan, J. Bacteriol., 119, 1000 (1974)

1406. Lactate Oxidase

SOURCE: Mycobacterium smegmatis
SOLVENT: a. 0.01 M Imidazole-HCl buffer, pH 7.0
b. 0.01 M Tris, 0.01 M Imidazole, 0.01 M piperazine, pH 7
COMMENTS: 1. Spectra of enzyme, enzyme plus tartronate, and photoirradiation of this complex
2. Spectra of L-lactate-reduced enzyme and photoirradiation-formed adduct with bromoacetate or bromopropionate
3. Spectral changes of 2-S-FMN-lactate oxidase (oxidized) on anaerobic reduction with L-lactate and then photoirradiation in presence of bromoacetic or bromopropionic acids
4. Spectral changes associated with the anaerobic reaction of α-hydroxybutynoate
REFERENCE: S. Ghisla, V. Massey, and Y. S. Choong, J. Biol. Chem., 254, 10602 (1979)

1407. Lactate Oxidase; Fluorodinitrobenzene Modified

SOURCE: Mycobacterium smegmatis
SOLVENT: 0.05 M Sodium phosphate buffer, pH 6.0
COMMENTS: 1. Spectral changes on anaerobic reduction of DNP-modified enzyme with L-lactate
2. Spectral changes on reaction of DNP-modified enzyme with sodium bisulfite
REFERENCE: Y. S. Choong, M. G. Shepherd, and P. A. Sullivan, Biochem. J., 173, 255 (1978)

1408. Lactate Oxidase

SOURCE: Mycobacterium smegmatis
SOLVENT: 0.01 M Imidazole-HCl, pH 7.0
COMMENTS: 1. Spectral changes during anaerobic reaction of reduced enzyme with 2-keto-3-butynoate
2. Spectral changes during anaerobic reaction of enzyme with 2-hydroxy-3-butynoate
3. Spectrum of 2-hydroxy-3-butynoate modified enzyme after sodium borohydride reduction
REFERENCE: S. Ghisla, H. Ogata, V. Massey, A. Schonbrunn, R. H. Abeles, and C. T. Walsh, J. Biol. Chem., 15, 1791 (1976); C. T. Walsh, A. Schonbrunn, O. Lockridge, V. Massey, and R. H. Abeles, J. Biol. Chem., 247, 6004 (1972)

1409. Lactate Oxidase

SOURCE: Mycobacterium smegmatis
SOLVENT: 0.1 M Tris-acetate buffer, pH 7.8
COMMENTS: Spectral changes during reconstitution of enzyme from apoenzyme and FMN
REFERENCE: Y. S. Choong, M. G. Shepherd, and P. M. Sullivan, Biochem. J., 145, 37 (1975)

1410. Lactate Oxidase

SOURCE: Mycobacterium smegmatis
SOLVENT: 0.01 M Imidazole-HCl buffer, pH 7.0
COMMENTS: Spectral changes due to illumination of enzyme-oxalate complex
REFERENCE: S. Ghisla and V. Massey, J. Biol. Chem., 250, 577 (1975)

1411. Lactate Oxidase

SOURCE: Mycobacterium smegmatis
SOLVENT: 0.01 M Imidazole-HCl, pH 7.0
COMMENTS:
1. Spectra of enzyme and reduced (L-lactate) forms
2. Spectra of enzyme and after reaction of reduced form with bromoacetic acid
3. Spectral changes due to phosphate
4. Spectral changes due to anaerobic reaction with pyruvate

REFERENCE: S. Ghisla, V. Massey, J. M. Lhoste, and S. G. Mayhew, Biochemistry, 13, 589 (1974); O. Lockridge, V. Massey, and P. A. Sullivan, J. Biol. Chem., 247, 8097 (1972)

1412. Lactate Oxygenase

SOURCE: Mycobacterium phlei
SOLVENT: 0.05 M Potassium phosphate buffer, pH 7.0
COMMENTS: Spectra of oxidized enzyme and anaerobically reduced enzyme (lactate)
REFERENCE: S. Takemori, Y. Nakai, K. Nakazawa, and J. Katagiri, Arch. Biochem. Biophys., 154, 137 (1973)

1413. Lactoferrin

SOURCE: Human milk
SOLVENT: 0.1 M Sodium bicarbonate
REFERENCE: W. F. Line, D. A. Sly, and A. Bezkorovainy, Int. J. Biochem., 7, 203 (1976)

1414. Lactoferrin, apo-, nitrated

SOURCE:	Human lactoferrin from human milk
COMMENTS:	Changes in spectrum due to different times of nitration
REFERENCE:	B. Teuwissen, P. L. Masson, P. Osinski and J. F. Heremans, Eur. J. Biochem., 35, 366 (1973)

1415. β-Lactoglobulin

SOURCE:	Human milk
SOLVENT:	0.066 M Phosphate buffer, pH 7.0
COMMENTS:	Spectral changes with time of protein-trans-retinylidene propylamine complex
REFERENCE:	J. Horowitz and J. Heller, J. Biol. Chem., 249, 4712 (1974)

1416. β-Lactoglobulin, deoxyribulose-

SOLVENT:	a. 0.01 N NaOH b. 0.01 N HCl
REFERENCE:	L. Berrens, Immunochemistry, 4, 37 (1967)

1417. Lactoperoxidase

SOURCE:	Commercial
SOLVENT:	pH 8.8
COMMENTS:	Spectra of enzyme, enzyme plus hydrogen peroxide, enzyme plus hydrogen peroxide plus dithiothreitol, and enzyme plus hydrogen peroxide plus dithiothreitol plus iodide
REFERENCE:	H. Edelhoch, G. Irace, M. L. Johnson, J. L. Michot, and J. Nunez, J. Biol. Chem., 254, 11822 (1979)

1418. Lactoperoxidase

SOURCE:	Commercial
SOLVENT:	10 mM and 50 mM Sodium phosphate, pH 7.2
COMMENTS:	1. Spectra of enzyme and of complexes with glutathione and with hydrogen peroxide 2. Spectra of enzyme-cysteine complex and after addition of iodide 3. Spectra of enzyme-cysteine complex and after addition of guaiacol
REFERENCE:	J. Pommier and H. J. Cahnmann, J. Biol. Chem., 254, 3006 (1979)

1419. Lactoperoxidase (Beef)

SOURCE:	Beef milk

SOLVENT: pH 7.10
COMMENTS: Spectra of free enzymes and of enzymes plus hydrogen peroxide
REFERENCE: S. Kimura and I. Yamazaki, Arch. Biochem. Biophys., 198, 580 (1979)

1420. Lactoperoxidase

SOURCE: Beef milk
COMMENTS: Spectral changes due to reaction of enzyme with 3-amino-1,2,4-triazole and hydrogen peroxide in presence and absence of ascorbic acid
REFERENCE: J. Y. Chang and W. A. Schroeder, Arch. Biochem. Biophys., 156, 475 (1973)

1421. Leghemoglobin

SOURCE: Lupinus lutenus L
COMMENTS: Spectra of high spin and low spin forms
REFERENCE: Y. V. Peive, B. P. Atanasov, G. Y. Zhiznevsky, and N. N. Krasnobaeva, Doklady Acad. USSR, 202, 482 (1972)

1422. Leghemoglobin

SOURCE: Lupinus lutens
COMMENTS: Spectra of high-spin and low-spin material
REFERENCE: Y. V. Pieve, B. P. Atanasov, G. Y. Zhiznevskaya, and N. N. Krasnobaeva, Doklady Akad. Nauk., USSR (Translation), 202, 39 (1972)

1423. Leghemoglobin I and II

SOURCE: Lupinus lutens
SOLVENT: 0.01 M Phosphate buffer, pH 6.3 and pH 6.6
COMMENTS: Spectra of leghemoglobin, reduced (dithionite) leghemoglobin, oxyleghemoglobin, and reduced (dithionite) oxyleghemoglobin
REFERENCE: S. S. Melik-Sarkisyan, V. V. Yarovenko, G. L. Shaposhnikov, L. P. Vladzievskaya, and V. L. Kretovich, Biokhimiya (Biochemistry), 39, 593 (1974)

1424. Leghemoglobin Derivatives

SOURCE: Lupinus luteus leghemoglobin
SOLVENT: 0.1 M Phosphate buffer, pH 6.5
COMMENTS: 1. Spectra of ferri-protein and following derivatives: H_2O, pyridine, imidazole, azide, hydroxide, thiocyanate and cyanide

2. Spectra of ferro-protein and following derivatives: thiocyanate, pyridine, imidazole, nicotinic acid, alkylisocyanide, azide, cyanide, and carbon monoxide

REFERENCE: Y. V. Peive, B. P. Atanasov, G. Y. Zhiznevskaya, and N. Krasnobaeva, Doklady Akad. Nauk. USSR (Translation), 202, 965 (1972)

1425. Leghemoglobin a

SOURCE: Soybeans
SOLVENT: Phosphate buffer, pH 7.0
COMMENTS: Spectra of mesoheme leghemoglobin a and mesoheme monoanhydride leghemoglobin a
REFERENCE: N. Ellfolk and U. Perttila, Biochim. Biophys. Acta, 534, 153 (1978)

1426. Leghemoglobin a

SOURCE: Soybeans
SOLVENT: pH 8.4
COMMENTS: Spectra of leghemoglobin a and its hydrogen peroxide compound
REFERENCE: G. Sievers and M. Ronnberg, Biochim. Biophys. Acta, 533, 293 (1978)

1427. Leghemoglobin IV and Ferric Leghemoglobin

SOURCE: Leghemoglobin plus hydrogen peroxide
SOLVENT:
a. Sodium borate, pH 8.5
b. 0.01 M Potassium phosphate buffer, pH 7.5
c. Glycine-NaOH, pH 10

COMMENTS:
1. Spectra of acid and alkaline form at 77°K of leghemoglobin (IV)
2. Spectra of ferric leghemoglobin at 77°K and room temperature

REFERENCE: J. B. Wittenberg, J. Biol. Chem., 253, 5690 (1978); I. Aviram, B. A. Wittenberg, and J. B. Wittenberg, J. Biol. Chem., 253, 5685 (1978)

1428. Leghemoglobin, Ferric

SOURCE: Soybeans
SOLVENT:
a. 0.1 M Potassium phosphate, pH 5.3
b. 0.1 M Sodium acetate buffer, pH 5.3

COMMENTS:
1. Spectra in solvents a and b at 20°C
2. Spectral changes at 77°K upon addition of acetate to protein

REFERENCE: C. A. Appleby, W. F. Blumberg, J. Peisach, B. A. Wittenberg, and J. B. Wittenberg, J. Biol. Chem., 251, 6090 (1976)

1429. Leghemoglobin a

SOURCE: Soybeans
SOLVENT: 0.02 M Phosphate buffer, pH 7.05
COMMENTS: Spectra of ferric, fluoroferric, cyanoferric, oxyferrous, and ferrous forms
REFERENCE: N. A. Nicola, E. Minasian, C. A. Appleby, and S. J. Leach, Biochemistry, 14, 5141 (1975)

1430. Leghemoglobin a and ferri-

SOURCE: Soybeans
SOLVENT: a. Sodium phosphate buffer, pH 6.0
b. 0.05 M Sodium tetraborate, pH 9.2
COMMENTS: 1. Spectra of deoxy-, oxy-, and carbonyl forms (solvent a)
2. Spectra of ferrileghemoglobin (solvent a) and ferrileghemoglobin a hydroxide (solvent b)
3. Spectra of ferrileghemoglobin a cyanide, azide and fluoride forms (solvent a)
REFERENCE: N. Ellfolk and G. Sievers, Biochim. Biophys. Acta, 405, 213 (1975)

1431. Leghemoglobin c_1 and c_2

SOURCE: Soybeans
SOLVENT: 0.1 M 2-(N-Morpholino) ethane sulfonic acid, pH 5.2
REFERENCE: C. A. Appleby, N. A. Nicola, J. G. R. Hurrell, and S. J. Leach, Biochemistry, 14, 4444 (1975)

1432. Leghemoglobin a, ferri-

SOURCE: Soybeans
SOLVENT: a. 0.05 M Sodium phosphate buffer, pH 6.5
b. 8.8 M Urea
COMMENTS: 1. Spectra in buffer, urea, and after renaturation
2. Spectra of reduced (dithionite) renatured form, oxygenated form, native ferro-leghemoglobin a and its oxygenated form
REFERENCE: N. Ellfolk, G. S. Sievers, and A. Harmoinen, Acta Chem. Scand., 22B, 1195 (1974)

1433. Leghemoglobin

SOURCE: Soybean root nodules
SOLVENT: 50 mM 2-(N-Morpholino) ethane sulfonic acid, pH 5.3
COMMENTS: 1. Spectra of ferrous and ferric leghemoglobins and their nicotinic acid complexes
2. Spectra of oxyleghemoglobin in presence and absence of nicotinic acid

REFERENCE: C. A. Appleby, B. A. Wittenberg, and J. B. Henberg, Proc. Nat. Acad. Sci., 70, 564 (1973)

1434. Leghemoglobin a, ferrous and ferric

SOURCE: Soybeans
SOLVENT: 0.1 M (2-N-Morpholino) ethane sulfonic acid buffer, pH 5.2
COMMENTS: Spectra of ferrous and ferric leghemoglobins and of their complexes with X ligand
REFERENCE: C. A. Appleby, B. A. Wittenberg, and J. B. Wittenberg, J. Biol. Chem., 248, 3183 (1973)

1435. Leghemoglobin

SOURCE: Soybean root nodules
COMMENTS: Spectra of oxyleghemoglobin and reduced (dithionite) leghemoglobin before and after reaction with 3,3'-diaminobenzidine oxide
REFERENCE: F. J. Bergersen and D. J. Goodchild, Aust. J. Biol. Sci., 26, 741 (1973)

1436. Leghemoglobin 1

SOURCE: Soybean root nodules
SOLVENT: 0.05 M Bis (2-hydroxyethyl) imino-tris (hydroxymethyl) methane, pH 6.5
COMMENTS: Spectra of deoxy-, oxy-, and carbonmonoxy forms
REFERENCE: T. Imamura, A. Riggs, and Q. H. Gibson, J. Biol. Chem., 247, 521 (1972)

1437. Leghemoglobin

SOURCE: Subclover (Trifolium subterraneum cultivar woogenellup) and soybeans cultivar Bethal
SOLVENT: 0.05 M Phosphate buffer, pH 7.1
REFERENCE: K. R. Thulborn, E. Minasian, and S. J. Leach, Biochim. Biophys. Acta, 578, 476 (1979)

1438. Leghemoglobin, Apo-, (Legoglobin)

SOURCE: Yellow field lupine
SOLVENT: 0.01 M Phosphate buffer, pH 6.7
COMMENTS: Spectra of deoxy-, carboxy- and oxy-legoglobins, and hydroxy, imidazole and cyanide complexes of ferrilegoglobin
REFERENCE: Y. A. Sharonov, V. V. Yaorovenko, S. S. Melik-Sarkisyan, and V. L. Kretovich, Biokhimiya (Biochemistry), 38, 65 (1973)

1439. Leghemoglobin I and II

SOURCE: Lupinus lutens L. (yellow lupine)
COMMENTS: Spectra of ferric and oxyferrous forms
REFERENCE: S. S. Melik-Sarkisyan, V. V. Yarovenko, G. L. Shaposhnikov, L. P. Vladzievskaya, and V. L. Kretovich, Biokhimiya (Biochemistry), 39, 593 (1974)

1440. Leucine Aminopeptidase

SOLVENT: H_2O and 1 M hydroxylamine
COMMENTS: Spectra of protein split off Sepharose conjugate with hydroxylamine, and native enzyme
REFERENCE: J. Lasch and R. Koelsch, Eur. J. Biochem., 82, 181 (1978)

1441. Leucine Aminopeptidase 1 and 2

SOURCE: Aspergillus oryzae
SOLVENT: pH 7.4
REFERENCE: J. P. Labbe and P. Rebeyrotte, Biochimie, 56, 842 (1974)

1442. Leucine Aminopeptidase

SOURCE: Beef eye lens
COMMENTS: Spectra of free protein and protein covalently bound to Sepharose 6B
REFERENCE: R. Koelsch, J. Lasch, I. Marquardt, and H. Hanson, Anal. Biochem., 66, 556 (1975)

1443. L-Leucine-α-ketoglutarate Transaminase

SOURCE: Acetobacter suboxydans
SOLVENT: 0.01 M Potassium phosphate buffer, pH 7.0
REFERENCE: T. Tachiki and T. Tochikura, Agric. Biol. Chem., 40, 2187 (1976)

1444. L-Leucine-pyruvate Transaminase

SOURCE: Acetobacter suboxydans
SOLVENT: 0.01 M Potassium phosphate buffer, pH 6.1 and pH 8.0
COMMENTS: Spectra of apoenzyme and apoenzyme plus pyridoxal 5'-phosphate
REFERENCE: T. Tachiki, M. Moriguchi, and T. Tochikura, Agric. Biol. Chem., 39, 43 (1975); T. Tachiki and T. Tochikura, Biochim. Biophys. Acta, 397, 342 (1975)

1445. Levansucrase

SOURCE: Bacillus subtilis
SOLVENT: 0.05 M Potassium phosphate buffer, pH 6
REFERENCE: G. Gonzy-Treboul, R. Chambert, and R. Dedonder, Biochimie, 57, 17 (1975)

1446. Ligamentum Nuchae

SOURCE: Beef ligamentum nuchae and trout fibrils
SOLVENT: a. 0.1 N HCl
b. 0.1 N NaOH
COMMENTS: Spectra of elastase digests in acid and alkali
REFERENCE: M. Spina, S. Garbisa, J. M. Field, and A. Serafini-Fracassini, Arch. Biochem. Biophys., 192, 430 (1979)

1447. Lipase

SOURCE: Pig pancreas
SOLVENT: 0.1 M Phosphate buffer, pH 6.0
COMMENTS: Spectra of native and diethyl pyrocarbonate modified lipase
REFERENCE: C. Chapus and M. Semeriva, Biochemistry, 15, 4988 (1976)

1448. Lipase II

SOURCE: Rice bran
REFERENCE: T. Aizono, M. Funatsu, Y. Fujiki, and M. Watanabe, Agric. Biol. Chem., 40, 317 (1976)

1449. Lipase

SOURCE: Rice bran
SOLVENT: 0.01 M Phosphate buffer containing 0.001 M $CaCl_2$, pH 6.2
REFERENCE: M. Funatsu, Y. Aizono, K. Hayashi, M. Watanabe, and M. Eto, Agric. Biol. Chem., 35, 734 (1971)

1450. Lipase

SOURCE: Candida paralipolytica
SOLVENT: Acetate buffer, pH 5.0
REFERENCE: Y. Ota, T. Nakamiya, and K. Yamada, Agric. Biol. Chem., 34, 1368 (1970)

1451. Lipoamide Dehydrogenase

SOURCE: Fungi (Phytophthora erythroseptica and Pythium ultimum)
SOLVENT: 0.03 M Phosphate buffer, pH 7.6
COMMENTS: Spectra of oxidized enzyme, half-reduced forms (NADH) and fully reduced forms (dithionite)
REFERENCE: H. Fehrmann and C. Veeger, Biochim. Biophys. Acta, 350, 292 (1974)

1452. Lipoamide Dehydrogenase

SOURCE: Azotobacter vinelandii
SOLVENT: 0.05 M Phosphate buffer, pH 7.0
COMMENTS: Spectra of enzyme and reduced (aerobic, NADPH) enzyme
REFERENCE: T. W. Brewster, R. A. DeAbrew, A. DeKok, J. Visser, and C. Veeger, Eur. J. Biochem., 59, 335 (1975)

1453. Lipoamide Dehydrogenase

SOURCE: Mackerel (Pneumatophorus japonicus japonicus) dark muscle
SOLVENT: 50 mM Phosphate buffer, pH 7.6
REFERENCE: T. Muramatsu, S. Kan, and M. Hirashi, Comp. Biochem. Physiol., 61B, 247 (1978)

1454. Lipoamide Dehydrogenase

SOURCE: Escherichia coli
SOLVENT: pH 5.8, pH 6.3, pH 7.1, and pH 7.6
COMMENTS:
1. Spectral changes during dithionite addition
2. Effect of guanidine HCl, NaCl, and 3-aminopyridine adenine nucleotide on enzyme spectrum at EH_2 level

REFERENCE: K. D. Wilkinson and C. H. Williams, Jr., J. Biol. Chem., 254, 852, 863 (1979)

1455. Lipoamide Dehydrogenase

SOURCE: Pig heart
SOLVENT: 0.1 M Phosphate buffer, pH 7.6
COMMENTS:
1. Spectral changes during anaerobic titration with NAD^+ of 2-electron reduced enzyme (EH_2)
2. Spectral changes upon addition of reduced acetylpyridine adenine dinucleotide at 1.5°C
3. Spectral changes upon addition of oxidized acetylpyridine adenine dinucleotide to EH_2 form of enzyme

4. Spectral changes upon addition of reduced nicotinamide hypoxanthine dinucleotide to oxidized enzyme
5. Spectral changes upon addition of oxidized nicotinamide hypoxanthine dinucleotide to EH_2 form of enzyme

REFERENCE: R. G. Mathews, D. P. Ballou, and C. H. Williams, Jr., J. Biol. Chem., 254, 4974 (1979)

1456. Lipoamide Dehydrogenase

SOURCE: Pig heart
SOLVENT: 0.1 M Phosphate buffer, pH 7.6 containing EDTA
COMMENTS:
1. Spectral changes during anaerobic reaction of EH_2 form with iodoacetamide
2. Spectra of native enzyme and carboxymethyl modified enzyme
3. Spectral changes during reduction of carboxymethyl enzyme-H_2 form with dithionite and NADH

REFERENCE: C. Thorpe and C. H. Williams, J. Biol. Chem., 251, 3553 (1976)

1457. Lipoamide Dehydrogenase

SOURCE: Pig heart
SOLVENT: 0.05 M Phosphate buffer, pH 6.30
COMMENTS: Spectral changes upon addition of dithionite and dehydrolipoamide
REFERENCE: R. G. Matthews and C. H. Williams, J. Biol. Chem., 251, 3956 (1976)

1458. Lipoamide Dehydrogenase

SOURCE: Pig heart
SOLVENT: pH 7.0
COMMENTS: Spectrum of enzyme in the absence of and presence of various guanidine HCl concentrations (0-2 M)
REFERENCE: K. Ogasahara, K. Kolke, M. Hamada, and T. Hiraoka, J. Biochem. Tokyo, 79, 819 (1976)

1459. Lipoamide Dehydrogenase, alkylated

SOURCE: Pig heart
SOLVENT: 0.1 M Phosphate buffer, pH 7.6, containing EDTA and sodium arsenite
COMMENTS: Spectra of carboxyamidomethyl modified enzyme and after anaerobic reaction with NADH
REFERENCE: R. G. Matthews and C. H. Williams, Jr., Biochim. Biophys. Acta, 370, 39 (1974)

1460. Lipoamide Dehydrogenase

SOURCE: Pig heart
SOLVENT: 0.1 M Phosphate buffer, pH 7.6, containing 0.3 mM EDTA
COMMENTS: Spectrum in presence of 1 M NaCl or 0.7 M guanidine HCl plus 0.3 M NaCl
REFERENCE: C. Thorpe and C. H. Williams, Jr., Biochemistry, 13, 3263 (1974)

1461. Lipoprotein C-III-A, apo- and Lipoprotein C-III-B, apo-

SOURCE: Human serum
REFERENCE: J. T. Sparrow, H. J. Pownall, F. J. Hsu, L. D. Blumenthal, A. R. Culwell, and A. M. Gotto, Biochemistry, 16, 5427 (1977)

1462. Lipoprotein, Low-density (LDL)

SOURCE: Human serum
COMMENTS: Spectra of LDL from subject on normal diet and from subject on β-carotene enriched diet
REFERENCE: G. C. Chen and J. P. Kane, Biochemistry, 13, 3330 (1974)

1463. Lipoprotein, apo-

SOURCE: Human and pig sera
SOLVENT: 0.05 M Tris-HCl buffer, pH 8.0
REFERENCE: R. L. Jackson, H. N. Baker, O. D. Taunton, L. C. Smith, C. W. Garner, and A. M. Gotto, Jr., J. Biol. Chem., 248, 2639 (1973)

1464. Lipoprotein-GLN-II, apo- (Apo A-II)

SOURCE: Human serum
SOLVENT: 0.005 M Ammonium bicarbonate, pH 8.0
REFERENCE: S. E. Lux, K. M. John, and H. B. Brewer, J. Biol. Chem., 247, 7510 (1972)

1465. Lipoprotein, Low-Density, apo-

SOURCE: Human serum
SOLVENT: 7.6 M Guanidine HCl
REFERENCE: R. Smith, J. R. Dawson, and C. Tanford, J. Biol. Chem., 247, 3376 (1972)

1466. Lipoprotein, High-Density

SOURCE: Human serum
COMMENTS: Spectral changes due to exposure to oxygen for 0-42 hours
REFERENCE: W. A. Robison and G. J. Nelson, Biochim. Biophys. Acta, 175, 448 (1969)

1467. Lipoprotein, High-Density, apo-, Fractions I, II and III

SOURCE: Beef serum
SOLVENT: 0.1 M Tris-HCl buffer, pH 8.0
REFERENCE: A. Jonas, Biochim. Biophys. Acta, 393, 460 (1975)

1468. Lipoprotein, High-Density, 1-Pyrene-butyrate (PB) and 1-Dimethylaminonaphthalenesulfonyl (DNS) Derivatives

SOURCE: Beef lipoprotein
SOLVENT: a. 0.1 M Sodium phosphate buffer, pH 7.0, plus 0.025% EDTA
b. 0.03 M Sodium phosphate buffer, pH 7.0 plus 0.3% sodium dodecylsulfate
COMMENTS: Spectra of both modified proteins in solvent a and of both delipidated derivatives in solvent b
REFERENCE: A. Jonas, J. Biol. Chem., 247, 7773 (1972)

1469. Lipoprotein

SOURCE: Locust (Locusta migratoria) hemolymph
SOLVENT: Phosphate-saline
REFERENCE: Y. Peled and A. Tietz, Insect Biochem., 5, 61 (1975)

1470. Lipoprotein, High-Density, apo-, Fractions III and IV

SOURCE: Monkey (Macacus rhesus) serum
SOLVENT: pH 6.5 and pH 12.0
REFERENCE: C. Edelstein, L. T. Lim, and A. M. Scanu, J. Biol. Chem., 248, 7653 (1973)

1471. Lipoprotein-Cytosol Retinylester Complex

SOURCE: Rat liver cytosol
COMMENTS: Spectra after gel filtration and hydroxyapatite treatment
REFERENCE: J. Heller, Arch. Biochem. Biophys., 198, 562 (1979)

1472. Lipoprotein-Pigment Complexes

SOURCE: a. Chromatium minutissimum
b. Rhodopseudomonas palustris
REFERENCE: Y. E. Erokhin, V. A. Chugunov, Z. K. Makhneva, I. M. Agrikova, and G. B. Vasilev, Biokhimiya (Biochemistry), 43, 669 (1978)

1473. Lipoprotein Lipase

SOURCE: Mucor javanicus IAM 6108
SOLVENT: 0.01 M Phosphate buffer, pH 6.5
REFERENCE: T. Saiki, Y. Takagi, T. Suzuki, T. Narasaki, G. Tamura, and K. Arima, Agric. Biol. Chem., 33, 414 (1969)

1474. Lipoxygenase-1

SOURCE: Soybeans
SOLVENT: 0.1 M Sodium borate, pH 9.0
REFERENCE: L. J. M. Spaapen, G. A. Veldink, T. J. Liefkens, J. F. G. Vliegenthart, and C. M. Kay, Biochim. Biophys. Acta, 574, 301 (1979)

1475. Lipoxygenase-1

SOURCE: Soybeans
SOLVENT: pH 9.0
COMMENTS: Spectral changes due to addition of 13-L-hydroperoxylinoleic acid under aerobic conditions
REFERENCE: J. J. M. C. DeGroot, G. J. Garssen, G. A. Veldink, J. F. G. Vliegenthart, J. Boldingh, and M. R. Egmond, FEBS Lett., 56, 50 (1975)

1476. Lipoxygenase

SOURCE: Soybeans
SOLVENT: 0.05 M Phosphate buffer, pH 7.5
COMMENTS: Spectral changes due to addition of L-cysteine plus o-phenanthroline
REFERENCE: M. Roza and A. Francke, Biochim. Biophys. Acta, 327, 24 (1973)

1477. Lipoxygenase

SOURCE: Horse bean (Vicia faba L.)
REFERENCE: Y. Beaux, R. Drapron, J. Nicolas, and J. M. Caillat, Biochimie, 55, 253 (1973)

1478. Luciferase

SOURCE: Earthworm (Diplocardia longa)
SOLVENT: 0.05 M Potassium phosphate, pH 7.5
REFERENCE: M. J. Cormier, R. Bellisario, and T. E. Spencer, Biochemistry, 11, 2256 (1972)

1479. Luciferase

SOURCE: Photobacterium phosphoreum
SOLVENT: 0.1 M Phosphate buffer, pH 7.0
COMMENTS: Spectra of native and renatured luciferase
REFERENCE: K. Yoshida and T. Nakamura, J. Biochem. Tokyo, 74, 915 (1973); K. Matsuda and T. Nakamura, J. Biochem. Tokyo, 72, 951 (1972)

1480. Luciferase

SOURCE: Renilla reniformis
REFERENCE: J. C. Matthews, K. Hori, and M. J. Cormier, Biochemistry, 16, 85 (1977)

1481. Luciferin

SOURCE: Pholas dactylus
SOLVENT: a. 0.01 M Acetate buffer, pH 4.8
b. 0.01 M Phosphate buffer, pH 7.0
COMMENTS: Spectrum in solvent b taken before and after dialysis against 5 M guanidine and then buffer
REFERENCE: J. P. Henry, M. F. Isambert, and A. M. Michelson, Biochimie, 55, 83 (1973)

1482. Lutropin, 2-nitro-4-azidophenyl, β-subunit

SOURCE: Sheep pituitary gland
SOLVENT: 0.1 M Ammonium bicarbonate
REFERENCE: B. D. Burleigh, W. K. Liu, and D. N. Ward, J. Biol. Chem., 253, 7179 (1978)

1483. Lysine Monooxygenase

SOURCE: Pseudomonas fluorescens
COMMENTS: Spectral changes due to anaerobic reaction with lysine
REFERENCE: S. Yamamoto, F. Hirata, T. Yamauchi, M. Nozaki, O. Hyaisihi, and K. Hiromi, Z. Naturforsch., 27b, 1056 (1972)

1484. Lysine Monooxygenase

SOURCE: Pseudomonas fluorescens
SOLVENT: a. 0.5 M Potassium phosphate buffer, pH 8.0
b. 0.1 M Sodium acetate buffer, pH 5.5
COMMENTS: Spectral changes due to reduction of enzyme with sodium dithionite in solvents a and b
REFERENCE: M. I. S. Flashner and V. Massey, J. Biol. Chem., 249, 2579 (1974)

1485. β-Lysine Mutase

SOURCE: Clostridium sticklandii
SOLVENT: pH 6.8 buffer containing EDTA and NaCl
REFERENCE: J. J. Baker, C. Van der Drift, and T. Stadtman, Biochemistry, 12, 1054 (1973)

1486. L-Lysine Oxidase

SOURCE: Trichoderma viride
SOLVENT: 0.02 M Potassium phosphate buffer, pH 7.4
REFERENCE: H. Kusakabe, K. Kodama, A. Kuniwaka, H. Yoshino, H. Misono, and K. Soda, J. Biol. Chem., 255, 976 (1980)

1487. Lysozyme

SOURCE: Human skin
REFERENCE: H. Ogawa, H. Miyazaki, and M. Kimura, J. Inves. Derm., 57, 111 (1971)

1488. Lysozyme

SOURCE: Mink spleen, kidneys, and liver
REFERENCE: G. M. Malinina, I. A. Cherkasov, N. A. Kravchenko, and V. A. Berestov, Biokhimiya (Biochemistry), 42, 877 (1977)

1489. Lysozyme

SOURCE: Rat urine
REFERENCE: T. Yuzuriha, K. Katayama, and T. Fujita, Biochim. Biophys. Acta, 490, 235 (1977)

1490. Lysozyme

SOURCE: Rat kidney
SOLVENT: 0.15 M Phosphate buffer, pH 6.5, containing 0.1 M NaCl
REFERENCE: R. Raghunathan and S. Gurnani, Ind. J. Biochem. Biophys., 11, 2731 (1974)

1491. Lysozyme

SOURCE: Rat, urine
SOLVENT: 6 M Guanidine HCl, pH 6.4 and pH 12.0
REFERENCE: R. S. Mulvey, R. J. Gualtieri, and S. Beychok, Biochemistry, 13, 782 (1974)

1492. Lysozyme

SOURCE: Phage WT, eR1-75 and eRRR T4
REFERENCE: M. L. Elwell and J. A. Schellman, Biochim. Biophys. Acta, 494, 367 (1977); M. Elwell and J. Schellman, Biochim. Biophys. Acta, 386, 309 (1975)

1493. Lysozyme

SOURCE: Commercial
SOLVENT: 0.1 M Phosphate buffer, pH 7.0
REFERENCE: T. Ichikawa and H. Terada, Biochim. Biophys. Acta, 580, 120 (1979)

1494. Lysozyme

SOURCE: Chicken egg white
SOLVENT: 8 M Urea
COMMENTS:
1. Spectral changes during reaction of aqueous hydrazine with reduced and carboxymethylated (RCM) lysozyme
2. Spectral changes upon cleavage of RCM-ozone-inactivated lysozyme

REFERENCE: F. Sakiyama, J. Biochem. Tokyo, 82, 365 (1977)

1495. Lysozyme

SOURCE: Chicken egg white
SOLVENT: Water
COMMENTS:
1. Spectra of native and gamma-irradiated protein
2. Spectra of native and gamma-irradiated protein at different temperatures

REFERENCE: S. M. Herbert and B. M. Tolbert, RAD. Res., 65, 268 (1976)

1496. Lysozyme, α,α'-dibromo-p-xylenesulfonyl-

SOURCE: Chicken egg white lysozyme
SOLVENT: 0.1 M Sesquicarbonate buffer, pH 10
REFERENCE: A. C. Chung and R. A. Day, Int. J. Pep. Prot. Res., 8, 357 (1976)

1497. Lysozyme

SOURCE: Chicken egg white
COMMENTS: Spectral changes during ozonization in water
REFERENCE: M. Kuroda, F. Sakiyama, and K. Narita, J. Biochem. Tokyo, 78, 641 (1975)

1498. Lysozyme

SOURCE: Chicken egg white
COMMENTS: Spectra of protein modified with 2,4-pentanedione and then treated with hydroxylamine
REFERENCE: H. F. Gilbert and M. H. O'Leary, Biochemistry, 14, 5194 (1975)

1499. Lysozyme

SOURCE: Chicken egg white
SOLVENT: 0.1 M Sodium acetate buffer, pH 4.5
COMMENTS: Spectra of native lysozyme and oxindole alanine-108 lysozyme
REFERENCE: T. Imoto and J. A. Rupley, J. Mol. Biol., 80, 657 (1973)

1500. Lysozyme, fluoresceinthiocarbamyl-

SOURCE: Chicken egg white lysozyme
SOLVENT: 0.01 M Acetate buffer, pH 3.4
COMMENTS: Spectra of native and modified protein
REFERENCE: M. Hiramatsu, N. Okabe, and K. Tomita, J. Biochem. Tokyo, 73, 971 (1973)

1501. Lysozyme and 4-nitrobenz-2-oxa-1,3-diazole Derivatives

SOURCE: Chicken egg white
SOLVENT:
a. 0.05 M Borate, pH 8.5-10.5
b. 0.05 M Phosphate, pH 7.2 and pH 11.5
c. 0.05 M Citrate, pH 2.3

COMMENTS:
1. Spectral changes during reaction of 7-chloro-4-nitrobenz-2-oxa-1,3-diazole (NBD) with lysozyme over period of 0-480 minutes
2. Spectra of O^{TYR}-NBD-lysozyme, O^{TYR}-N-NBD-lysozyme, and N-NBD-lysozyme

REFERENCE: A. A. Oboderin, E. Boedefeld, and P. L. Luisi, Biochim. Biophys. Acta, 328, 20 (1973); A. A. Oboderin, E. Boedefeld, and P. L. Luisi, Biochim. Biophys. Acta, 328, 31 (1973)

1502. Lysozyme

SOURCE: Chicken egg white
COMMENT: Spectral changes due to irradiation with 0.5 M_{RAD} and 1.0 M_{RAD}
REFERENCE: T. Kume, Y. Sato, and Y. Umemoto, Agric. Chem., 47, 545 (1973)

1503. Lysozyme

SOURCE: Chicken egg white
COMMENTS: Spectral changes due to C-acetylation of tryptophan residues of lysozyme
REFERENCE: A. Preiviero, G. Prota, and M. A. Coletti-Previero, Biochim. Biophys. Acta, 285, 269 (1972)

1504. Lysozyme

SOURCE: Chicken egg white
SOLVENT: 6 M Guanidine HCl, pH 6.0 and pH 11.7
REFERENCE: G. Barth, E. Bunnenberg, and C. Djerassi, Anal. Biochem., 48, 471 (1972)

1505. Lysozyme and 2-nitrophenylsulfonyl-Derivative

SOURCE: Chicken egg white
SOLVENT: Water, pH 7.0
REFERENCE: Y. Schechter, Y. Burstein, and A. Patchornik, Biochemistry, 11, 653 (1972)

1506. Lysozyme and several azo derivatives

SOURCE: Chicken egg white
SOLVENT: 0.01 M Borate-carbonate buffer, pH 9.0
COMMENTS: Spectra of the following azolysozymes: azophenylsulfonic, azo-p-benzoic acid, azotetrazole, and azophenyldiethylmethylammonium chloride
REFERENCE: M. Franek and Z. Pechan, Scripta Fac. Sci. Nat. Ujep Brunensis Chemia, 1, 67 (1971)

1507. Lysozyme

SOURCE: Chicken egg white
SOLVENT: 1/15 M Phosphate buffer, pH 6.20
COMMENTS: Spectral changes due to photooxidation
REFERENCE: N. A. Kravchenko and V. K. Lapuk, Biokhimiya (Biochemistry), 34, 672 (1969)

1508. Lysozyme and Lysozyme Methyl Esters

SOURCE: Chicken egg white
REFERENCE: N. A. Kravchenko, T. V. Chentsova, and E. D. Kaverzneva, Biokhimiya (Biochemistry), 33, 295 (1968)

1509. Lysozyme, S-Carboxymethyl-

SOURCE: Commercial lysozyme
SOLVENT: 6 M Guanidinium chloride
COMMENTS: Spectrum of S-carboxymethyl-lysozyme before and after irradiation
REFERENCE: K. J. Dilley, Biochem. J., 133, 821 (1973)

1510. Lysozyme-like Enzyme

SOURCE: Staphylococcus aureus
SOLVENT: a. Water
b. 0.7% NaCl
COMMENTS: Spectrum of lysozyme-like enzyme before defatting (solvent a) and after defatting (solvent b)
REFERENCE: V. A. Leonenko, N. A. Kravchenko, I. A. Cherkasov, T. I. Afanaseva, and Z. V. Ermoleva, Biokhimiya (Biochemistry), 39, 912 (1974)

1511. Lytic Enzyme

SOURCE: Fungi imperfecti mold
SOLVENT: a. 0.01 M Potassium phosphate buffer, pH 8.0
b. 0.1 N NaOH
REFERENCE: J. Nagasaki, J. Fukuyama, S. Yamamoto, and R. Kobayashi, Agric. Biol. Chem., 38, 349 (1974)

1512. Macromomycin

SOURCE: Streptomyces macromomyceticus
SOLVENT: a. H_2O
b. 0.1 N HCl
c. 0.1 N NaOH
REFERENCE: W. B. Im, C. K. Chiang, and R. Montgomery, J. Biol. Chem., 253, 3259 (1978)

1513. Malate Dehydrogenase

SOURCE: Pig heart cytoplasm
COMMENTS: Spectra of protein and protein-$PtCl_4^{2-}$ complex
REFERENCE: M. Wade, D. Tsernoglou, E. Hill, L. Webb, and L. Banaszak, Biochim. Biophys. Acta, 322, 124 (1973)

1514. Malate Dehydrogenase

SOURCE: Pig heart mitochondria
SOLVENT: 0.05 M Sodium phosphate buffer, pH 7.5
COMMENTS: Spectra of native enzyme, pyridoxal-5'-phosphate inactivated enzyme, and pyridoxal-5'-phosphate-sodium borohydride inactivated enzyme
REFERENCE: M. J. Wimmer, T. Mo, D. L. Sawyer, and J. H. Harrison, J. Biol. Chem., 250, 710 (1975)

1515. Malate Dehydrogenase

SOURCE: Rat liver and heart mitochondria
SOLVENT: 0.02 M Citrate, pH 5.5
REFERENCE: K. N. Kuan, G. L. Jones, and C. S. Vestling, Biochemistry, 18, 4366 (1979)

1516. Malate Dehydrogenase

SOURCE: Rat liver
SOLVENT: 0.1 M Potassium phosphate buffer, pH 7.4
REFERENCE: F. Wada, N. Numata, Y. Eguchi, and Y. Sakamoto, Biochim. Biophys. Acta, 410, 237 (1975)

1517. Malate Dehydrogenase

SOURCE: Halobacterium (Dead Sea)
SOLVENT: a. 0.01 M Sodium phosphate, pH 7.3
b. 0.01 M Sodium phosphate, pH 7.3, containing 4.26 M NaCl
REFERENCE: M. Mevarich, H. Eisenberg, and E. Neumann, Biochemistry, 16, 3781 (1977)

1518. Malate-Vitamin K Reductase

SOURCE: Mycobacterium phlei
SOLVENT: 0.3 M KCl and 10% glycerol in 0.02 M Tris HCl, pH 8.2
REFERENCE: K. Imai and A. F. Brodie, J. Biol. Chem., 248, 7489 (1973)

1519. Malic Enzyme

SOURCE: Bryophyllum tubiflorum
REFERENCE: P. C. Brandon and T. N. Van Boekel-Mol, Eur. J. Biochem., 35, 62 (1973)

1520. Maltose Phosphorylase

SOURCE: Lactobacillus brevis

SOLVENT: 0.005 M Citrate buffer, pH 6.6
REFERENCE: A. Kamogawa, K. Yokobayashi, and T. Fuku, Agric. Biol. Chem., 37, 2813 (1973)

1521. Mannanase

SOURCE: Bacillus subtilis K-50
SOLVENT: a. 0.01 M Phosphate buffer, pH 7.0
b. 0.1 N NaOH
REFERENCE: S. Emi, J. Fukumoto, and T. Yamamoto, Agric. Biol. Chem., 36, 991 (1972)

1522. α-Mannosidase

SOURCE: Phaseolus vulgaris
SOLVENT: 0.1 M Sodium acetate buffer, pH 4.0
COMMENTS: Spectral changes due to reaction with N-bromosuccinimide
REFERENCE: E. Paus, Biochim. Biophsy. Acta, 533, 446 (1978)

1523. Melitoate Hydroxylase

SOURCE: Pseudomonas sp.
SOLVENT: 0.1 M Potassium phosphate, pH 7.3
COMMENTS:
1. Spectrum of oxidized enzyme
2. Spectral changes during anaerobic light irradiation
3. Change in spectrum due to melitoate
4. Spectrum of photoreduced enzyme plus melitoate
5. Spectral changes due to anaerobic reduction by NADH
6. Spectral changes due to oxidation of photoreduced enzyme by NAD^+

REFERFNCE: S. Strickland and V. Massey, J. Biol. Chem., 248, 2944, 2953 (1973)

1524. Membranes

SOURCE: Acholeplasma laidlawii
SOLVENT: 0.02 M Sodium dodecylsulfate, 0.02 M Tris-HCl, pH 8.0
REFERENCE: G. Dresdner, Can. J. Microbiol., 24, 1093 (1978)

1525. Membrane, Purple

SOURCE: Halobacterium halobium
SOLVENT: pH 2.5, 0.01 M Cetyltrimethyl ammonium bromide, pH 3.0 and pH 8.0
COMMENTS: Spectra of native membrane, membrane in detergent, after red light exposure, after blue light exposure, and after 550 mM exposure

REFERENCE: D. Oesterhelt and W. Stoeckenius, Nature New Biology, 233, 149 (1971); K. R. Naqui, J. Gonzales-Rodriquez, R. J. Cherry, and D. Chapman, Nature New Biology, 245, 249 (1973)

1526. Membrane, Purple (Bacteriorhodopsin)

SOURCE: Halobacterium halobium
COMMENTS: Spectra of membrane, apomembrane, and reconstituted membrane
REFERENCE: D. Oesterhelt and L. Schuhmann, FEBS Lett., 44, 262 (1974)

1527. Membrane, Purple

SOURCE: Halobacterium halobium
SOLVENT: 0.01 M Potassium phosphate buffer, pH 7.0
REFERENCE: B. M. Becher and J. Y. Cassim, Prep. Biochem., 5, 161 (1975); Biophys. J., 16, 1183 (1976)

1528. Membrane, Purple

SOURCE: Halobacterium halobium
REFERENCE: N. Dencher, in Biochemistry of Sensory Functions (L. Jaenecke, ed.), 161 (1974), Springer Verlag, New York

1529. Membrane, Purple, sodium borohydride-reacted

SOURCE: Halobacterium halobium
SOLVENT: pH 7.0
COMMENTS:
1. Spectral changes on exposure to UV radiation
2. Spectral changes on adding 6.6 N HCl
3. Spectral changes on adding cetyltrimethyl ammonium bromide

REFERENCE: J. Peters, R. Peters, and W. Stockenius, FEBS Lett., 61, 128 (1978)

1530. Membrane, Purple

SOURCE: Halobacterium halobium
COMMENTS: Spectral changes due to reaction with sodium borohydride
REFERENCE: J. Peters, R. Peters, and W. Stockenius, FEBS Lett., 61, 128 (1976)

1531. Membrane, Purple

SOURCE: Halobacterium halobium
COMMENTS: Spectra of dark adapted and light adapted forms

REFERENCE: B. Becher and J. Y. Cassim, Biophys. J., 16, 1183 (1976)

1532. Membrane, Purple (Bacteriorhodopsin)

SOURCE: Halobacterium halobium
COMMENTS: Changes in the spectrum due to tyrosine modification with tetranitromethane
REFERENCE: M. Campos-Cavieres, T. A. Moore, and R. N. Perham, Biochem. J., 179, 233 (1979)

1533. Membrane, Purple (Bacteriorhodopsin)

SOURCE: Halobacterium halobium
SOLVENT: Aqueous suspension
REFERENCE: M. Stockburger, W. Klusmann, H. Gattermann, G. Massig, and R. Peters, Biochemistry, 18, 4886 (1979)

1534. Membrane, Purple and K, L, and M Intermediates

SOURCE: Halobacterium halobium R_1
SOLVENT:
a. 0.02 M Phosphate buffer, pH 7.0
b. 0.02 M Phosphate buffer, 67% glycerol, pH 7.0

COMMENTS:
1. Spectra of light-adapted, sonicated material in solvent a and in solvent a with 1% Ammonyx LO and 25% NaCl
2. Spectra of bleach-extracted membrane and after regeneration with all-trans-retinal 1 in solvent a
3. Spectral changes associated with formation of intermediates K, L and M and conversion back to purple membrane in solvent b

REFERENCE: B. Becher, F. Tokunaga, and T. G. Ebrey, Biochemistry, 17, 2293 (1978)

1535. Membrane, Purple and Membrane, Red

SOURCE: Halobacterium cutirubrum
SOLVENT: Water
REFERENCE: S. C. Kushwaha, M. Kates, and W. G. Martin, Can. J. Biochem., 53, 284 (1975)

1536. Membranes

SOURCE: Mitochondria of untreated and chloramphenicol-treated Neurospora crassa
REFERENCE: H. Weiss and H. J. Kolb, Eur. J. Biochem., 99, 139 (1979)

1537. Membranes

SOURCE: Escherichia coli
COMMENTS: Spectra of reduced and oxidized forms
REFERENCE: R. W. Hendler and R. I. Shrager, J. Biol. Chem., 254, 11288 (1979)

1538. Membrane

SOURCE: Beef eye, rod outer segment
SOLVENT: Cholic acid solution
COMMENTS: Spectra of solubilized material and ammonium sulfate fractions
REFERENCE: M. Makino, T. Hamanaka, Y. Orii, and Y. Kito, Biochim. Biophys. Acta, 495, 299 (1977)

1539. Membrane

SOURCE: Beef eye, rod outer segments
SOLVENT: 0.1 M Phosphate, pH 6.5
COMMENTS: Spectra of native membranes and methylmercuric-iodide-treated membranes before and after illumination and after addition of hydroxylamine
REFERENCE: F. J. M. Daemen, P. J. G. M. Van Braugel, P. A. A. Jansen, and S. L. Bonting, Biochim. Biophys. Acta, 453, 374 (1976)

1540. Membrane, Amidinated

SOURCE: Beef eye, rod outer segments
SOLVENT: 67 mM Phosphate buffer, pH 6.5, containing 2% digitonin and 50 mM hydroxylamine
COMMENTS: Spectra of non-amidinated and several times amidinated material
REFERENCE: W. J. De Grip, F. J. M. Daemen, and S. L. Bonting, Biochim. Biophys. Acta, 323, 125 (1973)

1541. Membrane

SOURCE: Beef erythrocytes
SOLVENT: Water, pH 9.3-9.5
REFERENCE: D. Mazia and A. Ruby, Proc. Nat. Acad. Sci., 61, 1005 (1978)

1542. Mesentericopeptidase

SOURCE: Bacillus mesentericus
SOLVENT: pH 7.0 and pH 13.0
REFERENCE: N. Genov, Int. J. Pept. Prot. Res., 7, 325 (1975)

1543. Metallothionein

SOURCE: Rat and chicken livers
SOLVENT: pH 6.6
REFERENCE: U. Weser, F. Donay, and H. Rupp, FEBS Lett., 32, 171 (1973)

1544. Metallothionein

SOURCE: Chicken liver
SOLVENT: 8.3 mM Sodium phosphate buffer, pH 7.0
COMMENTS: Spectral changes during titration with copper dissolved in acetonitrile-water [$Cu(CH_3CN)_4ClO_4$]
REFERENCE: H. Rupp and U. Weser, FEBS Lett., 44, 293 (1974); H. Rupp, W. Voelter and U. Weser, Hoppe-Seyler's Z. Physiol. Chem., 356, 755 (1975)

1545. Metallothionein

SOURCE: Rat liver
COMMENTS: Spectrum of the cadmium-binding protein before and after Sephadex G50 and Sephadex G-75 chromatography
REFERENCE: D. R. Winge and K. V. Rajagopalan, Arch. Biochem. Biophys., 153, 755 (1972)

1546. Metallothionein

SOURCE: Mouse liver
REFERENCE: M. Nordberg, G. F. Nordberg, and M. Piscator, Environ. Physiol. Biochem., 5, 396 (1975)

1547. Metallothionein

SOURCE: Rabbit renal tissue
COMMENTS: Spectra of cadmium metallothionein in presence of different amounts of mercury
REFERENCE: M. Nordberg, B. Trojanowska, and F. Nordberg, Environ. Physiol. Biochem., 4, 153 (1974)

1548. Metallothionein, Forms I and II

SOURCE: Rabbit liver
COMMENTS: Spectra after removal of ampholytes
REFERENCE: G. F. Nordberg, M. Nordberg, M. Piscator, and O. Vesterberg, Biochem. J., 126, 491 (1972)

1549. Methane Mono-oxygenase, Component C

SOURCE: Methylococcus capsulatus (Bath)
SOLVENT: 20 mM Sodium phosphate buffer, pH 7.0
COMMENTS: Spectral changes due to anaerobic photoreduction in the presence of EDTA or NADH
REFERENCE: J. Colby and H. Dalton, Biochem. J., 171, 461 (1978)

1550. Methanol Dehydrogenase

SOURCE: Hyphomicrobium X
SOLVENT: 0.02 M Potassium phosphate, pH 7.0
REFERENCE: J. A. Duine, J. Frank, and J. Westerling, Biochim. Biophys. Acta, 524, 277 (1978)

1551. Methanol Dehydrogenase

SOURCE: Methylomonas methanica
SOLVENT: 50 mM Phosphate buffer, pH 7.0
COMMENTS: Spectra of holo- and apo-enzyme
REFERENCE: R. N. Patel, C. To. Hou, and A. Felix, J. Bacteriol., 133, 641 (1978)

1552. Methanol Dehydrogenase

SOURCE: Methylosinus sporium
COMMENTS: 50 mM Phosphate buffer, pH 8.0
REFERENCE: R. N. Patel and A. Felix, J. Bacteriol., 128, 413 (1976)

1553. Methanol Dehydrogenase

SOURCE: Methylococcus capsulatus
COMMENTS: Spectra of soluble and solubilized enzymes
REFERENCE: A. M. Wadzinski and D. W. Ribbons, J. Bacteriol., 122, 1364 (1975)

1554. Methanol-oxidizing Enzyme

SOURCE: Candida boidinii
SOLVENT: 0.1 M Potassium phosphate buffer, pH 7.5
REFERENCE: H. Sahm and F. Wagner, Eur. J. Biochem., 36, 250 (1973)

1555. Methemalbumin

SOURCE: Human blood
COMMENTS: 1. Spectra of methemalbumin and reduced (dithionite) methemalbumin

2. Spectrum of serum before and after reduction

REFERENCE: C. B. Walberg, M. C. Geokas, and H. Rinderknecht, Clin. Chim. Acta, 48, 279 (1973)

1556. Methemoglobin Reductase

SOURCE: Blue-white dolphin (Stenella caeruleo-alba) erythrocytes
SOLVENT: 0.1 M Phosphate buffer, pH 7.0
REFERENCE: C. Shimizu and F. Matuura, Agric. Biol. Chem., 32, 587 (1968)

1557. Methemoglobin Reductase

SOURCE: Human erythrocytes
SOLVENT: 0.1 M Sodium phosphate buffer, pH 7.5
COMMENTS: Spectra of oxidized and reduced (hydrosulfite) enzyme
REFERENCE: F. Kuma and H. Inomata, J. Biol. Chem., 247, 556 (1972)

1558. Methioninase

SOURCE: Pseudomonas putida
SOLVENT: 0.1 M Potassium phosphate buffer, pH 7.5, containing 10 mM 2-mercaptoethanol
COMMENTS: Spectra of native enzyme and of enzyme after treatment with hydroxylamine
REFERENCE: S. Ito, T. Nakamura, and Y. Eguchi, J. Biochem. Tokyo, 79, 1263 (1976)

1559. Methioninase

SOURCE: Pseudomonas ovalis
SOLVENT: 0.01 M Potassium phosphate buffer, pH 7.2
COMMENTS: Spectra of holo- and apo-enzymes
REFERENCE: H. Tanaka, N. Eskai, T. Yamamoto, and K. Soda, FEBS Lett., 66, 307 (1976)

1560. L-Methionine γ-Lyase

SOURCE: Pseudomonas ovalis
SOLVENT: 0.01 M Potassium phosphate buffer, pH 7.2
COMMENTS: Spectra of holoenzyme, reduced (dithionite) holoenzyme, and apoenzyme
REFERENCE: H. Tanaka, N. Esaki, and K. Soda, Biochemistry, 16, 100 (1977)

1561. Methionine γ-Lyase

SOURCE: Pseudomonas ovalis
SOLVENT: 50 mM Potassium pyrophosphate buffer, pH 7.3
COMMENTS: 1. Spectral changes during reaction with L-propargylglycine
2. Spectral change on reaction of L-propargylglycine-inactivated enzyme with dithiothreitol
REFERENCE: M. Johnston, D. Jankowski, P. Marcotte, H. Tanaka, N. Esaki, K. Soda, and C. Walsh, Biochemistry, 18, 4690 (1979)

1562. L-Methionine γ-Lyase

SOURCE: Pseudomonas putida (IFO 3738)
SOLVENT: 0.2 M Potassium phosphate buffer, pH 8.0
COMMENTS: Spectral changes due to reaction of enzyme with DL-selenomethionine and DL-vinylglycine
REFERENCE: N. Esaki, H. Tanaka, S. Vemura, T. Suzuki, and K. Soda, Biochemistry, 18, 407 (1979)

1563. Methionine Synthetase, R Component and F Component

SOURCE: Escherichia coli
SOLVENT: 0.08 M Potassium phosphate buffer, pH 7.0
REFERENCE: K. Fujii and F. M. Huennekens, J. Biol. Chem., 249, 6745 (1974)

1564. Methylamine Dehydrogenase

SOURCE: Pseudomonas AM1
SOLVENT: 10 mM Sodium potassium phosphate buffer, pH 7.0
COMMENTS: Spectra of enzyme, light subunit, and heavy subunit
REFERENCE: S. Shirai, T. Matsumoto, and J. Tobari, J. Biochem. Tokyo, 83, 1599 (1978)

1565. Methylamine Dehydrogenase

SOURCE: Pseudomonas sp. J
SOLVENT: a. 50 mM Sodium potassium phosphate buffer, pH 7.0
b. 10 mM Potassium phosphate buffer, pH 7.0
COMMENTS: 1. Spectra of heavy subunit, light subunit, and light subunit plus methylamine hydrochloride (solvent a)
2. Spectra of enzyme, enzyme plus methylamine hydrochloride, and enzyme and methlyamine hydrochloride plus phenazine methosulfate (solvent b)

REFERENCE: T. Matsumoto, B. Y. Hirako, and J. T. Bari, Biochim. Biophys. Acta, 522, 303 (1978); T. Matsumoto, Biochim. Biophys. Acta, 522, 291 (1978)

1566. N-Methylglutamate Synthetase

SOURCE: Pseudomonas MA (ATCC 23819)
SOLVENT: a. 0.1 M Potassium phosphate buffer, pH 6.8
b. 110 mM Potassium N-tris-(hydroxymethyl) methylglycine, pH 8.3, containing potassium chloride, dithiothreitol, magnesium chloride, and EDTA MgNa
COMMENTS: 1. Spectrum of deaza FMN-reconstituted enzyme (solvent a)
2. Spectral changes during reaction of deaza FMN enzyme with glutamate (solvent b)
3. Spectral changes during reaction of substrate-reduced deaza FMN enzyme with ammonia
REFERENCE: M. S. Jorns and L. B. Hersh, J. Biol. Chem., 250, 3620 (1975)

1567. Metmyoglobin Reductase

SOURCE: Blue-white dolphin (Stenella caeruleo-alba)
SOLVENT: 0.1 M Phosphate buffer, pH 7.0
REFERENCE: C. Shimizu and F. Matsuura, Agric. Biol. Chem., 35, 468 (1971)

1568. α_1-Microglobulin

SOURCE: Human urine
SOLVENT: 0.1 M Sodium phosphate buffer, pH 7.0
REFERENCE: B. Ekstrom and I. Bergaard, J. Biol. Chem., 252, 8048 (1977)

1569. Microsomes

SOURCE: Peas
SOLVENT: 0.1 M Potassium phosphate, pH 7.2
REFERENCE: A. Ishimaru and I. Yamazaki, J. Biol. Chem., 252, 199 (1978)

1570. Mitochondria and Mitochondria-like Particles

SOURCE: Baker's yeast, respiring
REFERENCE: R. S. Criddle, D. Paltauf, H. Plattner, and G. Schatz, J. Gen. Physiol., 54, 575 (1969)

1571. Mitochondria and Mitoplasts

SOURCE: Rat liver
SOLVENT: 0.25 M Sucrose
REFERENCE: G. Gremel, A. Waksman, and P. Pajot, FEBS Lett., 74, 239 (1977)

1572. Monellin

SOURCE: Dioscoreophyllum cummsii
SOLVENT: 0.01 M Sodium phosphate buffer, pH 7.2
REFERENCE: J. A. Morris, R. Martenson, G. Diebler, and R. H. Cagan, J. Biol. Chem., 248, 534 (1973)

1573. Monoamine Oxidase

SOURCE: Beef liver mitochondria
SOLVENT: 50 mM Phosphate, pH 7.2, containing Triton X-100
COMMENTS: Spectra of the oxidized enzyme and the substrate-reduced enzyme
REFERENCE: J. I. Salach, Arch. Biochem. Biophys., 292, 128 (1979)

1574. Monoamine Oxidase

SOURCE: Beef kidney
SOLVENT: 0.05 M Phosphate, pH 7.6
COMMENTS: Spectral change due to the addition of pargyline
REFERENCE: H. Y. K. Chuang, D. R. Patek, and L. Hellerman, J. Biol. Chem., 249, 2381 (1974)

1575. Monoamine Oxidase

SOURCE: Beef liver
SOLVENT: 50 mM Sodium phosphate, pH 7.2, containing Triton X-100
COMMENTS: Spectral changes due to addition of 3-dimethylamino-1-propyne
REFERENCE: A. L. Maycock, R. H. Abeles, J. I. Salach, and T. P. Singer, Biochemistry, 15, 114 (1976)

1576. Monoamine Oxidase

SOURCE: Pig liver
COMMENTS: Spectra of protein, protein plus benzylamine, and protein plus dithionite
REFERENCE: W. R. Carper, D. D. Stoddard, and D. F. Martin, Biochim. Biophys. Acta, 334, 287 (1974)

1577. Monoamine Oxidase

SOURCE: Pig brain
SOLVENT: 0.1 M Potassium phosphate buffer, containing 0.1% cholate, pH 7.0
COMMENTS: Spectra of enzyme and reduced (dithionite) enzyme
REFERENCE: N. Minamura and K. T. Yasunobu, Biochem. Pharm., 27, 2737 (1978)

1578. Monoamine Oxidase

SOURCE: Human liver
SOLVENT: 0.05 M Phosphate buffer, pH 7.4
REFERENCE: I. F. Nostrand and M. D. Glantz, Arch. Biochem. Biophys., 158, 1 (1973)

1579. Monooxygenase of 4-Methoxybenzoate O-Demethylase

SOURCE: Pseudomonas putida
SOLVENT: 50 mM Potassium phosphate buffer, pH 8.0
COMMENTS: Spectra taken during stepwise anaerobic reduction with dithionite
REFERENCE: F.-H. Bernhardt, H. Pachowsky, and H. Staudinger, Eur. J. Biochem., 57, 241 (1975)

1580. Monooxygenase

SOURCE: Pig liver microsomes
SOLVENT: 0.1 M Potassium phosphate, pH 7.5
COMMENTS: Spectra of oxidized, reduced, and peroxyflavin forms
REFERENCE: L. L. Poulsen and D. M. Ziegler, J. Biol. Chem., 254, 6449 (1979)

1581. Mucin

SOURCE: Beef cervix
SOLVENT: Phosphate buffer, pH 7.4
REFERENCE: E. Van Roost, P. L. Masson, J. F. Heremans, and F. Andre, Biochim. Biophys. Acta, 317, 286 (1973)

1582. Mutarotase

SOURCE: Beef kidney
SOLVENT: 20 mM Barbitol buffer, pH 7.5
COMMENTS: Spectra of O-acetylated and deacetylated (hydroxylamine) enzyme
REFERENCE: P. H. Fishman, P. G. Pentchev, and J. M. Bailey, Biochemistry, 12, 2450 (1973)

1583. Mycodextranase

SOURCE: Penicillium melinii
SOLVENT: a. Water
b. 0.5 N KOH
REFERENCE: A. L. Rosenthal and J. H. Nordin, J. Biol. Chem., 250, 5295 (1975)

1584. Myeloperoxidase

SOURCE: Human leukocytes
SOLVENT: 100 mM Potassium phosphate, pH 7.3
COMMENTS: Spectra of oxidized and reduced (dithionite) forms
REFERENCE: A. R. J. Bakkenist, R. Wever, T. Vulsma, H. Plat, and B. F. Van Gelder, Biochim. Biophys. Acta, 524, 45 (1978)

1585. Myeloperoxidase

SOURCE: Human leukemic leukocytes
SOLVENT: 0.05 M Phosphate buffer, pH 5.4
COMMENTS: Spectra of enzyme, enzyme plus hypochlorite, enzyme plus ascorbate, and enzyme plus hydrogen peroxide
REFERENCE: J. W. Naskalski, Biochim. Biophys. Acta, 485, 291 (1977)

1586. Myeloperoxidase

SOURCE: Human neutrophilic granulocytes
SOLVENT: 50 mM Phosphate/citrate buffer, pH 4.6
COMMENTS:
1. Spectra of enzyme, enzyme plus cyanide, and enzyme, cyanide and hydrogen peroxide
2. Effect of NaCl on spectra in 1
3. Spectral changes of enzyme-CN^- complex due to addition of NaCl, 0-70.6 mM

REFERENCE: J. M. Zgliczynski and T. Stelmaszynska, Biochim. Biophys. Acta, 567, 309 (1979)

1587. Myeloperoxidase

SOURCE: Human neutrophilic granules
SOLVENT:
a. 0.15 M Citrate-phosphate buffer, pH 3.28, pH 5.23, pH 6.34
b. 50 mM Citrate-phosphate buffer, pH 3.17, pH 4.78, pH 6.28

COMMENTS:
1. Spectral changes due to the addition of NaCl (0-400 mM) at the pH values in a
2. The effect of pH on the myeloperoxidase spectrum using pH values in b

REFERENCE: T. Stelmaszynska and J. M. Zgliczynski, Eur. J. Biochem., 45, 305 (1974)

1588. Myeloperoxidase

SOURCE: Human leukocytes
SOLVENT: 0.1 M Potassium phosphate buffer, pH 7.0
COMMENTS: Spectra of the oxidized and reduced (hydrosulfite) components of the enzyme (isozymes)
REFERENCE: N. T. Felberg and J. Schultz, Arch. Biochem. Biophys., 148, 407 (1972)

1589. Myeloperoxidase

SOURCE: Human blood
SOLVENT: 0.1 M Potassium phosphate buffer, pH 7.0
COMMENTS: Spectra of oxidized and reduced enzyme prepared without trypsin
REFERENCE: N. T. Felberg, G. J. Putterman, and J. Schultz, Biochem. Biophys. Res. Comm., 37, 213 (1969)

1590. Myeloperoxidase

SOURCE: Pig leukocytes
SOLVENT: 0.1 M Phosphate buffer, pH 7.5
COMMENTS: Spectral changes due to superoxide ion in the presence and absence of superoxide dismutase
REFERENCE: T. Odajima & I.Yamazaki, Biochim. Biophys. Acta, 284, 355 (1972)

1591. Myeloperoxidase

SOURCE: Pig leukocytes
SOLVENT:
a. 0.01 M Acetate, pH 5.5
b. Water
c. 10 mM HCl (pH 2.6)
d. 30 mM HCl (pH 2)

COMMENTS:
1. Spectra of enzyme, enzyme-cyanide complex, and reduced (hydrosulfite) enzyme-cyanide complex (solvent a)
2. Spectra in solvents b and d and after dialysis against solvent a

REFERENCE: T. Odajima and I. Yamazaki, Biochim. Biophys. Acta, 284, 368 (1972)

1592. Myeloperoxidase

SOURCE: Pig leukocytes
SOLVENT: 0.1 M Phosphate buffer, pH 7.5 or pH 7.0
COMMENTS: Spectra of myeloperoxidase-hydroxylamine complex, reduced (dithionite) myeloperoxidase-azide complex, and myeloperoxidase-cyanide complex
REFERENCE: T. Odajima and I. Yamazaki, Biochim. Biophys. Acta, 284, 360 (1972)

1593. Myeloperoxidase, Ferro-

SOURCE: Dog pus
SOLVENT: a. 0.1 M Sodium potassium phosphate, pH 7.0
b. 0.1 M Sodium phosphate, pH 7.0, 5 M urea
c. 0.1 M Potassium phosphate, pH 6.6
COMMENTS: 1. Spectra of dithionite-reduced protein and of stable cyanide complexes (solvent a)
2. Spectral changes on dithionite reduction and formation of unstable cyanide complexes (solvent c)
3. Spectral changes on exposure to dodecyl sulfate of dithionite-reduced enzyme (solvent b)
REFERENCE: J. E. Harrison, J. Biol. Chem., 254, 4536 (1979)

1594. Myoglobin, met-

SOURCE: Commercial
SOLVENT: pH 6
COMMENTS: Spectra of protein, protein-tannic acid complex, and of complex after adding Tween 80
REFERENCE: A. Zanobi, P. Janni, and A. M. Firenzuoli, Experientia, 15, 1015 (1967)

1595. Myoglobin

SOURCE: Commercial
SOLVENT: 0.1 M Potassium phosphate buffer, pH 7.4
COMMENTS: Spectral changes due to varying flash intensities
REFERENCE: H. Shimada, T. Iizuka, R. Ueno, and Y. Ishimura, FEBS Lett., 98, 290 (1979)

1596. Myoglobin

SOURCE: Aplysia
SOLVENT: 2% Borate buffer, pH 9.1
COMMENTS: Spectra at 25°C, 50°C, 63.5°C, 66°C, 75°C, and 77°C
REFERENCE: M. Brunori, G. M. Giacometti, E. Antonini, and J. Wyman, J. Mol. Biol., 63, 139 (1972)

1597. Myoglobin

SOURCE: Aplysia
SOLVENT: a. 2% Borate buffer, pH 9.1
b. 2% Borate buffer, pH 9.1 plus 7%-n-butanol
COMMENTS: 1. Spectrum of myoglobin covalently bound to Sephadex G-100

2. Spectra of ferric, ferrous, and CO-forms resin bound in the native and denatured conformations

REFERENCE: G. M. Giacometti, A. Colosimo, S. Steffanini, B. Brunori, and E. Antonini, Biochim. Biophys. Acta, 285, 320 (1972)

1598. Myoglobin

SOURCE: Nassa mutablis
SOLVENT: 50 mM Potassium phosphate, pH 7.0
COMMENTS: Spectra of oxy-, CO-, and deoxy-forms
REFERENCE: G. Geraci, A. Sada, and C. Cirotto, Eur. J. Biochem., 77, 555 (1977)

1599. Myoglobin

SOURCE: Beef
SOLVENT: Phosphate buffer, pH 6.0, 0.2 ionic strength
COMMENTS: Spectra of met-, oxy-, and carboxy-forms
REFERENCE: S. K. Wolfe, D. A. Watts, and W. D. Brown, J. Agric. Food Chem., 26, 217 (1978); J. R. Quinn and A. M. Pearson, J. Food Sci., 29, 429 (1964)

1600. Myoglobin, met- and Derivatives

SOURCE: Beef heart
COMMENTS: Spectra of metmyoglobin, ferrimyoglobin peroxide and acid ferrimyoglobin peroxide
REFERENCE: J. B. Fox, Jr., R. A. Nickolas, S. A. Ackerman, and C. E. Swift, Biochemistry, 13, 5178 (1974)

1601. Myoglobin, met-

SOURCE: Beef heart
COMMENTS: Spectra of protein before and after reaction with metmyoglobin reductase
REFERENCE: L. Hagler, R. I. Coppes, Jr., and R. H. Herman, J. Biol. Chem., 254, 6505 (1979)

1602. Myoglobin

SOURCE: Reindeer (Rangifer tarandus), moose (Alces alces), and wapiti (Cervus elaphus XANTHOPYGUS)
COMMENTS: Spectra of met-, carboxy-, reduced, and oxy-forms
REFERENCE: B. F. Sukhomlinov, M. Z. Datskiv, and I. I. Maleeva, Zhurnal Evolyutsionni Biokhimii i Fiziologii (Trans.), 13, 157 (1977)

1603. Myoglobin

SOURCE: Red kangaroo (Megaleia rufa) and horse
SOLVENT: pH 6.0–pH 11.0 (0°C and 77°K)
COMMENTS:
1. Spectra of kangaroo myoglobin (IV): acid form, intermediate form, and alkaline form
2. Spectra of horse myoglobin (IV): acid and alkaline forms (20°C and 77°K)
3. Spectra of alkaline ferric myoglobin (kangaroo): 77°K and room temperature
4. Spectra of alkaline ferric myoglobin (horse) at 0°C and 77°K

REFERENCE: J. B. Wittenberg, J. Biol. Chem., 253, 5694 (1978)

1604. Myoglobin

SOURCE:
1. Salmon shark (Lamna ditropis)
2. Mackerel (Pneumatophorus japonicus)
3. Skipjack (Katswonus pelamis)
4. Bigeye tuna (Parathunnus sibi)
5. Turtle (Caretta caretta)
6. Sperm whale
7. Horse

COMMENTS: Spectra of oxy-, CO-, deoxy-metcyanide, and met-forms for all species
REFERENCE: H. Tomita and Y. Tsuchiga, Tohoku J. Agric. Res., 22, 228 (1971)

1605. Myoglobin

SOURCE: Bigeye tuna (Parathunnus sibi)
SOLVENT: M/20 Phosphate buffer, pH 7.0
COMMENTS: Spectra of oxy-, CO-, deoxy-, met-, and metcyanide forms
REFERENCE: F. Matsuur and K. Hashimoto, Bull, Jap. Soc. Sci. Fish., 20, 946 (1955)

1606. Myoglobin

SOURCE: Blue-white dolphin (Stenella caeruleo-alba)
COMMENTS: Spectra of ferri form
REFERENCE: T. Matsui, C. Shimizu, and F. Matsuura, Bull. Jap. Soc. Sci. Fish., 41, 771 (1975)

1607. Myoglobin

SOURCE: Pig muscle
SOLVENT: Potassium phosphate, pH 6.5
COMMENTS: Spectra of met-, CO-reduced, reduced (hydrosulfite) and cyanomet- forms

REFERENCE: T. Nakanishi and M. Izummoto, Agric. Biol. Chem., 36, 1505 (1972)

1608. Myoglobin, met-

SOURCE: Human muscle
COMMENTS: Spectra taken before and after enzymic reduction
REFERENCE: A. Hayashi, T. Suzuki, and M. Shin, Biochim. Biophys. Acta, 310, 309 (1973)

1609. Myoglobin

SOURCE: Horse
SOLVENT: 0.4 M Phosphate buffer, pH 7.0
COMMENTS: Spectra of carbonyl and nitrosyl derivatives
REFERENCE: R. W. Romberg and R. J. Kassner, Biochemistry, 18, 5387 (1979)

1610. Myoglobin

SOURCE: Horse heart
SOLVENT: Phosphate buffer, pH 8.0
COMMENTS: Spectra of Ru^{III}-myoglobin, Ru^{II}-myoglobin, and Ru^{II}(CO)-myoglobin
REFERENCE: D. R. Paulson, A. W. Addison, D. Dolphin, and B. R. James, J. Biol. Chem., 254, 7002 (1979)

1611. Myoglobin

SOURCE: Horse heart
SOLVENT: pH 5.9
COMMENTS: Spectra of native histidine-93-modified, and histidine-93 and 64-modified proteins
REFERENCE: M. Folin, G. Gennari, and G. Jori, Photochem. Photobiol., 20, 357 (1977)

1612. Myoglobin

SOURCE: Horse heart
SOLVENT: 0.1 M Potassium phosphate buffer, pH 7.0
COMMENTS: Spectral changes during addition of carbon monoxide to proto-oxymyoglobin or 2,4-diformyl-oxymyoglobin
REFERENCE: M. Sono, P. D. Smith, J. A. McCray, and T. Asakura, J. Biol. Chem., 251, 1418 (1976)

1613. Myoglobin

SOURCE: Horse heart

COMMENTS: 1. Spectra of oxymyoglobin at 25°K, 30°K, 60°K, 100°K, and 293°K
2. Spectra of deoxymyoglobin at 20°K, 45°K, 60°K and 293°K
3. Spectra of metmyoglobin fluoride at 25°K and 50°K

REFERENCE: J. Springall, M. J. Stillman, and A. J. Thomson, Biochim. Biophys. Acta, 453, 494 (1976)

1614. Myoglobin

SOURCE: Horse heart myoglobin

SOLVENT: a. 0.1 M Potassium phosphate buffer, pH 6.0
b. 0.1 M Glycine-NaOH, pH 10.0
c. 2 mM Potassium phosphate, pH 5.95-pH 10.0

COMMENTS: 1. The spectra of the acid forms (solvent a), the alkaline forms (solvent b), cyanide complexes (solvent a), azide complexes (solvent a), fluoride complexes (solvent a), of the following ferric formyl-myoglobins: proto-, 2-vinyl-4-formyl, 2-formyl-4-vinyl-, and 2,4-diformyl
2. Spectral changes during transition from alkaline form to acid form of ferric 2-vinyl-4-formylmyoglobin (solvent c)
3. Spectral changes during addition of azide (0-632 mM) to ferric 2-formyl-4-vinyl myoglobin (solvent a)

REFERENCE: M. Sono and T. Asakura, J. Biol. Chem., 251, 2664 (1976)

1615. Myoglobin

SOURCE: Horse heart mitochondria

SOLVENT: 0.1 M Potassium phosphate buffer, pH 7.0

COMMENTS: Spectra of 2-formyl-4-vinyl-, 2-vinyl-4-formyl-, and 2,4-diformyl- derivatives of oxy, deoxy, and carbonmonoxy myoglobins

REFERENCE: M. Sono and T. Asakura, J. Biol. Chem., 250, 5227 (1975)

1616. Myoglobin

SOURCE: Horse heart mitochondria

SOLVENT: a. 0.1 M Phosphate buffer, pH 6
b. pH 8
c. pH 10-pH 10.8
d. pH 9-pH 11

COMMENTS: 1. Spectra of meso-, deutero-, hemato-, proto-, protomonomethylester, and protodimethyl ester derivatives of met-, reduced (dithionite) and CO-reduced forms (solvents a, b)
2. Spectra of alkaline met forms (solvent c and solvent d)
3. Spectra of hydrogen peroxide oxidized forms (solvent a and solvent b)

REFERENCE: M. Tamura, T. Asakura, and T. Yonetani, Biochim. Biophys. Acta, 295, 467 (1973)

1617. Myoglobin

SOURCE: Horse heart mitochondria

SOLVENT: a. pH 6
b. 0.1 M Phosphate buffer, pH 7.5
c. pH 6.8-pH 11.1

COMMENTS: 1. Spectra of oxy-, deoxy-, and met- forms of meso-, deutero-, hemato-, protoheme-, protoheme monomethyl ester, and protoheme-dimethyl ester (solvent a)
2. Spectral changes during formation of oxy-protoheme-monomethyl ester myoglobin from met-protoheme-monomethyl ester myoglobin by enzyme reduction (solvent b)
3. Spectral changes during acid form to alkaline form transition due to pH change (solvent c)
4. Spectral changes due to changes in azide ion concentration with protoheme myoglobin

REFERENCE: M. Tamura, G. V. Woodrow, and T. Yonetani, Biochem. Biophys. Acta, 317, 34 (1973)

1618. Myoglobin and Myiglobin

SOURCE: Horse heart

COMMENTS: 1. Spectrum of CO-myoglobin
2. Spectral change on treating myiglobin with tetradecylsulfate in 0.02 M Sodium bicarbonate

REFERENCE: H. Krimphardt and D. Lubbers, Biochem. Z., 330, 342 (1958)

1619. Myoglobin, Ferric Chlorocruoro-

SOURCE: Sperm whale

SOLVENT: a. 0.2 M Tris-HCl, pH 7
b. 0.2 M Glycine-NaOH, pH 10.0
c. 0.2 M Sodium phosphate buffer, pH 7.0

COMMENTS: 1. Spectra of acidic form, imidazole complex, azide complex (solvent a) and alkaline form (solvent b) at 23°C and 89°K

2. Spectra of acidic form and azide complex (solvents a and c) at 89°K

REFERENCE: T. Iizuka and Y. Orii, Biochim. Biophys. Acta, 328, 239 (1973)

1620. Myoglobin, met-

SOURCE: Sperm whale
SOLVENT: Phosphate buffer, pH 7.2
COMMENTS: Spectral change due to aerobic incubation with L-dihydroxyphenylalanine
REFERENCE: S. S. Tate, J. Orlando, and A. Meister, Proc. Nat. Acad. Sci., 69, 2505 (1972)

1621. Myoglobin

SOURCE: Sperm whale
SOLVENT: 0.1 M Phosphate buffer, pH 7.0
COMMENTS: Spectra of protocobaltous, mesocobaltous, and deuterocobaltous derivatives of the oxy- and deoxy-myoglobins
REFERENCE: T. Yonetani, H. Yamamoto, and G. V. Woodrow, J. Biol. Chem., 249, 682 (1974)

1622. Myoglobin, meso, Cobalt

SOURCE: Sperm whale myoglobin
COMMENTS: Spectra of oxy-, deoxy-, and photolyzed oxymeso cobalt myoglobin
REFERENCE: T. Yonetani, H. Yamamoto, and T. Iizuka, J. Biol. Chem., 249, 2168 (1974)

1623. Myoglobin-ligand Complexes

SOURCE: Sperm whale myoglobin
SOLVENT: Ethylene glycol-water
COMMENTS:
1. Spectra of aquo Fe(III)-, Fe(III) azide-, Fe(III) imidazole-, and Fe(III) fluoride myoglobins after irradiation with electrons from cobalt-60 at 77°K
2. The effect of temperature on intermediates formed in 1; temperature 77°K-154°K

REFERENCE: Z. Gasyna, Biochim. Biophys. Acta, 577, 207 (1979)

1624. Myoglobin III Hydroxide

SOURCE: Sperm whale
REFERENCE: S. A. Asher and T. M. Schuster, Biochemistry, 18, 5377 (1979)

1625. Myoglobin, semisynthetic and derivatives

SOURCE: Sperm whale myoglobin
COMMENTS: Spectra of N^{ε}_{19}-acetimidomyoglobin, N^{ε}_{19}-acetimidoapomyoglobin, trifluoroacetic acid treated N^{ε}_{19}-acetimidoapomyoglobin, rehemed N^{ε}_{19}-acetimidomyoglobin, and des-Val^{1}, N^{ε}_{19}-acetimidomyoglobin
REFERENCE: R. D. DiMarchi, G. W. Neireiter, W. H. Garner, and F. R. N. Gurd, Biochemistry, 18, 3101 (1979)

1626. Myoglobin, Cu^{II}-

SOURCE: Sperm whale
SOLVENT: 0.5 M 2-(N-morpholino) ethane sulfonic acid and 0.05 M KCl, pH 7.1
REFERENCE: K. Alston and C. B. Storm, Biochemistry, 18, 4292 (1979)

1627. Myoglobin, Ruthenium Carbon Monoxide Complex

SOURCE: Sperm whale
SOLVENT: a. 0.1 M Phosphate buffer, pH 7
b. 2 M NaCl in 0.1 M phosphate buffer, pH 7
REFERENCE: T. S. Srivasta, Biochim. Biophys. Acta, 491, 599 (1977)

1628. Myoglobin, met-

SOURCE: Sperm whale
SOLVENT: N-ethyl morpholine-HCl buffer, pH 6.4
COMMENTS: Spectral changes due to changing concentration of cupric ion
REFERENCE: M. T. Lamy, P. C. Ribeiro, O. R. Nascimento, and G. Bemski, FEBS Lett., 71, 29 (1976)

1629. Myoglobin II and Carbonmonoxide Myoglobin II

SOURCE: Sperm whale
SOLVENT: a. pH 7.86 and pH 5.15 (0.1 M KCl)
b. pH 9.36 and pH 5.00 (0.1 M KCl)
COMMENTS: Spectra in the presence of excess dithionite
REFERENCE: Y. Hayashi, H. Yamada, and I. Yamazaki, Biochim. Biophys. Acta, 427, 608 (1976)

1630. Myoglobin, met-

SOURCE: Sperm whale
SOLVENT: a. 1/15 M Phosphate buffer, pH 7.0 in heavy water (D_2O)

b. pH 11.5 in D_2O
COMMENTS: Spectra of ferri met-myoglobin and of its complexes with fluoride, cyanide, imidazole, or azide (solvent a) and of the hydroxide complex (solvent b)
REFERENCE: T. Nozawa, T. Yamamoto, and M. Hatano, Biochim. Biophys. Acta, 427, 28 (1976)

1631. Myoglobin, met-

SOURCE: Sperm whale
SOLVENT: a. 65 mM Phosphate buffer, pH 6.86
b. 65 mM Phosphate buffer, pH 9.18
c. 0.2 M Sodium carbonate buffer, pH 11.45
COMMENTS: Spectra of complexes of met-myoglobin with the following substances:
1. 2-Mercaptoethanol, ethylmercaptan, n-propylmercaptan, isopropylmercaptan, and sodium sulfide (solvent a)
2. Cysteine, cysteine methylester, 2-mercaptoethanol, ethylmercaptan, n-propylmercaptan, isopropylmercaptan, and sodium sulfide (solvent b)
3. Ethylmercaptan, n-propylmercaptan, isopropylmercaptan, 2-mercaptoethanol, and sodium sulfide (solvent c)
4. Spectra of dithionite-reduced complexes of 1, 2, and 3 except for sodium sulfide complexes
REFERENCE: T. Shimizu, T. Nozawa, and M. Hatano, Biochim. Biophys. Acta, 434, 126 (1976); T. Shimizu, T. Nozawa, M. Hatano, Y. Imai, and R. Sato, Biochemistry, 14, 4172 (1975)

1632. Myoglobin, Ferri-

SOURCE: Sperm whale
SOLVENT: a. Tris-HCl buffers, pH 7-pH 9
b. Glycine-NaOH buffers, pH 9.4 and pH 10.4
COMMENTS: 1. Spectra at 83°K in solvents a and b over pH range indicated
2. Spectra at 83°K of ferri-myoglobin in water, pH 6.9; plus formate, pH 4.3; plus cyanate, pH 7.0; plus fluoride; plus methanol, pH 7.0
3. Spectral changes in Tris-HCl, pH 7.0, due to temperature changes: 79°K-197°K
REFERENCE: S. Yoshida, T. Iizuka, T. Nozawa, and M. Hatano, Biochim. Biophys. Acta, 405, 122 (1975)

1633. Myoglobin, Ferri-

SOURCE: Sperm whale
SOLVENT: a. pH 7.0-11.0 (water solutions)

b. pH 7.6-11.0 (deuterium oxide solutions)

COMMENTS: Spectral changes due to changes in pH and pD

REFERENCE: T. Iizuka, and I. Morishima, Biochim. Biophys. Acta, 400, 143 (1975)

1634. Myoglobin

SOURCE: Sperm whale

SOLVENT: pH 4.6

COMMENTS: Spectra before and after irradiation with 270 nm light

REFERENCE: C. C. C. Vidigal and G. Cilento, Biochem. Biophys. Res. Comm., 62, 184 (1975)

1635. Myoglobin, Ytterbium^{III} Mesoporphyrin IX-

SOURCE: Sperm whale apomyoglobin

SOLVENT: 0.05 M Tris-HCl, pH 8.0

REFERENCE: W. D. Horrocks, Jr., R. F. Venteicher, C. A. Spilburg, and B. L. Vallee, Biochem. Biophys. Res. Comm., 64, 317 (1975)

1636. Myoglobin

SOURCE: Sperm whale

SOLVENT: 0.1 M Potassium phosphate buffer, pH 7.0

COMMENTS:
1. Spectral changes occuring on photodissociation of carbon monoxide from carbon monoxide myoglobin and upon reassociation at 4.2°K
2. Spectra of deoxygenated and carbon monoxide forms of ferrous myoglobin

REFERENCE: T. Iizuka, H. Yamamoto, M. Kotani, and T. Yonetani, Biochim. Biophys. Acta, 371, 126 (1974)

1637. Myoglobin, Cobaltous

SOURCE: Sperm whale myoglobin

SOLVENT: 0.1 M Potassium phosphate buffer, pH 7.0

COMMENTS:
1. Spectral changes due to photodissociation of oxygenated mesoporphyrin IX cobaltous myoglobin and recombination with oxygen at 4.2°K
2. Spectra of the oxygenated forms and photolyzed forms at 50°K of cobaltous protoporphyrin IX myoglobin and cobaltous deutero porphyrin IX myoglobin

REFERENCE: T. Iizuka, H. Yamamoto, M. Kotani, and T. Yonetani, Biochim. Biophys. Acta, 351, 182 (1974)

1638. Myoglobin

SOURCE: Sperm whale
SOLVENT: 0.1 M Phosphate buffer, pH 7.0
COMMENTS: 1. Spectra of ethylisocyanide complexes of meso- and deutero-myoglobins
2. Spectra of chlorocruoro myoglobins: met-, oxy-, and ethylisocyanide complex
REFERENCE: R. Makino and I. Yamazaki, Arch. Biochem. Biophys., 165, 485 (1974)

1639. Myoglobin, carboxysulf-

SOURCE: Sperm whale myoglobin
SOLVENT: a. 0.03 M Potassium phosphate, pH 8.0
b. 0.1 M Phosphate buffer, pH 8.0
COMMENTS: 1. Spectra of carboxysulfmyoglobin at ice temperature and at liquid nitrogen temperature
2. Spectra of deoxysulfmyoglobin under argon and of carboxysulfmyoglobin under carbon monoxide
REFERENCE: J. A. Berzofsky, J. Peisach, and J. O. Alben, J. Biol. Chem., 247, 3774 (1972)

1640. Myoglobin

SOURCE: Sperm whale
SOLVENT: 0.05 M Phosphate buffer, pH 7.5
COMMENTS: Spectral changes during reaction of oxymyoglobin with ferricytochrome c
REFERENCE: C-S. C. Wu, P. Duffy, and W. D. Brown, J. Biol. Chem., 247, 1899 (1972)

1641. Myoglobin

SOURCE: Sperm whale myoglobin
SOLVENT: 0.01 M Phosphate buffer, pH 7.2, containing potassium cyanide
COMMENTS: Spectra of MbX (major chromatographic component), diglycine methylester myoglobin, and dihistidine methylester myoglobin
REFERENCE: M. Z. Atassi and R. P. Singhal, J. Biol. Chem., 247, 5980 (1972)

1642. Myoglobin

SOURCE: Sperm whale
COMMENTS: Spectra of the following myoglobins: myoglobin, oxy-, nitrous oxide-, metcyanide-, metazide, metnitrous oxide-, carbon monoxide, metfluoro, aquomet-, metthiocyanate-, metnitrite-, and methydroxide

REFERENCE: J. Bolard and A. Garnier, Biochim. Biophys. Acta, 263, 335 (1971)

1643. Myoglobin, Cobaltous

SOURCE: Sperm whale
SOLVENT: Phosphate buffer, pH 6.9
COMMENTS: Spectral changes during oxygenation of deoxycobaltous myoglobin
REFERENCE: C. A. Spillburg, B. M. Hoffman, and D. H. Petering, J. Biol. Chem., 247, 4219 (1972)

1644. Myoglobin, sulf-

SOURCE: Sperm whale myoglobin
SOLVENT: 0.033 M Phosphate buffer, pH 8.0
COMMENTS: Spectra of deoxy- and oxysulf- myoglobins
REFERENCE: J. Peisach, J. A. Berzofsky, and W. E. Blumberg, in International Symposium on Oxidases and Related Redox Systems (T. E. King, H. S. Mason, and M. Morrison, eds.), Vol. 1, 267 (1971), University Park Press, Baltimore

1645. Myoglobin, met- and guanidinated metmyoglobin

SOURCE: Sperm whale
SOLVENT: pH 7.0
COMMENTS: Spectra of metmyoglobin and guanidinated metmyoglobin
REFERENCE: L. J. Banaszak, E. H. Eylar, and F. R. N. Gurd, J. Biol. Chem., 238, 1989 (1963)

1646. Myosin, Tropo-

SOURCE: Rabbit muscle
SOLVENT: 0.5 M KCl, 0.5 mM dithiothreitol, 0.05 M potassium phosphate, pH 6.8 and pH 11.2
REFERENCE: B. Bullard, D. A. Mercola, and W. F. H. M. Mommaerts, Biochim. Biophys. Acta, 434, 90 (1976)

1647. Myosin, g_1, g_2, and g_3 Chains

SOURCE: Rabbit muscle
SOLVENT: 20 mM Tris-HCl, pH 8.0
REFERENCE: K. Yagi, Y. Okamoto, and Y. Yazawa, J. Biochem. Tokyo, 77, 333 (1975)

1648. Myosin, Light Chains, L_1, L_2, L_3, and L_4

SOURCE: Rabbit white skeletal muscle

SOLVENT: 20 mM Tris-HCl, pH 8.5
REFERENCE: W. T. Perrie, L. B. Smillie, and S. V. Perry, Biochem. J., 135, 151 (1973)

1649. Myosin, Light Chains

SOURCE: Rabbit muscle
SOLVENT: 0.05 M Potassium phosphate buffer, pH 6.8
REFERENCE: S. Lowey and J. C. Holt, Cold Spring Harbor Symp. Quant. Biol., 37, 19 (1972)

1650. Myosin

SOURCE: Rabbit skeletal muscle
SOLVENT: pH 7.3 and pH 13
REFERENCE: H. Finck, Biochim. Biophys. Acta, 111, 239 (1965)

1651. Myosin and Actomyosin

SOURCE: Rabbit muscle
SOLVENT: pH 7 and pH 13
REFERENCE: E. Milhalyi and A. J. Rowe, Biochem. Z., 345, 267 (1966)

1652. Myosin, Light Chains: EDTA, DTNB, and SH

SOURCE: Scallop
SOLVENT: 25 mM Tris-HCl, pH 7.6
REFERENCE: J. Kendrick-Jones, E. M. Szentkiralyi, and A. G. Szent-Gyorgyi, J. Mol. Biol., 104, 747 (1976)

1653. Myosin

SOURCE: Scallop (Patinopecten yessoensis) smooth adductor muscle
SOLVENT: 0.6 M KCl, 2 mM 2-mercaptoethanol, 10 mM potassium phosphate buffer, pH 7.0
REFERENCE: S. Watanabe and K. Hashimoto, Bull. Jap. Soc. Sci. Fish., 42, 1161 (1976)

1654. Myosin, Light Chains: EDTA and SH

SOURCE: Shellfish (Chlamys nipponensis akazara) striated adductor muscle
SOLVENT: 40 mM Potassium phosphate, pH 7.0
REFERENCE: K. Nishita, T. Ojima, and S. Watanabe, J. Biochem. Tokyo, 86, 663 (1979)

1655. Myosin, Light Chains: LC_1, LC_2, and LC_3

SOURCE: Carp white muscle
SOLVENT: Neutral pH
REFERENCE: F. Huriaux and B. Focant, Arch. Int. Physiol. Biochem., 85, 917 (1977)

1656. Myosin, Light Chains: LC_1, LC_2, and LC_3

SOURCE: Trout (Salmo trutta fario)
SOLVENT: 0.05 M Ammonium bicarbonate
REFERENCE: A. Perzanowska, C. Gerday, and B. Focant, Comp. Biochem. Physiol., 60B, 295 (1978)

1657. Myosin, Light Chains: Lb1 and Lb2

SOURCE: Lobster
REFERENCE: J. M. Regenstein, Comp. Biochem. Physiol., 56B, 239 (1977)

1658. Myosin, Light Chains: L_1, L_2, L_3

SOURCE: Pig heart muscle and human heart muscle
SOLVENT: 0.05 Ammonium bicarbonate
REFERENCE: C. Klotz, M. C. Aumont, J. J. Leger, and B. Swynghedauw, Biochim. Biophys. Acta, 386, 461 (1975)

1659. Myosin

SOURCE: Chicken pectoralis superficialis muscle
SOLVENT: 0.5 M NaCl, 20 mM Sodium phosphate buffer, pH 7.0
COMMENTS: Spectra of myosin, 5,5'-dithiobis-(2-nitrobenzoic acid)-treated (regenerated) myosin and non-regenerated myosin
REFERENCE: N. Katoh and S. Kubo, J. Biochem. Tokyo, 77, 949 (1975)

1660. Myosin, Light Chains

SOURCE: Adult chicken muscle
SOLVENT: 20 mM Sodium phosphate, pH 7.0
COMMENTS: Spectra of unfractionated and fractionated light chains
REFERENCE: N. Katoh, S. Kubo, and H. Takahashi, J. Biochem. Tokyo, 74, 771 (1973)

1661. Myosin, Low Molecular Weight Proteins (LMP-I, LMP-II, LMP-III)

SOURCE: Chicken white breast muscle
REFERENCE: S. L. Gaffin and S. Watanabe, J. Mechanochem. Cell Motility, 1, 139 (1972)

1662. Myosin, Light Chains

SOURCE: Chicken skeletal muscle
COMMENTS: Spectra of fractions I, II, and III from DEAE-Sephadex chromatography
REFERENCE: B. Z. Horvath and E. Gaetjens, Biochim. Biophys. Acta, 263, 779 (1972)

1663. Myosin and Actomyosin

SOURCE: Chicken skeletal muscle, heart muscle, and smooth (gizzard) muscle
SOLVENT: pH 7.3 and pH 13
REFERENCE: H. Finck, Biochim. Biophys. Acta, 111, 208, 221, and 231 (1965)

1664. Myrosinase

SOURCE: Yellow mustard powder (Brassica juncea)
SOLVENT: a. 50 mM Phosphate buffer, pH 8.5
b. 20 mM Carbonate buffer, pH 10
COMMENTS: Spectra of 2-methoxy-5-nitropone modified myrosinase (solvent a) and of 2-hydroxy-5-nitrobenzyl modified myrosinase (solvent b)
REFERENCE: M. Ohtsuru and T. Hata, Biochim. Biophys. Acta, 567, 384 (1979)

1665. Neuraminidase

SOURCE: Clostridium perfringens
SOLVENT: 0.1 M Phosphate buffer, pH 7
REFERENCE: S. Nees, R. W. Vek, and R. Schauer, Hoppe-Seyler's Z. Physiol. Chem., 356, 1027 (1975); S. Nees and R. Schauer, Behring Inst. Mitt., 55, 112 (1974)

1666. Neurophysin

SOURCE: Beef pituitary gland
SOLVENT: pH 5, pH 6.2, pH 9.0
COMMENTS: Spectra of protein and nitrated protein
REFERENCE: E. Breslow and J. Weiss, Biochemistry, 11, 2474 (1972)

1667. Neurotensin

SOURCE: Beef hypothalamus
SOLVENT: 0.1 M Acetic acid containing 0.001 M 2-mercaptoethanol
REFERENCE: R. Carraway and S. E. Leeman, J. Biol. Chem., 248, 6854 (1973)

1668. Nicotinamide Adenine Dinucleotide (NADH) Dehydrogenase

SOURCE: Peptostreptococcus elsdenii
REFERENCE: C. D. Whitfield and S. G. Mayhew, J. Biol. Chem., 247, 2811 (1974)

1669. Nicotinamide Adenine Dinucleotide (NADH) Dehydrogenase

SOURCE: Torulopsis utilis
COMMENTS: Spectra of enzyme before and after addition of NADH
REFERENCE: S. O. C. Tottmar and C. I. Ragan, Biochem. J., 124, 853 (1971)

1670. Nicotinamide Adenine Dinucleotide (NADH) Dehydrogenase

SOURCE: Photobacterium phosphoreum
SOLVENT: 40 mM Tris-HCl, pH 7.7
COMMENTS: Spectra taken before and after addition of NADH
REFERENCE: T. Imagawa and T. Nakamura, J. Biochem. Tokyo, 84, 547 (1978)

1671. Nicotinamide Adenine Dinucleotide Phosphate (NADPH) Dehydrogenase

SOURCE: Rat liver cytosol
SOLVENT: 2.5 mM Ammonium bicarbonate, pH 7.8
REFERENCE: R. Wallin, Biochem. J., 181, 127 (1979)

1672. Nicotinamide Adenine Dinucleotide Phosphate (NADPH) Dehydrogenase

SOURCE: Pigeon heart
COMMENTS: Spectra of native and photoinactivated, methylene-blue or rose bengal, enzyme
REFERENCE: S. E. Severin, V. I. Telepneva, and L. A. Tseitlin, Biokhimiya (Biochemistry), 36, 849 (1977)

1673. Nicotinamide Adenine Dinucleotide Phosphate (NADP) Reductase

SOURCE: Thiocapsa roseopersicina strain BBS
COMMENTS: Spectra of oxidized and reduced forms
REFERENCE: I. N. Gogotov and T. V. Laurinavichene, Biokhimiya (Biochemistry), 42, 1002 (1977)

1674. Nicotinamide Adenine Dinucleotide Phosphate ($NADP^+$) Reductase

SOURCE: Spinach chloroplasts
SOLVENT: a. pH 8.0
b. 0.05 M Tris buffer, pH 7.0

COMMENTS: 1. Spectral changes during reaction with dithionit in presence of methylviologen under argon in solvent a
2. Spectral changes during reaction with NADPH under argon in solvent a
3. Spectral changes during photoreduction with EDTA in vacuo in solvent b

REFERENCE: J. J. Keirns and J. H. Wang, J. Biol. Chem., 247, 7374 (1972)

1675. Nicotinamide Adenine Dinucleotide (NAD^+) Reductase

SOURCE: Prosthecochloris aestuarii
SOLVENT: 0.05 M Tris-HCl buffer, pH 7.8
COMMENTS: Spectra of oxidized and reduced (dithionite) forms
REFERENCE: Y. Shioi, K. Takamiya, and M. Nishimura, J. Biochem. Tokyo, 79, 361 (1976)

1676. Nicotinamide Adenine Nucleotide (Reduced) Dehydrogenase

SOURCE: Peptostreptococcus elsdenii
SOLVENT: 0.11 M Potassium phosphate, pH 7.0
COMMENTS: Spectral changes during the anaerobic titration with each of the following: dithionite and NADH
REFERENCE: C. D. Whitfield and S. G. Mayhew, J. Biol. Chem., 249, 2811 (1974)

1677. Nitrate Reductase

SOURCE: Bacillus licheniformis
SOLVENT: 0.07 M Phosphate buffer, pH 7.0
COMMENTS: Spectra of isolated and dithionite-treated enzyme
REFERENCE: T. Vant Riet, F. B. Wientjes, J. Van Doorn, and R. J. Planta, Biochim. Biophys. Acta, 576, 347 (1979)

1678. Nitrate Reductase

SOURCE: Chlorella vulgaris
SOLVENT: 0.1 M Potassium phosphate buffer, pH 6.9
COMMENTS: Spectra before and after addition of NADH
REFERENCE: L. Giri and C. S. Ramadoss, J. Biol. Chem., 254, 11703 (1979)

1679. Nitrate Reductase

SOURCE: Rhodopseudomonas capsulata AD2
SOLVENT: 0.1 M Potassium phosphate, pH 7
COMMENTS: Spectra of oxidized and dithionite-reduced enzyme
REFERENCE: K. Alef and J. H. Klemme, Z. Naturforsch, 34C, 33 (1979)

1680. Nitrate Reductase

SOURCE: Chlorella vulgaris
SOLVENT: 0.08 M Potassium phosphate buffer, pH 7.5
COMMENTS: Spectra taken before and after addition of NADH and dithionite
REFERENCE: L. P. Solomonson, G. H. Lorimer, R. L. Hall, R. Borchers, and J. L. Bailey, J. Biol. Chem., 250, 4120 (1975)

1681. Nitrate Reductase

SOURCE: Clostridium perfringens
REFERENCE: S. Seki-Chiba and M. Ishimoto, J. Biochem. Tokyo, 82, 1663 (1977)

1682. Nitrate Reductase I and II

SOURCE: Klebsiella aerogenes
SOLVENT: 0.065 M Phosphate buffer, pH 7.8
REFERENCE: J. Van Riet and R. J. Planta, Biochim. Biophys. Acta, 379, 81 (1975)

1683. Nitrate Reductase

SOURCE: Rhodotorula glutinis
SOLVENT: 0.1 M Sodium phosphate buffer, pH 7.5
COMMENTS: Spectra before and after addition of NADPH
REFERENCE: M. G. Guerro and M. Gutierrez, Biochim. Biophys. Acta, 482, 272 (1977)

1684. Nitrite Reductase

SOURCE: Neurospora crassa
SOLVENT: a. Standard potassium phosphate buffer
b. 0.1 M Potassium phosphate buffer, pH 7.5
COMMENTS: 1. Spectrum of enzyme in solvents a and b
2. Spectral changes during formation of enzyme-carbon monoxide complexes caused by FAD, FAD and nitrite, and reducing agents (solvent b)
REFERENCE: M. A. Lafferty and R. H. Garrett, J. Biol. Chem., 249, 7555 (1974); J. M. Vega, R. H. Garrett, and L. M. Siegel, J. Biol. Chem., 250, 7980 (1975)

1685. Nitrite Reductase

SOURCE: Spinach
SOLVENT: a. 50 mM Phosphate buffer, pH 7.5, containing 10% glycerol

b. 50 mM Tris buffer, pH 7.5, containing 0.02 M NaCl
c. 100 mM Potassium phosphate buffer, pH 7.7
d. 300 mM Potassium phosphate buffer, pH 8.5
e. 50 mM Potassium phosphate buffer, pH 7.7

COMMENTS: 1. Spectra before and after addition of nitrite (solvent b)
2. Spectra of oxidized and reduced (dithionite) forms

REFERENCE: S. Ida, J. Biochem. Tokyo, 82, 915 (1977); P. J. Aparicio, D. R. Knaff, and R. Malkin, Arch. Biochem. Biophys., 169, 102 (1975); J. Shimizu and G. Tamura, J. Biochem. Tokyo, 75, 999 (1974); M. J. Murphy, L. M. Siegel, S. R. Tove, and H. Kamin, Proc. Nat. Acad. Sci., 71, 612 (1974); J. Cardenas, J. L. Berea, J. Rivas, and C. G. Moreno, FEBS Lett., 23, 131 (1972)

1686. Nitrite Reductase

SOURCE: Spinach
SOLVENT: 0.1 M Potassium phosphate buffer, pH 7.7 or pH 7.0
COMMENTS: 1. Spectra of oxidized and reduced (dithionite) forms
2. Spectra of enzyme before and after reaction with p-chloromercuribenzoate (pH 7.0)
3. Spectra of the following enzyme complexes: nitrite, nitrite plus dithionite, hydroxylamine, hydroxylamine plus dithionite, cyanide, cyanide plus dithionite, carbon monoxide, methylviologen plus carbon monoxide, carbon monoxide plus dithionite, and dithionite, methylviologen and carbon monoxide

REFERENCE: J. M. Vega and H. Kamin, J. Biol. Chem., 252, 896 (1977)

1687. Nitrite Reductase

SOURCE: Achromobacter cycloclastes
COMMENTS: Spectra of reduced and oxidized forms
REFERENCE: H. Iwasaki and T. Matsubara, J. Biochem. Tokyo, 71, 645 (1972)

1688. Nitrite Reductase

SOURCE: Marrow plants (Cucurbita pepo)
COMMENTS: Spectra of the following enzyme complexes and reactions: dithionite, dithionite and nitrite, dithionite and carbon monoxide, and dithionite and cyanide

REFERENCE: D. P. Hucklesby, D. M. James, M. T. Banwell, and E. J. Hewitt, Phytochemistry, 15, 599 (1976); D. P. Hucklesby, D. M. James, and E. J. Hewitt, Biochem. Soc. Trans., 2, 436 (1974)

1689. Nitrite Reductase

SOURCE: Achromobacter fishei
SOLVENT: 0.05 M Phosphate buffer, pH 7.5
COMMENTS: Spectra of oxidized, reduced (dithionite), nitrate and dithionite, and hydroxylamine and dithionite forms
REFERENCE: O. Prakash and J. C. Sadana, Arch. Biochem. Biophys., 148, 614 (1972)

1690. Nitrite Reductase

SOURCE: Chlorella fusca
SOLVENT: 10 mM Tris, pH 7.5
COMMENTS: Spectra of oxidized and reduced (dithionite) forms
REFERENCE: W. G. Zumft, Biochim. Biophys. Acta, 276, 363 (1972)

1691. Nitrite Reductase

SOURCE: Pseudomonas aeruginosa
SOLVENT: a. Citrate-phosphate buffer, pH 2-13
b. pH 3-pH 10
c. pH 11.5
d. pH 5.0-9.1
COMMENTS: 1. Spectra of oxidized enzyme under NO (solvent a)
2. Spectra of reduced enzyme under NO (solvent b)
3. Spectral changes after reaction with NO (solvent c)
4. Spectral changes due to pH changes (solvent d)
REFERENCE: H. Shimada and Y. Orii, J. Biol. Chem. (Tokyo), 84, 1553 (1978)

1692. Nitrite Reductase

SOURCE: Pseudomonas aeroginosa
SOLVENT: 0.2 M Sodium phosphate buffer, pH 6
COMMENTS: Spectra of oxidized-NO form and of reduced-NO form (NADH and phenazinemethosulfate)
REFERENCE: H. Shimada and Y. Orii, FEBS Lett., 54, 237 (1975)

1693. Nitroalkane-oxidizing Enzyme

SOURCE: Hansenula mrakii
SOLVENT: 0.01 M Potassium phosphate buffer, pH 7.0

REFERENCE: T. Kido, T. Yamamoto, and K. Soda, J. Bacteriol., 126, 1261 (1976)

1694. Nitrogenase; Component I and Component II

SOURCE: Azotobacter vinelandii
COMMENTS: Spectra of oxidized and reduced (dithionite) forms of both components
REFERENCE: V. K. Shah and W. J. Brill, Biochim. Biophys. Acta, 305, 445 (1973)

1695. Nitrogenase, Ac2 Protein

SOURCE: Azotobacter chroococcum
SOLVENT: 25 mM Tris buffer, pH 7.4, containing 50 mM $MgCl_2$ (degassed)
COMMENTS:
1. Spectra of oxidized (phenazinemethosulfate) and reduced (acetylene) forms
2. Spectra of oxygen-inactivated form and reduced (dithionite) form

REFERENCE: M. G. Yates and K. Planque, Eur. J. Biochem., 60, 467 (1975)

1696. Nitrogenase, Kp1 Protein and Kp2 Protein

SOURCE: Klebsiella pneumoniae strain 5a1
SOLVENT: 25 mM Tris-HCl buffer, pH 7.4, containing DTT
COMMENTS: Spectra of oxidized and reduced (dithionite) forms
REFERENCE: R. R. Eady, B. E. Smith, K. A. Cook, and J. R. Postgate, Biochem. J., 128, 655 (1972)

1697. Nitrogenase

SOURCE: Clostridium pasteurianum
SOLVENT: 7 M-8 M Urea solution
COMMENTS: Spectra of alkaline and acidic solutions
REFERENCE: W. G. Zumft, Eur. J. Biochem., 91, 345 (1978)

1698. Nitrogenase

SOURCE: Rhodospirillum rubrum
REFERENCE: P. W. Ludden and R. H. Burris, Biochem. J., 175, 251 (1978)

1699. 2-Nitropropane Dioxygenase

SOURCE: Hansenula mrakii
SOLVENT: 10 mM Potassium phosphate buffers, pH 7.0 and pH 8.0

COMMENTS: 1. Spectra of native enzyme, reduced (2-nitropropane) and reoxygenated enzyme, pH 8.0
2. Spectra of native enzyme, reduced (nitromethane), and reoxygenated enzyme, pH 8.0
3. Spectra of oxidized and reduced (dithionite) forms, pH 7.0

REFERENCE: T. Kido, K. Soda, T. Suzuki, and K. Asada, J. Biol. Chem., 251, 6994 (1976)

1700. Non-Histone Proteins

SOURCE: Rat liver
SOLVENT: 10 mM Tris-HCl, pH 7, containing 0.4 M NaCl
REFERENCE: H. W. J. Van den Broek, L. D. Nooden, J. S. Sevall, and J. Bonner, Biochemistry, 12, 229 (1973)

1701. Non-Histone Proteins, HMG1 and HMG2 (High Mobility Group)

SOURCE: Calf thymus
SOLVENT: 0.01 N HCl containing 0.0001 M DTT
REFERENCE: C. Baker, I. Isenberg, G. H. Goodwin, and E. W. Johns, Biochemistry, 15, 1645 (1976)

1702. Nuclease and Nitrotyrosine^{115}-Nuclease

SOURCE: Staphylococcus aureus
SOLVENT: 0.05 M Tris-acetate, pH 5
REFERENCE: C. D. Bello and J. H. Griffin, J. Biol. Chem., 250, 1445 (1975)

1703. Nuclease Peptide-T-(6-48), Nitrotyrosine^{27}-

SOLVENT: 0.1 M Ammonium bicarbonate, pH 8.0
REFERENCE: I. M. Chaiken, J. Biol. Chem., 247, 1999 (1972)

1704. Nuclease O

SOURCE: Aspergillus oryzae
REFERENCE: T. Uozumi, G. Tamura, and K. Arima, Agric. Biol. Chem., 33, 645 (1969)

1705. Nuclei

SOURCE: Rat liver
SOLVENT: a. 5 mM $MgCl_2$, pH 7.5
b. 0.02 M KCl, pH 1.88-pH 10.0
c. 5 mM $MgCl_2$, pH 4.87-pH 8.50
d. 0.2 M KCl, pH 7.5

REFERENCE: D. E. Olins and A. L. Olins, J. Cell. Biol., 53, 715 (1972)

1706. Nucleocapsid

SOURCE: Measles infected cells
SOLVENT: 0.02 M Sodium phosphate buffer, pH 6.8
REFERENCE: D. J. Walters and P. H. Bussell, Virology, 61, 64 (1974)

1707. Nucleohistone IV, Ca^{2+}

SOURCE: Calf thymus tissue
SOLVENT: 0.001 M Tris, pH 7.0
COMMENTS: Spectra taken at 0.15 M, 0.05 M, and 0.005 M $CaCl_2$
REFERENCE: T. E. Wagner and V. Vandegrift, Biochemistry, 11, 1431 (1972)

1708. Nucleoprotein, Acidic

SOURCE: Pig lens
SOLVENT: 0.01 N HCl
REFERENCE: D. Glasser, FEBS Lett., 17, 185 (1971)

1709. Nucleotide Pyrophosphatase

SOURCE: Mung bean
COMMENTS: Spectra of native and AMP-modified enzyme
REFERENCE: C. V. Balakrishnan, S. D. Ravindranath, and N. Appaji Rao, Arch. Biochem. Biophys., 168, 163 (1975)

1710. Nucleotide Pyrophosphotransferase, ATP-

SOURCE: Streptococcus adephospholyticus
SOLVENT: 0.01 M Acetate buffer, pH 5.5, containing 0.1 M NaCl
REFERENCE: T. Nishino and S. Murao, Agric. Biol. Chem., 38, 2491 (1975)

1711. Old Yellow Enzyme

SOURCE: Gluconobacter suboxydans
SOLVENT: 0.05 M Potassium phosphate, pH 6, containing 2-mercaptoethanol
COMMENTS: Spectra of oxidized and reduced (NADPH or dithionite) forms
REFERENCE: O. Adachi, K. Matsushita, E. Shinagawa, and M. Ameyama, J. Biochem. Tokyo, 86, 699 (1979)

1712. Old Yellow Enzyme

SOURCE: Brewer's bottom yeast
SOLVENT: a. 0.1 M Tris-HCl containing 0.1 M ammonium sulfate and 1 μM phenylmethylsulfonyl-fluoride, pH 8.0
b. 0.1 M Phosphate, pH 7.0
COMMENTS: 1. Spectral changes due to the addition of 4-hydroxy-N-n-butylbenzamide, solvent a
2. Spectral changes due to addition of p-methoxybenzaldehyde, solvent b, plus phenylmethylsulfonyl fluoride
3. Spectral changes of 5-deaza-FMN-enzyme on addition of pentafluorophenol, solvent b
4. Spectral changes of 2-thio-FMN-enzyme on addition of pentafluorophenol, or p-chlorophenol in solvent b
5. Spectral changes of 7,8-dichloro-FMN-enzyme on the addition of pentafluorophenol, or p-chlorophenol in solvent b
REFERENCE: A. S. Abromovitz, and V. Massey, J. Biol. Chem., 251, 5327 (1976)

1713. Old Yellow Enzyme

SOURCE: Brewer's yeast
SOLVENT: 0.1 M Pyrophosphate buffer, pH 8.5
COMMENTS: 1. Spectral changes due to addition of p-hydroxybenzaldehyde
2. Spectral changes due to addition of dithionite to enzyme-p-hydroxybenzaldehyde complex
REFERENCE: R. G. Matthews, V. Massey, and C. C. Sweeley, J. Biol. Chem., 250, 9294 (1975)

1714. Orcinol Hydroxylase

SOURCE: Pseudomonas putida
SOLVENT: 20 mM Potassium phosphate buffer, pH 6.86
COMMENTS: 1. Spectra of enzyme with and without orcinol
2. Spectra of enzyme before and after reduction with NADH
REFERENCE: Y. Ohta and D. W. Ribbons, Eur. J. Biochem., 61, 259 (1976)

1715. Orcinol Hydroxylase

SOURCE: Pseudomonas putida
SOLVENT: 20 mM Potassium phosphate buffer, pH 6.8
COMMENTS: 1. Spectral changes during anaerobic reduction of enzymes by NADH

2. Spectral changes during anaerobic reaction with NADH in presence of either orcinol, resorcinol, or m-cresol
3. Spectral changes during reaction of reduced enzyme with oxygen in presence of orcinol or m-cresol

REFERENCE: Y. Ohta, I. J. Higgins, and D. W. Ribbons, J. Biol. Chem., 250, 3814 (1975)

1716. Ornithine Cyclase

SOURCE: Clostridium sporogenes
SOLVENT: 0.1 M Tris-Cl buffer, pH 8.0, containing 20% glycerol, 10 mM ornithine, and 2 mM DTT
REFERENCE: W. L. Muth and R. N. Costilon, J. Biol. Chem., 249, 7457 (1974)

1717. Ornithine Decarboxylase

SOURCE: Mouse liver
SOLVENT: 0.05 M Sodium phosphate buffer, pH 7.2
COMMENTS: Spectra of enzyme, enzyme plus ornithine, and enzyme plus dithiothreitol
REFERENCE: C. G. D. Morley and H. Ho, Biochim. Biophys. Acta, 438, 551 (1976)

1718. Ornithine Decarboxylase

SOURCE: Escherichia coli
SOLVENT: pH 6.0, pH 7.1, pH 8.1, and pH 6.7
COMMENTS: Spectra of holoenzyme, pH 6.0, 7.1, 8.1, and apoenzyme, pH 6.7
REFERENCE: D. Applebaum, D. L. Sabo, E. H. Fischer, and D. R. Morris, Biochemistry, 14, 3675 (1975)

1719. Ovalbumin

SOURCE: Chicken egg
SOLVENT: 0.06 M Sodium phosphate buffer, pH 7.0
COMMENTS: Spectra of native, 35% maleylated, and 64% maleylated proteins
REFERENCE: M. A. Quasin and A. Salahuddin, J. Biochem. Tokyo, 85, 1029 (1979)

1720. Ovalbumin

SOURCE: Chicken egg white
SOLVENT: 0.15 M Phosphate buffer, pH 7.0
COMMENTS: Spectra of protein and of protein succinylated to the extent of 31%, 50%, 57%, 66%, and 82%

REFERENCE: S. A. Kidwai, A. A. Ansari, and A. Salahuddin, Biochem. J., 155, 171 (1976)

1721. Ovalbumin

SOURCE: Chicken egg white
SOLVENT: 0.15 M Sodium phosphate buffer, pH 7.0
COMMENTS: Spectra of protein and of protein acetylated to the extent of 21%, 28%, 62%, 66%, 93%, and 98%
REFERENCE: A. A. Ansari, S. A. Kidwai, and A. Salahuddin, J. Biol. Chem., 250, 1625 (1975)

1722. Ovalbumin

SOURCE: Chicken egg white
SOLVENT: 0.05 N NaOH
COMMENTS: Spectral changes caused by methylene blue-sensitized photo-oxidation
REFERENCE: T. Gomyo and T. Sakurai, Agric. Biol. Chem., 32, 94 (1968)

1723. Ovalbumin, dinitrophenyl- and D-azo-Ovalbumin

SOURCE: Chicken egg white
SOLVENT: 0.1 N NaOH
REFERENCE: H. N. Eisen, M. E. Carsten, and S. Belman, J. Immunol., 73, 296 (1954)

1724. Ovomucoid

SOURCE: Turkey egg white
SOLVENT: 0.006 M Tris, pH 8.3
COMMENTS: Spectral changes associated with reoxidation of reduced protein
REFERENCE: L. B. Sjoberg and R. L. Feeney, Biochim. Biophys. Acta, 168, 79 (1968)

1725. Ovomucoid

SOURCE: Chicken egg white
SOLVENT: a. 50 mM Sodium phosphate, pH 7.5
b. 100 mM Sodium phosphate, 50% glycerol, pH 7.5
COMMENTS: 1. Spectra of native and acetylated protein after treatment with 7 M guanidine HCl (solvent a)
2. Spectra of native, half-acetylated and fully acetylated protein in solvent b at 77°K
REFERENCE: E. Kay, E. H. Strickland, and C. Billups, J. Biol. Chem., 249, 797 (1974)

1726. Ovomucoid and Derivatives

SOURCE: Chicken egg white
SOLVENT: pH 7.0 and 12.0
REFERENCE: A. Waheed and A. Salahuddin, Biochem. Biophys. Acta, 379, 147 (1975)

1727. Ovomucoid

SOURCE: Chicken egg white
COMMENTS: Spectra of native ovomucoid, carboxymethylated ovomucoid and periodate-oxidized ovomucoid
REFERENCE: M. A. Krysteva and I. D. Dobrev, Eur. J. Biochem., 74, 501 (1977)

1728. Ovotransferrin, Iron-Bicarbonate

REFERENCE: P. R. Carey and N. M. Young, Can. J. Biochem., 52, 273 (1974)

1729. Ovotransferrin and Iodinated Derivatives

SOURCE: Chicken egg white
REFERENCE: J. L. Phillips and P. Azari, Arch. Biochem. Biophys., 151, 445 (1972)

1730. Ovotransferrin

SOURCE: Chicken egg white
COMMENTS: Spectra of protein and succinyl derivatives
REFERENCE: H. Buttkus, J. R. Clark, and R. E. Feeney, Biochemistry, 4, 998 (1965)

1731. Ovotransferrin

SOURCE: Chicken (commercial)
REFERENCE: Y. Tominatsu, S. Kint, and J. R. Scherer, Biochem. Biophys. Res. Comm., 54, 1067 (1973)

1732. Oxidase, Terminal

SOURCE: Photobacterium phosphoreum
SOLVENT: 20 mM Potassium phosphate buffer, pH 7.0, containing Triton X-100
COMMENTS: Spectra of oxidized, reduced (dithionite), and reduced (ascorbic acid plus tetramethylphenylenediamine) plus potassium nitrite
REFERENCE: H. Watanabe, Y. Kamita, T. Nakamura, A. Takimoto, and Y. Yamanaka, Biochim. Biophys. Acta, 547, 70 (1979)

1733. Oxidase, Tryptophan Side-Chain

SOURCE: Pseudomonas (ATCC 29574)
SOLVENT: 50 mM Potassium phosphate buffer, pH 6.0
COMMENTS: Spectra of the oxidized and reduced (dithionite) forms
REFERENCE: H. Ushiro, K. Takai, S. Narumiya, S. Ito, and O. Hayaishi, J. Biol. Chem., 254, 11794 (1979)

1734. Oxygenase

SOURCE: Guinea pig liver
REFERENCE: K. Prema and K. P. Gopinathan, Biochem. J., 143, 613 (1974)

1735. Oxynitrilase

SOURCE: Sweet almonds
SOLVENT: 0.1 M Sodium citrate buffer, pH 5.5
COMMENTS: 1. Spectral changes due to the addition of thiocyanate or azide ions
2. Spectra of deaza-FAD oxynitrilase and the peroxide modified substance
REFERENCE: M. S. Jorns, J. Biol. Chem., 254, 12145 (1979)

1736. Oxytocin and Oxytoceine

SOLVENT: 0.05 M Tris-Cl, pH 2.5 and pH 9.0
COMMENTS: Spectra of oxytocin, oxytoceine, and oxytocin-1 Hg and oxytocin-2 Hg (pH 2.5)
REFERENCE: R. Sperling and M. Gorecki, Biochemistry, 13, 2347 (1974)

1737. Pantothenase

SOURCE: Pseudomonas fluorescens
SOLVENT: a. 50 mM Potassium phosphate, pH 7.0
b. pH 12, solvent a plus KOH
REFERENCE: R. K. Airas, E. A. Hietanen, and V. T. Nurmikko, Biochem. J., 157, 409 (1976)

1738. Papain, Hydroxynitrobenzyl-

SOLVENT: pH 4.04-pH 9.03, and 5 N NaOH
REFERENCE: J. E. Mole and H. R. Horton, Biochemistry, 12, 5278 (1973)

1739. Papain, N-4-nitrobenzo-2-oxa-1,3-diazole-

SOLVENT: a. 1 mM EDTA, pH 5.0
b. 1 mM EDTA, pH 5.0, containing 6 M guanidine HCl
c. Phosphate buffer, pH 6.6
COMMENTS: In addition to NBD derivative the spectrum of the mixed disulfide of papain and 7-(2'-mercaptoethylamino)-NBD is also given in solvents b and c
REFERENCE: G. Allen and G. Lowe, Biochem. J., 133, 679 (1973); T. Stuchbury, M. Shipton, R. Norris, J. P. G. Malthouse, K. Brocklehurst, J. A. L. Herbert, and H. Suschitzky, Biochem. J., 151, 417 (1975)

1740. Papain

SOLVENT: 50 mM Sodium acetate buffer, pH 4.5
COMMENTS: Effect of N-bromosuccinimide on the spectrum of hydroxyethyldisulfide derivative
REFERENCE: G. Lowe and A. S. Whitworth, Biochem. J., 141, 503 (1974)

1741. Particles, Naked Intranuclear

SOURCE: Human liver infected with hepatitis B virus
SOLVENT: 0.05 M Tris-HCl, 40 mM KCl, 10 mM $MgCl_2$, 0.1% 2-mercaptoethanol, pH 7.5
REFERENCE: S. Z. Hirschman, M. Gerber, and E. Garfinkel, Proc. Nat. Acad. Sci., 71, 3345 (1974)

1742. Parvalbumins III and V

SOURCE: Lung fish (Protopterus dolloi)
COMMENTS: Spectra of parvalbumins III and V
REFERENCE: C. Gerday, B. Joris, N. Gerardin-Otthiers, S. Collin, and G. Hamoir, Biochimie, 61, 589 (1979)

1743. Parvalbumin

SOURCE: Hake (Merluccius merluccius) muscle
SOLVENT: a. 0.5% Ammonium bicarbonate, pH 8.0
b. 0.05 M Phosphate buffer, pH 7.0
REFERENCE: J. Haiech, J. Derancourt, J. F. Pechere, and J. G. Demaille, Biochimie, 61, 583 (1979); J. Parello and J. F. Pechere, Biochimie, 53, 1079 (1971)

1744. Parvalbumin

SOURCE: Thornback ray (Raja clavata) muscle
SOLVENT: a. 0.05 M Phosphate buffer, pH 7.0
b. 0.1 M NaOH
c. Water

REFERENCE: J. Parello and J. F. Pechere, Biochimie, 53, 1079 (1971); C. Gerday and J. C. Teuwis, Biochim. Biophys. Acta, 271, 320 (1972)

1745. Parvalbumin

SOURCE: Turbot (Psetta maxima) muscle
REFERENCE: J. F. Pechere, J. Demaille, and J. P. Capony, Biochim. Biophys. Acta, 236, 391 (1971)

1746. Parvalbumin

SOURCE: Dogfish (Squalus acanthias) muscle
SOLVENT: a. 20 mM Tris, pH 7.5
b. 20 mM Sodium glycerophosphate, 1 mM EDTA, pH 6.8
c. 2 mM Tris, pH 8.0
REFERENCE: C. Heizmann, D. Malencik, and E. Fischer, Biochem. Biophys. Res. Comm., 57, 162 (1974); H. E. Blum, S. Pocinwong, and E. H. Fischer, Proc. Nat. Acad. Sci., 71, 2198 (1974); D. Malencik, C. Heizmann, and E. Fischer, Biochemistry, 14, 715 (1975)

1747. Parvalbumin

SOURCE: Chondrostoma nasus L. white muscle
SOLVENT: a. Water
b. 0.1 M NaOH
REFERENCE: A. Piront and C. Gerday, Comp. Biochem. Physiol., 46B, 349 (1973)

1748. Parvalbumin

SOURCE: Carp (Cyprinus carpio) muscle
COMMENTS: Spectra of native and phthalylated protein
REFERENCE: R. Bertrand, P. Pantel, and J. F. Pechere, Biochimie, 56, 515 (1974)

1749. Parvalbumin

SOURCE: Carp (Cyprinus carpio L) red muscle (supracarinalis)
SOLVENT: 0.05 M Ammonium bicarbonate, pH 7.9
REFERENCE: C. Gosselin-Rey, A. Piront, and C. Gerday, Biochem. Biophys. Acta, 532, 294 (1978)

1750. Parvalbumin

SOURCE: Whiting (Gadus merlangus L.), Haddock (Gadus aeglefinus L.)
SOLVENT: Phosphate-NaCl buffer, pH 7.5
REFERENCE: J. I. Cosset and C. Gerday, Comp. Biochem. Physiol., 55B, 537 (1976)

1751. Parvalbumin

SOURCE: Rabbit muscle
SOLVENT: a. 0.005 M Ammonium acetate, pH 8
b. 2 mM Tris-HCl, pH 7
REFERENCE: J. F. Pechere and M. Mousseron, Compte. Rend. Acad. Sci. Paris, 278D, 2577 (1974); J. P. Capony, C. Pina, and J. F. Pechere, Eur. J. Biochem., 70, 123 (1976); E. E. Strehler, H. M. Eppenberger, and C. W. Heizmann, FEBS Lett., 78, 127 (1977)

1752. Parvalbumins IVa and IVb

SOURCE: Frog (Rana temporaria) muscle
REFERENCE: C. Gosselin-Rey and C. Gerday, Biochim. Biophys. Acta, 492, 53 (1977)

1753. Parvalbumin

SOURCE: Chicken leg muscle
SOLVENT: 2 mM Tris-HCl, pH 7
REFERENCE: E. E. Strehler, H. M. Eppenberger, and C. W. Heizmann, FEBS Lett., 78, 127 (1977)

1754. Parvalbumin-like Protein

SOURCE: Chicken muscles
SOLVENT: 5 mM Tris-HCl buffer, pH 7
REFERENCE: C. W. Heizmann, M. T. Hauptle, and H. M. Eppenberger, Eur. J. Biochem., 80, 433 (1977)

1755. Pectinesterase

SOURCE: Tomato fruit (Lycopersicum esculentum var. Hikari)
SOLVENT: 100 mM Potassium phosphate, pH 7.5
REFERENCE: H. Nakagawa, Y. Yanagawa, and H. Takehana, Agric. Biol. Chem., 34, 991 (1970)

1756. Pectin Lyase

SOURCE: Aspergillus japonicus
SOLVENT: 0.1 M Acetate buffer, pH 5.0
REFERENCE: S. Ishii and T. Yokosuka, Agric. Biol. Chem., 39, 313 (1975)

1757. Penicillinase

SOURCE: Escherichia coli strain G11a1 (amp A)
SOLVENT: a. 0.01 M Potassium phosphate buffer, pH 6.9

b. pH 11.7 (solvent a plus KOH)
c. pH 2.0 (solvent a plus HCl)

REFERENCE: E. B. Lindstrom, H. G. Boman, and B. B. Steele, J. Bacteriol., 101, 218 (1970)

1758. Penicillinase

SOURCE: Staphylococcus aureus PC1
SOLVENT: 0.1 M KCl, 0.01 M Sodium phosphate buffer, pH 7.0
REFERENCE: B. Robinson and R. H. Pain, Biochem. J., 155, 331 (1976)

1759. Penicillopepsin, 1,2-epoxy-3-p-(nitrophenoxy) propane (EPNP)

SOURCE: Penicillium janthinellum
COMMENTS: Spectra of enzyme and EPNP-enzyme
REFERENCE: G. Mains and T. Hoffman, Can. J. Biochem., 52, 1018 (1974)

1760. Pepsin

SOURCE: Commercial
COMMENTS: Spectra of pepsin before and after tyrosinase treatment
REFERENCE: I. W. Sizer, J. Biol. Chem., 163, 145 (1946); I. W. Sizer and J. F. Fennessey, J. Biol. Chem., 188, 351 (1951); W. J. Haas, I. W. Sizer, and J. R. Loofbourow, Biochim. Biophys. Acta, 6, 589 (1951)

1761. Pepsin

SOURCE: Commercial
SOLVENT: M/20 KCl buffer, pH 2.0
COMMENTS: Effect of potassium permanganate on the spectrum of pepsin
REFERENCE: W. J. Haas, I. W. Sizer, and J. R. Loofbourow, Biochim. Biophys. Acta, 6, 601 (1951)

1762. Pepsin

SOURCE: Pig
COMMENTS: Spectra of native enzyme and dinitrophenyl hexamethylenediamine derivative
REFERENCE: L. F. Matyash, O. G. Ogloblina, and V. M. Stepanov, Biokhimiya (Biochemistry), 37, 893 (1972); Eur. J. Biochem., 35, 539 (1973)

1763. Pepsin

SOLVENT: Water
COMMENTS: Spectra of a solution of protein and of the solid protein
REFERENCE: G. M. Androes, H. R. Gloria, and R. F. Reinisch, Photochem. Photobiol., 15, 375 (1972)

1764. Pepsin and Biacetyl-Pepsin

SOURCE: Pig
SOLVENT: Water
REFERENCE: W. Y. Huang and J. Tang, J. Biol. Chem., 247, 2704 (1972)

1765. Pepsin and Derivative

SOURCE: Chicken
SOLVENT: 0.2 M HCl
COMMENTS: Spectra of enzyme and of disulfide with 2-thionitrobenzoate
REFERENCE: Y. Schechter, M. Rubinstein, R. Becker, and Z. Bohak, Eur. J. Biochem., 58, 123 (1975)

1766. Pepsinogen, Pyridoxal Phosphate-

SOURCE: Pig
SOLVENT: a. pH 2.8
b. pH 7.8
REFERENCE: T. H. Finlay, J. Biol. Chem., 249, 7476 (1974)

1767. Pepsinogen

SOURCE: Pig
SOLVENT: 0.032 M Disodium phosphate-0.004 M monosodium phosphate, pH 7.7
COMMENTS: Spectra of pepsinogen, citraconyl-, maleyl-, and itaconyl pepsinogens
REFERENCE: Y. Nakagawa and G. Perlmann, Arch. Biochem. Biophys., 149, 476 (1972)

1768. Pepsinogen

SOURCE: Chicken
SOLVENT: 0.1 M Phosphate buffer, pH 7.1
REFERENCE: A. Grinvald and I. Z. Steinberg, Biochemistry, 13, 5170 (1974)

1769. Peptidase

SOURCE: Bacillus subtilis
REFERENCE: N. Minimiura, T. Yamamoto, and J. Fukumoto, Agric. Biol. Chem., 30, 186 (1966)

1770. Peptidoglutaminase I and II

SOURCE: Bacillus circulans
SOLVENT: 0.01 M Phosphate buffer, pH 8.0, containing 0.1 M KCl
REFERENCE: M. Kikuchi and K. Sakaguchi, Agric. Biol. Chem., 37, 827 (1973)

1771. Peroxidase

SOURCE: Pig small intestines
SOLVENT: 20 mM KCl: pH 4.73, 4.95, 6.61, and 7.04
REFERENCE: S. Kimura and I. Yamazaki, Arch. Biochem. Biophys., 189, 14 (1978)

1772. Peroxidase

SOURCE: Pig small intestine
SOLVENT:
a. pH 7.10
b. pH 5.59
c. pH 7.83

COMMENTS:
1. Spectral changes upon the addition of hydrogen peroxide (solvent a)
2. Spectrum of peroxidase-benzhydroxamic acid complex (solvent b)
3. Changes in spectrum of 2 upon addition of hydrogen peroxide

REFERENCE: S. Kimura and I. Yamazaki, Arch. Biochem. Biophys., 198, 580 (1979)

1773. Peroxidase

SOURCE: Japanese radish root
SOLVENT: 10 mM Phosphate buffer, pH 7.0 and pH 5
COMMENTS: Spectra of proteins heated to 100°C for 10 minutes
REFERENCE: Y. Tamura and Y. Morita, J. Biochem. Tokyo, 78, 561 (1975)

1774. Peroxidase

SOURCE: Goat submaxillary gland
SOLVENT: Water
COMMENTS: 1. Spectra of enzyme and apoenzyme

2. Changes in the spectrum due to cyanide, azide, fluoride, dithionite, and dithionite plus carbon monoxide

REFERENCE: M. Bal, G. Sen, U. Mahajani, and A. G. Datta, Arch. Biochem. Biophys., 151, 216 (1972)

1775. Peroxidase

SOURCE: Guinea pig bone marrow and leucocytes (heterophil and eosinophil)
COMMENTS: Spectra of oxidized and reduced (dithionite) forms
REFERENCE: R. K. Dresser, S. R. Himmelhoch, W. H. Evans, M. Januska, M. Mage, and E. Shelton, Arch. Biochem. Biophys., 148, 452 (1972)

1776. Peroxidase 556

SOURCE: Rice (Oryzae sativa L.) embryos
COMMENTS: Spectra of reduced and oxidized forms
REFERENCE: Y. Morita and S. Ida, Agric. Chem. Biol., 32, 441 (1968)

1777. Peroxidase A and Peroxidase B

SOURCE: Wheat
COMMENTS: Spectra of oxidized and reduced forms
REFERENCE: M. F. Jeanjean, K. Kobrehel, and P. Feillet, Biochimie, 57, 145 (1975)

1778. Peroxidase P1 and Peroxidase P7

SOURCE: Turnip (Brassica napus var. blanc dur d'hiver)
SOLVENT:
a. Phosphate buffer, pH 7
b. Sodium bicarbonate, NaOH buffer, pH 11.30
c. Glycine-NaOH buffer, pH 10.0

COMMENTS:
1. Spectra in solvents a, b, and c
2. Spectral changes of reduced enzymes upon oxidation with oxygen in solvent a

REFERENCE: J. Ricard, G. Mazza, and R. J. P. Williams, Eur. J. Biochem., 28, 566 (1972)

1779. Peroxidase, Micro-

SOURCE: Commercial (horse heart cytochrome c)
SOLVENT: 0.1 M Phosphate buffer, pH 7.0
COMMENTS: Spectra of reduced and CO-microperoxidase
REFERENCE: V. S. Sharma, M. R. Schmidt, and H. M. Ranney, J. Biol. Chem., 251, 4267 (1976)

1780. Peroxidase

SOURCE: Commercial
SOLVENT: Phosphate, pH 7.1
COMMENTS: 1. Spectra of peroxidase, galactose oxidase, and mixture
2. Changes in spectra upon addition of galactose
REFERENCE: L. Cleveland, R. E. Coffman, P. Coon, and L. Davis, Biochemistry, 14, 1108 (1975)

1781. Peroxidase

SOURCE: Wild horseradish roots
SOLVENT: 0.05 M Phosphate
COMMENTS: Spectral changes caused by the addition of DL-tryptophan to ferriperoxidase and to ferroperoxidase in the presence of hydrosulfite.
REFERENCE: I. Yamazaki, R. Nakajima, K. Miyoshi, R. Makino, and M. Tamura in International Symposium on Oxidases and Related Redox Systems (T. E. King, H. S. Mason, and M. Morrison, eds.), Vol. I, 407 (1971), University Park Press, Baltimore, Maryland

1782. Peroxidase

SOURCE: Wild horseradish roots
SOLVENT: a. 0.1 M Acetate buffer, pH 5
b. 0.1 M Phosphate buffer, pH 4.5, pH 6, pH 7
c. Glycine-NaOH, pH 9.5, pH 11, and pH 11.5
COMMENTS: Spectra taken at 293°K and 77°K and in solvent b from 197°K-77°K
REFERENCE: M. Tamura and H. Hori, Biochim. Biophys. Acta, 284, 20 (1972)

1783. Peroxidases (Synthetic-HRP)

SOURCE: Wild horseradish root apoperoxidase plus various hemes
COMMENTS: 1. pH 6.1-pH 12.0: spectra of meso-HRP, deutero-HRP, proto-monoester-HRP, and protodiester-HRP
2. Spectra of Fe^{3+} forms, compound I, compound II, and compound III at pH 6.0 of deutero-HRP, meso-HRP, and proto-monoester-HRP
3. Spectra at pH 6 of fluoride and cyanide complexes of meso-HRP, deutero-HRP, proto-monoester-HRP, and proto-diester-HRP
4. Spectra at pH 6 of the reduced (dithionite) and reduced (dithionite) plus CO of meso-HRP, deutero-HRP, proto-HRP, proto-monoester-HRP, and proto-diester heme-HRP

REFERENCE: M. Tamura, T. Asakura, and T. Yonetani, Biochim. Biophys. Acta, 268, 292 (1972)

1784. Peroxidase

SOURCE: Horseradish
COMMENTS: Spectra of native form, compound I, and form made by reacting peroxidase with sodium chlorite
REFERENCE: L. Hager, D. Doubek and P. Hollenberg, in Molecular Basis of Electron Transport (J. Schultz and B. Cameron, eds.), 361 (1972), Academic Press, New York

1785. Peroxidase

SOURCE: Horseradish
SOLVENT: pH 7
COMMENTS: Spectra of the following forms of peroxidase: native, reduced (dithionite), reduced (dithionite) plus NO or sodium nitrite, plus NO and plus sodium nitrite
REFERENCE: J. S. Leigh, Jr., and G. H. Reed, J. Biol. Chem., 247, 2447 (1972)

1786. Peroxidase

SOURCE: Horseradish
SOLVENT: Phosphate buffer, pH 5.0, pH 7.0, pH 7.6
COMMENTS: Effect of hydrogen peroxide on spectrum
REFERENCE: R. Roman and U. B. Dunford, Biochemistry, 11, 2076 (1972); M. Tamura and I. Yamazaki, J. Biochem. Tokyo, 71, 311 (1972)

1787. Peroxidase (Synthetic-HRP)

SOURCE: Wild horseradish roots
SOLVENT: 0.1 M Potassium phosphate, pH 6.0
COMMENTS:
1. Spectra of following synthetic HRP's and their peroxide compounds: deutero-HRP-(B+C), meso-HRP-(B+C), hemato-HRP-(B+C), and 2,4-diacetyldeutero-HRP-(B+C)
2. Spectra of cyanide complex and hydrosulfite reduced form with and without CO

REFERENCE: R. Makino and I. Yamazaki, J. Biochem. Tokyo, 72, 655 (1972)

1788. Peroxidase (Reconstituted-HRP)

SOURCE: Apoperoxidase of horseradish
SOLVENT: 0.1 M Phosphate, pH 7.0

COMMENTS: Spectral changes during conversion of CO-ferro-peroxidase to ferriperoxidase in presence of air: proto-HRP and meso-HRP
REFERENCE: M. Tamura and T. Yonetani, Biochim. Biophys. Acta, 322, 211 (1973)

1789. Peroxidase (A_1 + A_2) and Synthetic Peroxidase (B+C)

SOURCE: Apoperoxidase of horseradish
SOLVENT: a. 0.1 M Sodium acetate buffer, pH 4.0
b. 0.1 M Potassium phosphate buffer, pH 7.0
COMMENTS: 1. Spectra of chlorocruoro-HRP (B+C) and its azide complex in solvent a
2. Spectra of cyanide complexes of chlorocruoro-HRP (B+C) plus sodium hydrosulfite in solvent b
3. Spectra of cyanide complexes of diacetyldeutero-HRP (B+C) in solvent b
4. Spectra of fluoride complexes of meso-HRP (B+C) and diacetyldeutero-HRP (B+C) in solvent a
REFERENCE: R. Makino and I. Yamazaki, Arch. Biochem. Biophys., 157, 356 (1973)

1790. Peroxidase (Synthetic-HRP)

SOURCE: Apoperoxidase of horseradish
SOLVENT: 0.2 M Phosphate buffer, pH 7.0
COMMENTS: The spectra of cyanide complexes in the presence of dithionite, of proto-HRP, meso-HRP, and deutero-HRP
REFERENCE: M. Tamura and T. Yonetani, Biochim. Biophys. Acta, 322, 211 (1973)

1791. Peroxidase (Synthetic-HRP)

SOURCE: Apoperoxidase of horseradish
SOLVENT: 0.01 M Sodium phosphate, pH 7.0
COMMENTS: 1. Spectral changes upon reaction of mesohaematin peroxidase with ascorbic acid and hydrogen peroxide
2. Spectra of diacetyl deuterohaematin peroxidase at pH 6.0 (10°C and 40°C) and pH 11.5 (25°C)
3. Spectra of oxidized and reduced (dithionite) forms of proto-, meso-, and diacetyl-deutero-haematin peroxidases
REFERENCE: P. I. Ohlsson and K. G. Paul, Biochim. Biophys. Acta, 315. 293 (1973)

1792. Peroxidase

SOURCE: Horseradish
SOLVENT: 0.02 M Potassium phosphate buffer, pH 6.8
COMMENTS: Spectra of peroxidase and benzhydroxamic acid derivative
REFERENCE: G. R. Schonbaum, J. Biol. Chem., 248, 502 (1973)

1793. Peroxidase (B+C) (HRP)

SOURCE: Wild horseradish roots
SOLVENT: a. 0.05 M Sodium acetate buffer, pH 5.0
b. 0.05 M Potassium phosphate, pH 7.0
COMMENTS: 1. Spectrum obtained on reaction of HRP with indole 3-acetic acid, then the addition of ascorbate in solvent a
2. Spectra of ferric HRP, after addition of hydrogen peroxide and ascorbic acid (Compound II) after further addition of hydrogen peroxide (Compound III) in solvent b
REFERENCE: H. Yamazaki and I. Yamazaki, Arch. Biochem. Biophys., 154, 147 (1973)

1794. Peroxidase

SOURCE: Horseradish
SOLVENT: 50 mM Sodium acetate, pH 4.25 containing ascorbate
COMMENTS: Spectral changes upon addition of hydroxymethyl hydroperoxide
REFERENCE: S. Marklund, Arch. Biochem. Biophys., 154, 614 (1973)

1795. Peroxidase (B+C) (HRP)

SOURCE: Horseradish
SOLVENT: pH 6.8 and pH 9.2
COMMENTS: Effect of pH on the spectrum of ferrous (dithionite) HRP
REFERENCE: H. Yamada and I. Yamazaki, Arch. Biochem. Biophys., 165, 728 (1974)

1796. Peroxidase (HRP)

SOURCE: Horseradish
SOLVENT: 0.05 M Sodium formate, pH 7.2
COMMENTS: 1. Spectra of HRP and Compound I in presence and absence of sodium formate
2. Spectrum of HRP in sodium formate, pH 7.2
3. Spectrum obtained after radiation of #2 with γ-rays

4. Spectrum of #2 plus hydrogen peroxide
5. Spectrum of #4 after radiation with γ-rays
6. Spectra obtained after pulse irradiation of HRP, 2 millisec-5 min. after pulse

REFERENCE: B. H. J. Bielski, D. A. Comstock, A. Haber, and P. C. Chan, Biochim. Biophys. Acta, 350, 113 (1974)

1797. Peroxidase and Chloroperoxidase

SOURCE: Horseradish
SOLVENT: 0.2 M Potassium phosphate-citrate buffer, pH 6.0
COMMENTS:
1. Spectral changes seen on addition of ethyl hydrogen peroxide
2. Spectral changes seen on addition of chlorite

REFERENCE: P. F. Hollenberg, T. Rand-Meir, and L. P. Hager, J. Biol. Chem., 249, 5816 (1974)

1798. Peroxidase A, Diacetyldeutero- (DDP)

SOURCE: Horseradish peroxidase
COMMENTS:
1. Spectra of ferrous DDP-complexes: HCN complex, pH 7.44 and pH 5.00; CO-complex, pH 8.95 and pH 4.95; NO-complex, pH 8.55 and pH 5.20; Anaerobic conditions in presence of 0.1 M KCl
2. pH effect on spectra of ferrous-O_2, pH 10.0 and pH 5.7 and ferric-cyanide complexes of DDP, pH 9.45, and 5.97

REFERENCE: H. Yamada and I. Yamazaki, Arch. Biochem. Biophys., 171, 737 (1975)

1799. Peroxidase

SOURCE: Horseradish
SOLVENT: 0.1 M Phosphate buffer, pH 7.0
COMMENTS:
1. Spectrum of peroxidase plus hydrogen peroxide (Compound I)
2. Spectrum of peroxidase plus guaiacol (Compound II)
3. Spectra made at -50° in methanol-water (50/50)

REFERENCE: M. Santimone, Can. J. Biochem., 53, 649 (1975)

1800. Peroxidase

SOURCE: Horseradish
SOLVENT:
a. 0.1 M Potassium phosphate, pH 7.01
b. pH 11.90

COMMENTS:
1. Spectra of Fe III-form and its fluoro-, and cyano-derivatives in solvent a
2. Spectra of Fe II-form and its CO- and cyanide derivatives in solvent a

3. Spectrum of alkaline Fe III-form in solvent b
REFERENCE: T. Nozawa, N. Kobayashi, and M. Hatano, Biochim. Biophys. Acta, 427, 652 (1976)

1801. Peroxidase

SOURCE: Horseradish
SOLVENT: 0.1 M Phosphate buffer, pH 7.0
COMMENTS: Spectrum of peroxidase plus dithiothreitol at -77°C
REFERENCE: J. Olsen and L. Davis, Biochim. Biophys. Acta, 445, 324 (1976)

1802. Peroxidase (HRP)

SOURCE: Horseradish
SOLVENT: a. Carbonate buffer, pH 10.51
b. Phosphate buffer, pH 7.0
COMMENTS: 1. Spectra of HRP and HRP-I
2. Spectra of HRP-I immediately after addition of 4a, 9b-dihydro-8,9b-dimethyl-3(4H)-dibenzofuranone (Pummerer's Ketone) and 4 and 8 min. later
REFERENCE: W. D. Hewson and H. B. Dunford, J. Biol. Chem., 251, 6043 (1976)

1803. Peroxidase, Apo (AHRP), and Derivatives

SOURCE: Horseradish peroxidase
SOLVENT: 0.1 M Sodium phosphate buffer, pH 6.0
COMMENTS: 1. Spectral changes during reaction of ferric protoporphyrin IX with AHRP
2. Spectra of following metallo-protoporphyrin IX-AHRP complexes: ferric, manganese, zinc and copper
3. Spectra of following metal-free porphyrin IX-AHRP complexes: protoporphyrin IX, hematoporphyrin-IX, and deuteroporphyrin-IX
REFERENCE: Y. J. Kang and J. D. Spikes, Arch. Biochem. Biophys., 172, 565 (1976)

1804. Peroxidase (HRP)

SOURCE: Horseradish
COMMENTS: Spectrum of HRP purified via Con-A chromatography
REFERENCE: M. G. Brattain, M. E. Marks, and T. G. Pretlow II, Anal. Biochem., 72, 346 (1976)

1805. Peroxidase C

SOURCE: Horseradish
SOLVENT: pH 7.21 and pH 9.85 in 0.1 M KCl

COMMENTS: Spectral changes due to pH of reduced (dithionite) CO-form under anaerobic conditions
REFERENCE: Y. Hayashi, Y. Yamada, and I. Yamazaki, Biochim. Biophys. Acta, 427, 668 (1976)

1806. Peroxidase, Artificial

SOURCE: Horseradish
COMMENTS: Spectra of hematin-peroxidase, and ethyl-, hydroxyethyl-, acetyl-, vinyl-, 1-propenyl-, and 1-butenyl-hematin peroxidases
REFERENCE: P. I. Ohlsson, K. G. Paul, and I. Sjoholm, J. Biol. Chem., 252, 8222 (1977)

1807. Peroxidase

SOURCE: Horseradish
COMMENTS: Spectra of five forms: ferroperoxidase; ferroperoxidase (dithionite); ferriperoxidase plus hydrogen peroxide, Compound I; ferriperoxidase plus hydrogen peroxide and ascorbic acid, Compound II; oxyperoxidase
REFERENCE: K. N. Yokota and I. Yamazaki, Biochemistry, 16, 1913 (1977)

1808. Peroxidase

SOURCE: Horseradish
SOLVENT: a. 1/15 M Deuterium oxide phosphate buffer, p^2H 6.86
b. Deuterium oxide p^2H 12.01
COMMENTS: Spectra of ferriperoxidase and its cyanide complex
REFERENCE: N. Kobayashi, T. Nozawa, and M. Hatano, Biochim. Biophys. Acta, 493, 340 (1977)

1809. Peroxidase (B,C), Cobalt- (HRP)

SOURCE: Horseradish
SOLVENT: 0.05 M Tris buffer, pH 8.2
COMMENTS: Spectra of Co^{+3}-HRP, Co HRP, Co HRP (O_2), and Co^{3+}HRP (CN^-)
REFERENCE: M.Y.R. Wang, B. M. Hoffman, and P. F. Hollenberg, J. Biol. Chem., 252, 6268 (1977)

1810. Peroxidase (HRP)

SOURCE: Horseradish
SOLVENT: 0.2 M Phosphate buffer, pH 7.0
COMMENTS: Spectra of ferri-HRP, ferro-HRP, and their nitric oxide complexes

REFERENCE: M. Tamura, K. Kobayashi, and K. Hayashi, FEBS Lett., 88, 124 (1978)

1811. Peroxidase, Reduced

SOURCE: Horseradish
SOLVENT: 0.2 M Sodium phosphate buffer, pH 6.8 plus glycerol (1:1)
COMMENTS: Spectra at room and liquid nitrogen temperatures
REFERENCE: Y. A. Sharonov, A. P. Mineyev, M. A. Livshitz, N. A. Sharonova, V. B. Zurkin, and Y. P. Lysov, Biophys. Str. Mech., 4, 139 (1970)

1812. Peroxidase (HRP)

SOURCE: Horseradish
SOLVENT: 0.05 M NaCl
COMMENTS: Spectra of HRP, butyric anhydride modified, succinic anhydride modified, and modified with trinitrobenzene
REFERENCE: N. N. Ugarova, E. D. Rozhkova, and I. V. Berezin, Biokhimiya (Biochemistry), 43, 983 (1978)

1813. Peroxidase C (HRP-C)

SOURCE: Horseradish
COMMENTS: Spectra of HRP-C, HRP-C plus hydrogen peroxide, and HRP-C plus hydrogen peroxide plus ascorbate
REFERENCE: Y. Hayashi and I. Yamazaki, Arch. Biochem. Biophys., 190, 446 (1978)

1814. Peroxidase (HRP)

SOURCE: Horseradish
SOLVENT: 0.2 M Citric acid-phosphate, pH 7.8, 6.0, and 4.9
COMMENTS: Spectra of HRP and HRP-azide complex
REFERENCE: I. Morishima, S. Ogawa, and T. Yonezawa, Biochim. Biophys. Acta, 537, 293 (1978)

1815. Peroxidase

SOURCE: Horseradish
SOLVENT:
a. 100 mM Cacodylate buffer, pH 7.1
b. 7% (V/V) buffer in 9% (W/V) AOT/silicone oil (AOT = sodium diisooctyl sulfosuccinate)
3. 1% (V/V) buffer, in 2.5% (W/V) AOT/n-heptane

REFERENCE: C. Balny and P. Douzou, Biochimie, 61, 445 (1979)

1816. Peroxidase (HRP)

SOURCE: Horseradish
SOLVENT: a. 3.7 mM Sodium carbonate buffer, pH 10.70
b. 4.6 mM Phosphate buffer, pH 7.09
c. 83 mM Acetate buffer, pH 3.80
COMMENTS: 1. Spectra of HRP, HRP plus H_2O_2 and HRP plus H_2O_2 plus ferrocyanide (solvent a)
2. Spectra of HRP plus H_2O_2 plus ferrocyanide (solvents a, b, and c)
3. Spectra of HRP plus chlorite (solvents a, b, and c)
4. Spectra of HRP plus chlorine dioxide (solvents a, b, and c)
REFERENCE: W. D. Hewson and L. P. Hager, J. Biol. Chem., 254, 3182 (1979)

1817. Peroxidase

SOURCE: Mycobacterium tuberculosis H37Rv
SOLVENT: a. 0.045% Potassium phosphate buffer, pH 7.0
b. 0.18% Potassium phosphate buffer, pH 7.0
REFERENCE: B. Gayathri-Devi, M. S. Shaila, T. Ramakrishnan, and K. P. Gopinathan, Biochem. J., 149, 187 (1975)

1818. Peroxygenase

SOURCE: Pea (Pisum sativum) seed microsomes
SOLVENT: 0.1 M Sodium phosphate buffer, pH 7.2, containing 0.05 M NaCl, 0.5 mM EDTA, 0.2% cholate, and 20% (V/V) glycerol
COMMENTS: 1. Spectra of partially purified oxidized and reduced (dithionite-methylviologen) CO-complex
2. Spectrum of oxidized form plus H_2O_2 plus CO
REFERENCE: A. Ishimaru, J. Biol. Chem., 254, 18427 (1979)

1819. Phalloidin, N-Nethyl

SOURCE: Amantia family
REFERENCE: H. Faulstich and T. Wieland, Eur. J. Biochem., 22, 79 (1971)

1820. Phenol Hydroxylase

SOURCE: Yeast (Trichosporon cutaneum)
SOLVENT: 0.1 M Potassium phosphate buffer, pH 7.6
COMMENTS: Spectra of protein, protein reduced with sodium borohydride, and protein reduced with dithionite
REFERENCE: H. Y. Neujahr and A. Gaal, Eur. J. Biochem., 35, 386 (1973)

1821. Phenol Hydroxylase, apo- and derivatives

SOURCE: Yeast (Trichosporon cutaneum)
COMMENTS: Spectra of apo-enzyme and following derivatives: p-mercuribenzoate and 3-carboxy-4-nitrophenyl sulfide
REFERENCE: H. Y. Neujahr and A. Gaal, Eur. J. Biochem., 58, 351 (1975)

1822. Phenol Hydroxylase

SOURCE: Yeast (Trichosporon cutaneum)
SOLVENT: 0.1 M Tris-sulfate, pH 7.6
COMMENTS: Spectra of enzyme and of enzyme with each of following: phenol, p-fluorophenol, and o-methylphenol
REFERENCE: H. Y. Neujahr and K. G. K. Jellen, J. Biol. Chem., 253, 8835 (1978)

1823. Phenylalanine Hydroxylase

SOURCE: Rat liver
SOLVENT: 0.01 M Tris-HCl, pH 6.8
REFERENCE: D. B. Fisher, R. Kirkwood, and S. Kaufman, J. Biol. Chem., 247, 5161 (1972)

1824. Phormidium Fragile

REFERENCE: C. Lemasson, N. Tandeau de Marsac, and G. Cohen-Bazire, Proc. Nat. Acad. Sci., 70, 3130 (1973)

1825. Phosphatase, Acid

SOURCE: Escherichia freundii K-1
SOLVENT: 0.006 M Potassium phosphate buffer, pH 7.0
REFERENCE: Y. Tani, T. Tochikura, H. Yamada, and K. Ogata, Agric. Biol. Chem., 32, 1220 (1968)

1826. Phosphatase, Acid

SOURCE: Sweet potato
SOLVENT: a. 0.1 M Phosphate buffer, pH 6.0
b. 0.1 N NaOH
REFERENCE: K. Uehara, S. Fujimoto, and T. Taniguchi, J. Biochem. Tokyo, 75, 627 (1974)

1827. Phosphatase, Acid

SOURCE: Spinach (Spinacea oleracea)
SOLVENT: 0.01 M Phosphate buffer, pH 6.5, containing 0.1 N NaCl

REFERENCE: S. Fujimoto, T. Nakagawa, S. Ishimitsu, and A. Ohara, Chem. Pharm. Bull., 25, 1459 (1977)

1828. Phosphatase, Alkaline

SOURCE: Bacillus subtilis
SOLVENT: 0.01 M Tris-HCl-0.01 M sodium acetate-0.2 M magnesium acetate, pH 7.4
REFERENCE: F. K. Yoshizumi and J. E. Coleman, Arch. Biochem. Biophys., 160, 255 (1974)

1829. Phosphatase, Alkaline

SOURCE: Escherichia coli
SOLVENT: pH 8.0
COMMENTS: Spectra of native and denatured proteins
REFERENCE: D. Gerard, G. Laustriat, and H. Lami, Biochim. Biophys. Acta, 263, 482 (1972)

1830. Phosphatase, Alkaline, Co(II)

SOURCE: Escherichia coli
SOLVENT: 0.01 M Tris, pH 5-pH 9
COMMENTS:
1. Spectra of dimer-Co_2(II) at several pH values
2. Spectra of dimer-Co_4(II)-phosphate complex at several pH values
3. Spectra of enzyme at pH 5 plus 2-4 atoms of Co(II) per mole of dimer

REFERENCE: J. S. Taylor, C. Y. Lau, M. L. Applebury, and J. E. Coleman, J. Biol. Chem., 248, 6216 (1973)

1831. Phosphatase, Alkaline, Co^{II}

SOURCE: Escherichia coli
SOLVENT: 10 mM Tris-HCl, pH 8.0
COMMENTS: Spectra in the presence and absence of hydrogen peroxide, and magnesium
REFERENCE: R. A. Anderson and B. L. VAllee, Proc. Nat. Acad. Sci., 72, 394 (1975); R. A. Anderson, W. F. Bosron, F. W. Kennedy, and B. L. Vallee, Proc. Nat. Acad. Sci., 72, 2989 (1975)

1832. Phosphatase, Alkaline, Cobalt

SOURCE: Escherichia coli
SOLVENT: 20 mM N-2-hydroxyethylpiperazine-N'-2-ethane sulfonic acid, pH 8.0
COMMENTS:
1. Spectral changes upon addition of cobalt sulfate to apoenzyme

2. Spectral changes of cobalt-phosphatases due to magnesium
3. Effect of phosphate on spectra of Co(II)-phosphatase in presence and absence of magnesium

REFERENCE: R. A. Anderson, F. S. Kennedy, and B. L. Vallee, Biochemistry, 15, 3710 (1976)

1833. Phosphoenolpyruvate Carboxykinase

SOURCE: Rat liver cytosol
SOLVENT: 10 mM Triethanolamine, pH 7.2, containing 10% glycerol, 0.1 mM EDTA and 1 mM glutathione
REFERENCE: G. Colombo, G. M. Carlson, and H. A. Lardy, Biochemistry, 17, 5321 (1978)

1834. Phosphoenolpyruvate Carboxylase

SOURCE: Escherichia coli
SOLVENT: 0.1 M Tris-Acetate buffer, pH 8.0
REFERENCE: T. Yoshinaga, H. Teraoka, K. Izui, and H. Katsuki, J. Biochem. Tokyo, 75, 913 (1974)

1835. Phosphofructokinase

SOURCE: Clostridium pasteurianum
SOLVENT: a. 0.05 M Potassium phosphate, pH 7
b. 0.1 N NaOH
REFERENCE: K. Uyeda and S. Kurooka, J. Biol. Chem., 245, 3315 (1970)

1836. Phosphofructokinase

SOURCE: Saccharomyces carlsbergensis
SOLVENT: pH 7.0 and 0.1 N NaOH
REFERENCE: N. Tamaki and B. Hess, Hoppe-Seyler's Z. Physiol. Chem., 356, 399 (1975)

1837. Phosphofructokinase

SOURCE: Rabbit muscle
SOLVENT: 25 mM β-glycerophosphate, 1 mM EDTA, pH 7.2
COMMENTS: Spectrum of phosphofructokinase which had been modified by α-bromo-4-hydroxy-3-nitroacetophenone
REFERENCE: K. J. Schwartz, Y. Nakagawa, and E. T. Kaiser, J. Am. Chem. Soc., 98, 6369 (1976)

1838. Phosphofructokinase

SOURCE: Rabbit muscle
SOLVENT: 0.1 M Phosphate, pH 8.0

COMMENTS: Spectra of phosphofructokinase modified with 7-chloro-4-nitrobenzo-2-oxa-1,3-diazole and unmodified enzyme
REFERENCE: P. M. Lad, N. M. Wolfman, and G. G. Hammes, Biochemistry, 16, 4802 (1977)

1839. Phosphofructokinase

SOURCE: Yeast
SOLVENT: 0.1 M Phosphate buffer, pH 7.0
REFERENCE: G. Kopperschlager, I. Lorenz, W. Diezel, J. Marquardt, and E. Hofmann, Acta Biol. Med. Germ., 29, 561 (1972)

1840. Phosphoglucomutase (PGM)

SOURCE: Rabbit muscle
SOLVENT: a. 20 mM Tris-HCl, pH 8.5
b. 20 mM Sodium acetate, pH 5.1
c. 50 mM Tris-HCl, pH 7.5
COMMENTS: 1. Spectrum of PGM-Co(II) complex (solvent a)
2. Effect of pH on spectrum (solvents a and b)
3. Effect of temperature, 28° and 3°C (solvent a)
4. Spectrum of PGM-Ni(II) complex (solvent c)
5. Effect of temperature, 1°-24.4°C (solvent c)
REFERENCE: W. J. Ray, Jr., and J. S. Multani, Biochemistry, 11, 2805 (1972)

1841. 6-Phosphogluconate Dehydrogenase

SOURCE: Candida utilis
COMMENTS: Spectra of enzyme containing 1.1 and 90 atom percent C^{13}
REFERENCE: R. T. Eakin, Biochim. Biophys. Acta, 377, 9 (1975)

1842. Phosphoglucose Isomerase

SOURCE: Rabbit muscle
SOLVENT: 50 mM N,N-Bis (2-hydroxyethyl) glycine buffer, pH 8.0, containing 0.9 mM pyridoxal-phosphate
COMMENTS: Spectra of native and photosensitized enzyme
REFERENCE: G. M. Hathaway and E. A. Noltmann, Arch. Biochem. Biophys., 179, 24 (1977)

1843. Phosphoglycerate Kinase

SOURCE: Beef liver
SOLVENT: a. 50 mM Triethanolamine-HCl buffer, pH 7.5
b. Solvent a containing 8 M urea

c. 10 mM Triethanolamine-HCl buffer, 100 mM NaCl, pH 7.0
d. 0.1 N NaOH

REFERENCE: K. D. Kulbe, M. Bojanovski, and W. Lamprecht, Eur. J. Biochem., 52, 239 (1975)

1844. Phospholipase A_2, dimethylaminonaphthylsulfonyl aminotyrosine-69-

SOURCE: Horse pancreas
SOLVENT: 50 mM Sodium acetate, pH 6.0
REFERENCE: H. Meyer, W. C. Puijk, R. Dijkman, M. M. E. L. Foda-Van der Hoorn, F. Pattus, A. J. Slotboom, and G. H. De Haas, Biochemistry, 18, 3589 (1979)

1845. Phospholipase A

SOURCE: Snake (Crotalus terrificus) venom
SOLVENT: a. 0.1 M HCl, pH 1.0
b. 0.1 M Sodium bicarbonate, pH 8.5
c. 0.1 M NaOH, pH 12.9
REFERENCE: H. Breithaupt, K. Rubsamen, and E. Habermann, Eur. J. Biochem., 49, 333 (1974)

1846. Phospholipase B

SOURCE: Vibrio el-tor
REFERENCE: G. C. Chatterjee and S. K. Das, Enzymologia, 28, 346 (1965)

1847. Phosphorylase A

SOURCE: Rabbit muscle
SOLVENT: a. pH 7.0
b. pH 12.5
COMMENTS: Spectra of native and acetylated enzyme
REFERENCE: E. S. Severin, N. B. Kozlova and P. L. Vulfson, Biokhimiya (Biochemistry), 36, 1054 (1971)

1848. Phosphorylase B

SOURCE: Rabbit skeletal muscle
SOLVENT: 30 mM β-glycerophosphate buffer, plus 1 mM EDTA, pH 6.8
COMMENTS: Spectra of native enzyme and enzyme-adenosine-5'-chloromethyl phosphonate complex
REFERENCE: L. K. Skloysheva, P. L. Vulfson, N. N. Gulyaev, and E. S. Severin, Biokhimiya (Biochemistry), 43, 1509 (1978)

1849. Phosphorylase B

SOURCE: Rabbit skeletal muscle
COMMENTS: Spectra of native and diethylpyrocarbonate modified enzyme
REFERENCE: L. K. Skolysheva, P. L. Vulfson, and E. S. Severin, Biokhimiya (Biochemistry), 42, 187 (1977)

1850. Phosphorylase B

SOURCE: Rabbit skeletal muscle
COMMENTS: Spectra of enzyme and enzyme treated with formaldehyde
REFERENCE: N. B. Livanova, A. D. Morozkin, and J. D. Shpikiter, in Pyridoxal Catalysis: Enzyme and Model Systems (E. E. Snell, A. E. Braunstein, E. S. Severin, and Y. M. Torchinsky, eds.), 569 (1968), Interscience, New York

1851. Phosphorylase b, glycogen

SOURCE: Rabbit muscle
SOLVENT: pH 1.1-pH 10
COMMENTS: Spectra of sodium borohydride-reduced enzyme at pH 1.1-pH 10 in glycerophosphate
REFERENCE: M. Cortijo and S. Shaltiel, Eur. J. Biochem., 29, 131 (1972)

1852. Phosphorylase b, Pyridoxal phosphate-

SOURCE: Rabbit muscle
REFERENCE: O. Avramovic-Zikic, K. Welinder, S. Shechosky, and J. Sodek, Can. J. Biochem., 51, 21 (1973)

1853. Phosphorylase b, Glycogen

SOURCE: Pig skeletal muscle
SOLVENT: 1.0 mM EDTA/50 mM glycerol phosphate, pH 7.0
REFERENCE: S. J. Tzartos and Q. E. Evangelopoulos, Eur. J. Biochem., 74, 233 (1977)

1854. Phosphorylase I, Starch

SOURCE: Victory freezer pea seeds
SOLVENT: a. pH 7
b. pH 13
REFERENCE: N. K. Matheson and R. H. Richardson, Phytochemistry, 17, 195 (1978)

1855. Phosphorylase Kinase

SOURCE: Rabbit muscle
SOLVENT: 5 mM Sodium glycerol phosphate buffer, pH 7.0, containing 0.2 mM EDTA
REFERENCE: T. Hayakawa, J. P. Perkins, D. A. Walsh, and E. G. Krebs, Biochemistry, 12, 567 (1973)

1856. Phosphorylase Kinase, δ-subunit

SOURCE: Rabbit skeletal muscle
REFERENCE: P. Cohen, A. Burchell, J. G. Foulkes, P. T. W. Cohen, T. C. Vanaman and A. C. Nairn, FEBS Lett., 92, 287 (1978)

1857. Phosphovitin, Green-Iron (III)-

SOURCE: Egg yolk
SOLVENT: a. 0.005 M Tris buffer, pH 7.5
b. Solvent a plus ethylene glycol (1/1:V/V)
REFERENCE: J. Webb, J. S. Multani, P. Saltman, N. A. Beach, and H. B. Gray, Biochemistry, 12, 1797 (1973)

1858. Phosphovitin

SOURCE: Chicken egg
SOLVENT: Cacodylate buffer, pH 6.8
COMMENTS: Spectra of protein and of protein-metal complexes of magnesium, calcium, manganese, and cobalt
REFERENCE: K. Grizzuti and G. E. Perlman, Biochemistry, 14, 2171 (1975)

1859. Phosphovitin Kinase

SOURCE: Calf brain
SOLVENT: 0.01 M Potassium phosphate buffer, pH 7.25
REFERENCE: O. Walinder, Biochim. Biophys. Acta, 258, 411 (1972)

1860. Photolyase, DNA-

SOURCE: Yeast
REFERENCE: H. Werbin and J. J. Madden, Photochem. Photobiol., 25, 421 (1977)

1861. Photoreaction Center

SOURCE: Rhodospirillum rubrum strain S1 and strain G9
SOLVENT: 10 mM Tris-HCl, pH 8.0, containing 0.03% dodecyl dimethylamine N-oxide

REFERENCE: C. Vadebroncoeur, H. Noel, L. Poirier, Y. Cloutier, and G. Gingras, Biochemistry, 18, 4301 (1979)

1862. Photoreaction Center

SOURCE: Rhodospirillum rubrum
COMMENTS: Spectra of reduced and oxidized centers at room temperature and at 77°K
REFERENCE: T. Mar and G. Gingras, Biochim. Biophys. Acta, 460, 239 (1977)

1863. Photoreactivating Enzyme

SOURCE: Escherichia coli
REFERENCE: K. L. Wun, A. Gih, and J. C. Sutherland, Biochemistry, 16, 921 (1977); R. M. Snapka and C. O. Fuselier, Photochem. Photobiol., 25, 415 (1977)

1864. Photosystem I and II Membrane Particles

SOURCE: Synechococcus cedrorum
REFERENCE: P. J. Newman and L. A. Sherman, Biochim. Biophys. Acta, 503, 343 (1978)

1865. Phycobilisomes

SOURCE: Rhodella violacea
SOLVENT: 0.65 M Phosphate buffer, pH 7.0
COMMENTS: Spectra of phycobilisomes and isolated tripartite units
REFERENCE: K. P. Koller, W. Wehrmeyer, and E. Morschel, Eur. J. Biochem., 91, 57-63 (1978)

1866. Phycobilisome, Phycocyanin, Allophycocyanin, and Phycoerythrin

SOURCE: Blue-green alga (Nostoc species)
SOLVENT:
a. 100 mM Sodium phosphate buffer, pH 7.0, containing 0.02% sodium azide
b. 50 mM Sodium phosphate buffer, pH 7.0 containing 0.01% sodium azide and 50% glycerol
c. 0.75 M Phosphate buffer, pH 7.0
d. 10 mM Phosphate buffer, pH 7.0

COMMENTS:
1. Spectra at 23°C in solvent a
2. Spectra at -196°C in solvent b
3. Spectra of phycobilisomes in solvent c
4. Spectra of dissociated phycobilisomes in solvent d

REFERENCE: B. H. Gray and E. Gantt, Photochem. Photobiol., 21, 121 (1975)

1867. C-Phycocyanin (PC), Allophycocyanin (APC), and C-Phycoerythrin (PE)

SOURCE: a. Spirulina platensis (PC and APC)
b. Fremyella diplosiphon (PE)
SOLVENT: 50 mM Phosphate buffer, pH 7.5
COMMENTS: 1. Spectral changes upon titration of PC with sodium dithionite
2. Spectral changes upon titration of urea-denatured PC with sodium dithionite
3. Spectra of denatured APC in the presence and absence of dithionite
4. Spectra of denatured PE in the presence and absence of dithionite
REFERENCE: W. Kufer and H. Scheer, Hoppe-Seyler's Z. Physiol. Chem., 360, 935 (1979)

1868. R-Phycocyanin (PC) and C-Phycoerythrin (PE)

SOURCE: a. Alga (Porphyridium cruentum) (PC)
b. Alga (Aphanocapsa species) (PE)
SOLVENT: a. 0.05 M Ammonium acetate buffer, pH 6.8
b. 8 M Urea-10 mM β-mercaptoethanol, pH 1.9
c. 0.02 M Ammonium acetate buffer, pH 7.0
d. 0.015 M Ammonium acetate buffer, pH 7.0, containing 1% (W/V) sodium dodecylsulfate
e. 7.8 M Urea, pH 3.0
f. 9.0 M Urea, pH 3.0
COMMENTS: 1. PE in solvents a and b
2. PC in solvents b, c, and d
3. β-Subunit of PC in solvents e and f
REFERENCE: A. N. Glazer and C. S. Hixson, J. Biol. Chem., 250, 5489 (1975)

1869. Phycocyanin and Phycoerythrin

SOURCE: Alga (Anacystis nidulans)
SOLVENT: 0.1 M Phosphate buffer, pH 7.0-ethylene glycol, 1:1 (V/V)
COMMENTS: Spectra taken at 77°K, 195°K and 293°K
REFERENCE: D. Frackowiak, K. Fiksinski, and J. Grabowski, Photosynthetica, 9, 185 (1975)

1870. Phycocyanin

SOURCE: Synechococcus species
SOLVENT: a. 0.2 M Sodium acetate, 0.001 M β-mercaptoethanol, pH 5.5
b. 0.05 M Sodium phosphate, 0.001 M β-mercaptoethanol, pH 7.0

COMMENTS: 1. Spectrum of hexameric protein in solvent a
2. Spectra of protein and its α- and β-subunits in solvent b
REFERENCE: A. N. Glazer, S. Fang and D. M. Brown, J. Biol. Chem., 248, 5679 (1973)

1871. Phycocyanin and α- and β-subunits

SOURCE: Cyanidium caldarium
SOLVENT: a. 8 M and 9 M Urea
b. Phosphate buffer, pH 6.0
COMMENTS: Spectra of protein and its α- and β-subunits
REFERENCE: L. F. Troxler, J. A. Foster, A. S. Brown, and C. Franzblau, Biochemistry, 14, 268 (1975); O. H. W. Kao, M. R. Edwards, and D. S. Berns, Biochem. J., 147, 63 (1975)

1872. C-Phycocyanin

SOURCE: Blue-green alga (Coccochloris elabens and Lyngbya species)
SOLVENT: 0.1 I Sodium phosphate buffer, pH 6.0
COMMENTS: 1. Spectra of whole cell suspensions and of purified protein of Coccochloris elabens
2. Spectrum of whole cell suspension of Lyngbya species
REFERENCE: O. H. W. Kao, D. S. Berns, and W. R. Town, Biochem. J., 131, 39 (1973)

1873. Phycocyanin (PC) and Chlorophyllin (CP)

SOURCE: Alga (Anacystis nidulans)
SOLVENT: 0.1 M Phosphate buffer, pH 7, plus glycerine, 1:1 (V/V)
COMMENTS: Spectra of PC and CP and of mixtures of the two proteins at 77°K and 293°K
REFERENCE: J. Grabowski, Photosynthetica, 6, 291 (1972)

1874. Phycocyanin

SOURCE: Alga (Chroomonas species)
SOLVENT: a. 9 M Urea, pH 2.0
b. 10 mM Phosphate buffer, .2 M NaCl, pH 2.0
COMMENTS: 1. Spectra of four components in solvent a
2. Spectra of native protein (solvent b) and denatured protein (solvent a)
REFERENCE: J. Jung, P. S. Song, R. J. Paxton, M. S. Edelstein, R. Swanson, and E. E. Hazen, Jr., Biochemistry, 19, 24 (1980)

1875. Phycocyanin and Hybrids (PC)

SOURCE: a. Synechococcus species, strain 6301
b. Aphanocapsa species, strain 6701
c. Anabaena species, strain 6411
SOLVENT: a. 0.20 M Phosphate buffer, 0.001 M β-mercapto-ethanol, pH 7.0
b. 8 M Urea, pH 3.0
COMMENTS: 1. Spectra of native proteins, PC, from sources a, b, and c in solvent a
2. Spectra of hybrids prepared from α- and β-chains of different PC's in solvent a
3. Spectra of PC and α- and β-subunits of source a in solvent b
REFERENCE: A. N. Glazer and S. Fang, J. Biol. Chem., 248, 663 (1973)

1876. C-Phycocyanin and α- and β-Subunits

SOURCE: Blue-green alga (Oscillatoria agardhii)
SOLVENT: 2 M NaCl, pH 7.0
REFERENCE: P. A. Torjesen and K. Sletten, Biochim. Biophys. Acta, 263, 258 (1972)

1877. C-Phycocyanin and α- and β-Subunits

SOURCE: Blue-green alga (Mastigocladus laminosus)
SOLVENT: 0.05 M Phosphate buffer, pH 8.0
REFERENCE: A. Binder, K. Wilson, and H. Zuber, FEBS Lett., 20, 111 (1972)

1878. Phycoerythrin I and II

SOURCE: a. Alga (Acrochaetium virgatulum)
b. Red Alga (Hypoglossum woodwardii)
COMMENTS: Spectral changes due to exposure to white light in presence of N-allylthiourea (Source b proteins)
REFERENCE: H. H. Van der Velde, Biochim. Biophys. Acta, 303, 246 (1973)

1879. Phycoerythrin (PE) and Phycocyanin (PC)

SOURCE: a. Alga (Rhodomonas lens) (PE)
b. Alga (Chroomonas species) (PC)
COMMENTS: Spectra of α- and β-subunits taken on unstained bands in SDS gels
REFERENCE: R. Maccoll, D. S. Berns, and O. Gibbons, Arch. Biochem. Biophys., 177, 265 (1976)

1880. Phycoerythrin

SOURCE: a. Alga (Cryptomonas species)
b. Blue green alga (Aphanocapsa species)
SOLVENT: 0.2 M Sodium acetate, pH 5.50
COMMENTS: 1. Spectra of cell suspension of a
2. Spectra of cell lysate of a
3. Spectra of phycoerythrin of a
4. Spectra of crude extract of b
5. All spectra made before and after illumination at 4°C
REFERENCE: A. N. Glazer, G. Cohen-Bazire, and R. Y. Stanier, Arch. Mikrobiol., 80, 1 (1971)

1881. Phycoerythrin I (PEI) and III (PEIII)

SOURCE: a. Alga (Rhodomonas species strain 3-C)
b. Alga (Cryptomonas ovata varipalustris)
SOLVENT: Citrate-phosphate buffer, pH 5.5
COMMENTS: PEI from source a and PEIII from source b
REFERENCE: C. Brooks and E. Gantt, Arch. Mikrobiol., 88, 195 (1977)

1882. B-Phycoerythrin (BPE), b-Phycoerythrin (bPE), α, β, and γ Subunits of B-Phycoerythrin

SOURCE: Alga (Porphyridium cruentum)
SOLVENT: a. 0.01 M Ammonium acetate containing 0.001 M sodium azide and 0.001 M 2-mercaptoethanol, pH 5.5
b. 8 M Urea, 10 mM β-mercaptoethanol, pH 1.9
COMMENTS: 1. Spectra of BEP and bPE in solvent a, and solvent b
2. Spectra of α, β, and γ-subunits of BPE in solvent b
REFERENCE: A. N. Glazer and C. S. Hixson, J. Biol. Chem., 252, 32 (1977)

1883. Phytochrome, Small and Large

SOURCE: a. Oat (Avena sativa L. cv. Garry)
b. Rye (Secale cereale L. cv. Balbo)
c. Pea (Pisum sativum L. cv. Alaska)
SOLVENT: 60 mM Potassium phosphate, pH 7.8
COMMENTS: 1. All spectra recorded after saturating far red light, saturating red light, and after 5 min. darkness at 3°C after saturating red light
2. Spectra of large oat protein in presence of 2-mercaptoethanol, ferric chloride, or copper chloride in the presence and absence of EDTA

REFERENCE: L. H. Pratt and S. C. Cundiff, Photochem. Photobiol., 21, 91 (1975); L. H. Pratt, Photochem. Photobiol., 21, 99 (1975)

1884. Phytoflavin

SOURCE: Blue-green alga (Anacystis nidulans)
SOLVENT: 5 mM Sodium phosphate, pH 7.5
COMMENTS: Spectra of quinone, semiquinone, and hydroquinone forms
REFERENCE: B. Entsch and R. M. Smillie, Arch. Biochem. Biophys., 151, 378 (1972); R. M. Smillie, Plant Physiol., 40, 1124 (1965)

1885. Pigment, CO-binding, P582

SOURCE: Desulfotomaculum nigrificans
SOLVENT: 0.1 M Tris-HCl buffer, pH 8.0, containing 0.5 M KCl and 1 mM EDTA
COMMENTS:
1. Spectrum of oxidized P582
2. Spectra of reduced (borohydride and sulfite) and CO-reduced forms
3. Effects of dithionite and dithionite-reduced methyl viologen on spectrum of P582
4. Spectrum of reduced (methyl viologen) P582 plus potassium nitrite

REFERENCE: P. A. Trudinger and L. A. Chambers, Biochim. Biophys. Acta, 293, 26 (1973)

1886. Pigments, Visual

SOURCE: a. Eledone moschata
b. Ascalaphus macaronius
SOLVENT: 2% Digitonin in 0.067 M sodium phosphate buffer, pH 6.0
COMMENTS: Spectra of P470 from source a and P345 from source b before and after irradiation
REFERENCE: R. Paulsen and J. Schwemer, Eur. J. Biochem., 40, 577 (1973)

1887. Pigments, Visual

SOURCE: Gekko gekko
SOLVENT: 2% digitonin, borate buffer, pH 8.5
COMMENTS: Spectral changes due to p-mercuribenzoate and reversal by dithiothreitol
REFERENCE: F. Crescitelli, Vision Res., 15, 743 (1975)

1888. Pigments, Visual (Porphyropsins)

SOURCE: a. Channel catfish (Ictalurus punctatus)
b. Redear sunfish (Lepomis microlophus)
c. Bluegill sunfish (Lepomis macrochirus)
SOLVENT: 2% Digitonin
REFERENCE: J. G. Parkes, J. H. Rockey, and P. A. Leibman, Biochim. Biophys. Acta, 428, 1 (1976)

1889. Plantacyanin, Oxidized

SOURCE: Spinach (Spinacea oleracea)
REFERENCE: V. T. Aikazyan and R. M. Nalbandyan, FEBS Lett., 55, 272 (1975)

1890. Plastocyanin

SOURCE: Wheat leaves
REFERENCE: A. A. Mutuskin, K. V. Pshenova, S. K. Alekhina, and P. A. Kolesnikov, Biokhimiya (Biochemistry), 36, 236 (1971)

1891. Plastocyanin

SOURCE: Parsley
COMMENTS: Spectra of holoplastocyanin, apoplastocyanin, and reconstituted plastocyanin
REFERENCE: M. T. Graziani, A. F. Agro, G. Rotilio, D. Barra, and B. Mondovi, Biochemistry, 13, 804 (1974)

1892. Plastocyanin, Co (II), Stellacyanin, and Azurin

SOURCE: Bean leaves (Phaseolus vulgaris)
COMMENTS: Spectra of cobalt (II) derivatives of plastocyanin, stellacyanin and azurin
REFERENCE: D. R. McMillin, R. C. Rosenberg, and H. B. Gray, Proc. Nat. Acad. Sci., 71, 4760 (1974)

1893. Plastocyanin

SOURCE: Spinach
SOLVENT: pH 7.5
COMMENTS: Spectra of holoplastocyanin, apoplastocyanin and reconstituted plastocyanin
REFERENCE: M. H. Siegelman, I. Rasched, and P. Boger, Biochem. Biophys. Res. Comm., 65, 1456 (1975)

1894. Plastocyanin

SOURCE: Alga (Scenedesmus)
COMMENTS: Spectra of holoplastocyanin, apoplastocyanin and reconstituted plastocyanin
REFERENCE: M. H. Siegelman, I. R. Rasched, K. J. Kunert, P. Kroneck, and P. Boger, Eur. J. Biochem., 64, 131 (1976)

1895. Plastocyanin and Ferredoxin

SOURCE: Ceratophylum demersum
REFERENCE: E. P. Oganezova and R. M. Nalbandyan, Biokhimiya (Biochemistry), 41, 651 (1976)

1896. Plastocyanin

SOURCE: Goosefoot (Chenopodium album)
SOLVENT: a. Water
b. Dimethylsulfoxide
REFERENCE: E. P. Oganezova and R. M. Nalbandyan, FEBS Lett., 82, 147 (1977)

1897. Poly (ADP-ribose) Synthetase

SOURCE: Calf thymus
SOLVENT: 50 mM Tris-Cl (pH 8.0) containing 10% glycerol, 0.2 M NaCl, and 10 mM 2-mercaptoethanol
REFERENCE: S. Ito, Y. Shizuta, and O. Hayaishi, J. Biol. Chem., 254, 3647 (1979)

1898. Polynucleotide Kinase

SOURCE: Bacteriophage T_4
SOLVENT: 20 mM Tris-HCl, pH 8.0, containing 5 mM 2-mercaptoethanol and saturated with nitrogen
REFERENCE: J. R. Lillehaug, Eur. J. Biochem., 73, 499 (1977)

1899. Polypeptides

SOURCE: n-Butanol soluble material from Streptomyces albus membranes
REFERENCE: V. Larraga, M. Nieto, and E. Munuz, Biochim. Biophys. Acta, 363, 26 (1974)

1900. Polypeptide, Fluorescent

SOURCE: Human lens
SOLVENT: pH 7.0
REFERENCE: A. Spector, D. Roy, and J. Stauffer, Exp. Eye Res., 21, 9 (1975)

1901. Polyvinyl Alcohol-degrading Enzyme

SOURCE: Pseudomonas strain
SOLVENT: Water
REFERENCE: Y. Watanabe, Y. Tsujisaka, N. Hamada, and M. Morita, Arch. Biochem. Biophys., 174, 575 (1976)

1902. Porphyringlobin

SOURCE: Human
SOLVENT: 0.05 M Potassium phosphate buffer, pH 7.0
COMMENTS: Effect of illumination on spectrum (500 W Xenon arc: 0-285 minutes)
REFERENCE: A. Treffry and S. Ainsworth, Biochem. J., 137, 319 (1974)

1903. Prealbumin

SOURCE: Rat brain
REFERENCE: A. N. Krichevskaya, A. A. Sinichkin, V. N. Prokofev, V. V. Kholobystov, and N. A. Kucherenko, Biokhimiya (Biochemistry), 42, 177 (1977)

1904. Prealbumin, 2-Hydroxy-5-nitrobenzyl Modified

SOURCE: Human plasma
REFERENCE: J. Horowitz and J. Heller, J. Biol. Chem., 249, 7181 (1974)

1905. Procarboxypeptidase, Cobalt and Carboxypeptidase, Cobalt

SOURCE: Beef
SOLVENT: 5 mM Tris-HCl-1.0 M NaCl, pH 7.1
REFERENCE: W. D. Behnske and B. L. Vallee, Proc. Nat. Acad. Sci., 69, 2442 (1972)

1906. Progesterone Receptor Subunit B

SOURCE: Chick oviduct
SOLVENT: 0.01 M Potassium phosphate, pH 7.4, containing 0.1 M KCl
COMMENTS: Spectra of B-subunit-progesterone complex after isolation, after dialysis against buffer, and buffer plus guanidine
REFERENCE: R. W. Kuhn, W. T. Schrader, W. A. Coty, P. M. Conn, and B. W. O'Malley, J. Biol. Chem., 252, 308 (1977)

1907. Proline Reductase

SOURCE: Clostridium sticklandii
SOLVENT: 0.05 M N-Tris (hydroxymethyl) methylglycine-KOH buffer, pH 8.6
REFERENCE: B. Seto and T. C. Stadtman, J. Biol. Chem., 251, 2435 (1976)

1908. Protein A

SOURCE: Staphylococcus aureus
SOLVENT: a. Water
b. pH 7.3 in 0.1 M KCl
REFERENCE: J. Sjoquist, B. Meloun, and H. Hjelm, Eur. J. Biochem., 29, 572 (1972); I. Sjoholm, Eur. J. Biochem., 51, 55 (1975)

1909. Protein A and Protein B

SOURCE: Bacillus megaterium spores
SOLVENT: 10 mM Potassium phosphate, pH 7.4
REFERENCE: P. Setlow, J. Biol. Chem., 250, 8168 (1975)

1910. Protein B and Protein B*

SOURCE: Escherichia coli outer membrane
SOLVENT: 1 mM Tris-acetate buffer, pH 7.0 containing 0.2% sodium dodecylsulfate
REFERENCE: R. A. F. Reithmeier and P. D. Bragg, Arch. Biochem. Biophys., 178, 527 (1977)

1911. C-Protein

SOURCE: Rabbit leg muscle
SOLVENT: 0.1 M NaF, 10 mM Potassium phosphate, pH 7.0
REFERENCE: G. Offer, C. Moos, and R. Starr, J. Mol. Biol., 74, 653 (1973)

1912. HU-Protein

SOURCE: Escherichia coli
SOLVENT: 20 mM Potassium phosphate buffer, pH 7.0, containing 100 mM KCl
COMMENTS: Spectrum of aspartokinase homoserine I dehydrogenase I also given
REFERENCE: J. Rouviere-Yaniv and F. Gros, Proc. Nat. Acad. Sci., 72, 3428 (1975)

1913. I-Protein

SOURCE: Chicken breast muscle
SOLVENT: 0.1 M KCl
COMMENTS: Spectral change due to treatment with ribonuclease
REFERENCE: K. Ohashi, S. Kimura, K. Deguchi, and K. Maruyama, J. Biochem. Tokyo, 81, 233 (1977)

1914. Protein L 7/L12

SOURCE: Escherichia coli
SOLVENT: 10 mM Tris-HCl, 20 mM $MgCl_2$, 50 mM KCl, pH 7.6
REFERENCE: C. A. Luer and K. P. Wong, Biochemistry, 18, 2019 (1979)

1915. Protein, M Component

SOURCE: Escherichia coli
SOLVENT: 0.25 M Phosphate buffer, pH 7.0
COMMENTS: Spectra of M component and M component plus cyanide; component M is methionine synthetase
REFERENCE: K. Fujii and F. M. Huennekens, J. Biol. Chem., 249, 6745 (1974)

1916. R-Protein

SOURCE: Ehrlich ascites tumor
SOLVENT: pH 7.0
REFERENCE: S. Natori, K. Takeuchi, and D. Mizuno, J. Biochem., 77, 1319 (1975)

1917. Protein S1

SOURCE: Escherichia coli MRE600
REFERENCE: R. Linde, N. Q. Khanh and H. G. Gassen, Meth. Enzymol., 60, 417 (1979)

1918. TsnC Protein

SOURCE: Escherichia coli
SOLVENT: 0.1 M Potassium phosphate, pH 7.4
REFERENCE: P. Modrich and C. Richardson, J. Biol. Chem., 250, 5508 (1975)

1919. Protein TU, Protein TS, Protein TU-Protein TS Complex, and Protein TU-Guanosine Diphosphate Complex

SOURCE: Escherichia coli
SOLVENT: 0.02 M Tris-HCl buffer, pH 7.5, 0.01 M magnesium acetate, and 5 mM 2-mercaptoethanol

REFERENCE: K. I. Arai, M. Kawakita, Y. Kaziro, T. Kondo, and N. Ui, J. Biochem. Tokyo, 73, 1095 (1973)

1920. Protein Z

SOURCE: Rabbit and rat liver
SOLVENT: 10 mM Phosphate buffer, pH 7.4
REFERENCE: Y. Matsushita, H. Umeyama, and I. Moriguchi, Chem. Pharm. Bull., 25, 647 (1977)

1921. Protein Component

SOURCE: Rat liver
SOLVENT: 0.005 M Tris-HCl buffer, pH 7.5
REFERENCE: I. Ishiguro, J. Naito, R. Shinohara, A. Shikura, and T. Terai, Biochim. Biophys. Acta, 365, 148 (1974)

1922. Protein Component 0.62

SOURCE: Merino sheep wool
SOLVENT: a. 8 M Urea, pH 6.0
b. 0.01 M NaOH, pH 12
REFERENCE: M. J. Frenkel, J. M. Gillespie, and E. F. Woods, Eur. J. Biochem., 34, 114 (1978)

1923. Protein, Fraction I, Large and Small Subunits

SOURCE: Tobacco (Nicotiana tabacum)
SOLVENT: pH 11.7
COMMENTS: Spectra of large and small subunits
REFERENCE: N. Kawashima, Y. Tanabe, and S. Iwai, Biochim. Biophys. Acta, 371, 417 (1974)

1924. Protein, Fraction I, Small Subunits

SOURCE: Tobacco (Nicotiana tabacum cv. turkish samsun)
SOLVENT: 0.1 M NaOH
COMMENTS: Spectra of S1 and S2
REFERENCE: J. C. Gray, S. D. Kung, and S. G. Wildman, Arch. Biochem. Biophys., 185, 272 (1978)

1925. Protein, Storage, Fraction I

SOURCE: Momordica charantia
REFERENCE: S. S. L. Li, Experientia, 33, 895 (1977)

1926. Protein, 2.8S

SOURCE: Soybean seeds

REFERENCE: I. A. Vaintraub, and A. D. Shutov, Biokhimiya (Biochemistry), 34, 795 (1969)

1927. Protein S4

SOURCE: Escherichia coli ribosomes
SOLVENT: 10 mM Tris, 20 mM $MgCl_2$, 360 mM KCl, pH 7.6
REFERENCE: S. H. Allen and K. P. Wong, Biophys. J., 23, 359 (1978)

1928. Protein, 22S

SOURCE: Sea urchin (Strongylocentrotus droebachiensis, Strongylocentrotus purpuratus, and Arbacia punctulata)
REFERENCE: R. E. Stephens, J. Cell. Biol., 32, 255 (1967)

1929. Protein, Component 2a

SOURCE: Fish (Tilapia macrochir Boulenger) white muscles
SOLVENT: a. Phosphate-NaCl buffer, pH 6.2
b. 0.1 N NaOH
REFERENCE: A. Piront, G. Hamoir, and R. Crokaert, Arch. Int. Physiol. Biochim., 76 (1968)

1930. Protein, Component 2 and Component 3

SOURCE: Carp, red and white muscles
SOLVENT: a. Phosphate-NaCl buffer, pH 7.3
b. 0.1 N NaOH
REFERENCE: S. Konosu, G. Hamoir, and J. F. Pechere, Biochem. J., 96, 98 (1965)

1931. Protein, Acetylcholine Receptor

SOURCE: Torpedo nobiliana
SOLVENT: 50 mM Potassium phosphate buffer, pH 7.0, in 0.1% Emulphogene BC-720
REFERENCE: W. M. Moore, L. A. Holladay, D. Puett, and R. N. Brady, FEBS Lett., 45, 145 (1974)

1932. Protein, Acidic and Protein, Basic

SOURCE: Cucumber (Cucumis sativus)
REFERENCE: K. A. Markossian, V. T. Aikazayan, and R. M. Nalbandyan, Biochim. Biophys. Acta, 359, 47 (1974)

1933. Protein, Acidic I and Protein, Acidic II

SOURCE: Chick brain
SOLVENT: 0.1 M Potassium phosphate buffer, pH 7.1, containing 1 mM EDTA
REFERENCE: Y. Kato, T. Kato, H. Kasai, T. Okuyama, and K. Uyemura, J. Biochem. Tokyo, 82, 43 (1977)

1934. Protein, Acidic

SOURCE: Beef liver
SOLVENT: 0.1 M Potassium phosphate buffer, pH 7.1, containing 1 mM EDTA
REFERENCE: T. Kato, Y. Kato, H. Kasei, and T. Okuyama, J. Biochem. Tokyo, 82, 1297 (1977)

1935. Protein, Activator of cAMP Diesterase

SOURCE: Beef brain
SOLVENT: a. 0.05 M NaCl, pH 7.5
b. 0.1 M KCl, pH 5.9 and pH 10.3
COMMENTS: Spectra of native protein and nitro derivatives
REFERENCE: P. G. Richman and C. B. Klee, Biochemistry, 17, 928 (1978)

1936. Protein, Activator of cAMP Diesterase

SOURCE: Beef brain
SOLVENT: 0.02 M N-2-Hydroxyethylpiperazine-N'-2-ethane sulfonic acid, buffer, pH 6.7
COMMENTS: Spectra of protein acetylated with N-acetylimidazole in the presence of Ca^{2+} or (EGTA) ethylenebis (oxyethylenenitrolo) tetraacetic acid
REFERENCE: P. G. Richman, Biochemistry, 17, 3005 (1978)

1937. Protein, Activator of Calcium-dependent cGMP Phosphodiesterase

SOURCE: Beef pancreas
SOLVENT: a. 20 mM Tris-HCl, pH 7.5
b. 0.1 M NaOH
REFERENCE: A. Vandermeers, M-C. Vandermeers-Piret, J. Rathe, R. Kutzner, A. Delforge, and J. Christophe, Eur. J. Biochem., 81, 379 (1977)

1938. Protein, Activator of Cyclic 3':5'-Nucleotide Phosphodiesterase

SOURCE: Beef brain
SOLVENT: 20 mM Tris-HCl, pH 8
REFERENCE: Y. P. Lin and W. Y. Cheng, J. Biol. Chem., 251, 4193 (1976)

1938a. Protein, L-Arabinose-binding, Mercurinitrophenol Labeled

SOURCE: Escherichia coli
SOLVENT: 0.1 M Tris-HCl, pH 7.85
REFERENCE: D. M. Miller III, M. E. Newcomer, and F. A. Quiocho, J. Biol. Chem., 254, 7521 (1979)

1939. Protein, azo-, h_2-5S

SOURCE: Rat liver
SOLVENT: a. 0.1 M NaCl plus 0.01 M sodium phosphate, pH 7.4
b. 88% formic acid
REFERENCE: S. Sorof, B. P. Sani, V. M. Kish, and H. P. Meloche, Biochemistry, 13, 2612 (1974)

1940. Protein, 4-azoquinoline-1-oxide Derivatives

SOURCE: 1. Rabbit serum protein
2. α-Amylase of bacteria
3. Beef serum albumin
SOLVENT: Phosphate buffered saline, pH 7.2
REFERENCE: S. Nakashima, K. Sugimura, and S. Ono, J. Biochem. Tokyo, 71, 849 (1972)

1941. Protein, Base-plate Parts

SOURCE: Bacteriophage P22
SOLVENT: 0.01 M Tris-HCl, 0.1 M NaCl, pH 7.4
REFERENCE: S. Iwashita and S. Kanegasaki, Eur. J. Biochem., 65, 87 (1976)

1942. Protein, basic

SOURCE: Pig spinal cord and peripheral nerve
SOLVENT: a. 0.1 M Sodium phosphate, pH 6.8
b. 0.1 M NaOH
REFERENCE: K. Uyemura, C. Tobari, and S. Hirano, Biochem. Biophys. Acta, 214, 190 (1970)

1943. Protein, Chromo- (Blue)

SOURCE: Ovary of a Turban shell (Turbo cornutus)
SOLVENT: 0.05 M Acetate-0.5 M NaCl buffer, pH 5.0
REFERENCE: K. Yamaguchi and T. Ogata, Bull. Jap. Soc. Sci. Fish., 44, 631 (1978)

1944. Protein, Blue

SOURCE: Rice Bran (Oryzae sativa L.)
COMMENTS: Spectra of reduced and oxidized forms
REFERENCE: S. Ida and Y. Morita, Agric. Biol. Chem., 33, 10 (1969)

1945. Protein, Blue Fluorescence

SOURCE: Photobacterium phosphoreum
SOLVENT: 50 mM Phosphate, pH 7.0
REFERENCE: R. Gast and J. Lee, Proc. Nat. Acad. Sci., 75, 833 (1978)

1946. Protein, Cadmium-binding

SOURCE: Mussels (Mytilus edulis L.)
SOLVENT: a. 0.002 M Tris, pH 8.6
b. 0.002 M Tris-HCl, pH 4
REFERENCE: S. G. George, E. Carpene, T. L. Coombs, J. Overnell, and A. Youngson, Biochim. Biophys. Acta, 580, 225 (1979)

1947. Protein, Calcium-binding

SOURCE: Earthworm (Lumbricus terrestris)
SOLVENT: 40 mM Tris, 1 mM magnesium acetate, 50 mM calcium chloride, pH 7.5
REFERENCE: D. M. Waisman, F. C. Stevens, and J. H. Wang, J. Biol. Chem., 253, 1106 (1978)

1948. Protein, Calcium-binding

SOURCE: Crayfish muscle (Astacus pontastacus leptodactylus leptodactylus)
SOLVENT: 50 mM Tris-HCl buffer, pH 8.0
REFERENCE: J. A. Cox, W. Wnuk, and E. A. Stein, Biochemistry, 15, 2613 (1976)

1949. Protein, Calcium-binding

SOURCE: a. Muscle (sarcoplasm) of Amphioxus
b. Sandworm
c. Crayfish
REFERENCE: L. Kohler, J. A. Cox, and E. A. Stein, Mol. Cell. Biochem., 20, 85 (1978)

1950. Protein, Calcium-binding

SOURCE: The intestinal mucosa of beef, pig, chick, horse, and guinea pig
REFERENCE: C. S. Fullemer and R. H. Wasserman, Biochim. Biophys. Acta, 393, 134 (1975)

1951. Protein, Calcium-binding Component of Troponin

SOURCE: Rabbit skeletal muscle
SOLVENT: 0.3 mM Sodium bicarbonate
COMMENTS: Same spectrum in presence and absence of calcium ion
REFERENCE: J. P. Van Eerd and Y. Kawasaki, Biochem. Biophys. Res. Comm., 47, 859 (1972)

1952. Protein, Calcium-binding Component of Troponin

SOURCE: Rabbit white skeletal muscle
SOLVENT: 0.1 M KCl-20 mM Tris-15 mM HCl, pH 7.6-0.1 mM calcium chloride
REFERENCE: J. F. Head and S. V. Perry, Biochem. J., 137, 145 (1974)

1953. Protein, Calcium-binding

SOURCE: Pig intestinal mucosa
SOLVENT: 0.02 M Tris-HCl, pH 8.0; 1.5 mM calcium chloride
REFERENCE: K. J. Dorrington, A. Hui, T. Hofmann, A. J. W. Hitchman, and J. E. Harrison, J. Biol. Chem., 249, 199 (1974)

1954. Protein, Caroteno-

SOURCE: Salps (Salpa cylindrica)
SOLVENT: a. 0.2 M Phosphate buffer
b. 5 M Urea and 7 M urea
REFERENCE: P. J. Herring, Comp. Biochem. Physiol., 613, 391 (1978)

1955. Protein - Chlorophyll Complex

SOURCE: Alga (Phormidium luridum var. olivaceae UTEX 426), Maize, and Barley Chlarina F2
REFERENCE: S. Reinman and J. P. Thornber, Biochim. Biophys. Acta, 547, 188 (1979)

1956. Protein - Chlorophyll Complex

SOURCE: Tobacco (Nicotiana tabacum L.)
REFERENCE: F. Henriques and R. B. Park, Plant Physiol., 62, 856 (1978)

1957. α-Protein, P-700-Chlorophyll-

SOURCE: Tobacco chloroplast lamellae
REFERENCE: J. A. Shiozawa, R. S. Alberte, and J. P. Thornber, Arch. Biochem. Biophys., 165, 888 (1974)

1958. Protein, Peridinin-Chlorophyll a-

SOURCE: Gleodinium species and Gonyaulax polyedra
COMMENTS: Spectra of native and denatured material
REFERENCE: P. S. Song, P. Koka, B. B. Prezelin, and F. T. Haxo, Biochemistry, 15, 4422 (1976)

1959. Protein-Chlorophyll Complex

SOURCE: Spinach and barley thylakoids
COMMENTS: Spectra of chlorophyll-protein complex 1 (also named P700-chlorophyll a-protein), an oligomer of this, and two oligomers of the light-harvesting chlorophyll a/b-protein complex
REFERENCE: J. M. Anderson, J. C. Waldron, and S. Thorne, FEBS Lett., 92, 227 (1978)

1960. Protein-Chlorophyll a/b Complex

SOURCE: Pea seeds (Pisum sativum var. Morse's Progress No. 9)
REFERENCE: J. J. Burke, C. L. Ditto, and C. J. Arntzen, Arch. Biochem. Biophys., 187, 252 (1978)

1961. Protein-Bacteriochlorophyll Complex

SOURCE: Rhodopseudomonas sphaeroides, strain 241 and strain R-26, and Rhodospirillum rubrum
COMMENTS: Spectra of chromatophores, reaction center particles and light-harvesting complex
REFERENCE: K. Sauer and L. A. Austin, Biochemistry, 17, 2011 (1978)

1962. Protein-Chlorophyll Complex

SOURCE: Pea leaves (Pisum sativum)
REFERENCE: B. Scott and R. P. F. Gregory, Biochem. J., 149, 341 (1975)

1963. Protein-Chlorophyll Complex

SOURCE: Rhodopseudomonas spheroides
REFERENCE: P. J. Fraker and S. Kaplan, J. Biol. Chem., 247, 2732 (1972)

1964. Protein, α-Clot-, and β-Clot-

SOURCE: Limulus lysate
REFERENCE: S. L. Gaffin, Biorheology, 13, 273 (1976)

1965. Protein, Contractile

SOURCE: Rat liver mitochondria
COMMENTS: Spectra of protein from rats of different ages
REFERENCE: A. N. Razumovich and T. G. Lasfovskaya, Biokhimiya (Biochemistry), 36, 44 (1971)

1966. Protein, Copper-

SOURCE: Alcaligenes faecalis
SOLVENT: 0.05 M Phosphate buffer, pH 7.0
COMMENTS: Spectra of oxidized and reduced (dithionite) forms
REFERENCE: T. Matsubara and H. Iwasaki, J. Biochem. Tokyo, 71, 747 (1972)

1967. Protein, Copper-, Acidic

SOURCE: Cucumber skin
REFERENCE: V. T. Aikazyan, K. A. Markosyan, and R. M. Nalbandyan, Biokhimiya (Biochemistry), 39, 982 (1974)

1968. Protein, Copper-binding Immunoglobulin, Fab Piece

SOURCE: Human serum
SOLVENT: 0.01 M Sodium phosphate, pH 7.2, containing 0.04 M NaCl
COMMENTS: Spectra of copper-containing protein and apo-protein
REFERENCE: B. L. Baker and D. E. Hultquist, J. Biol. Chem., 253, 84-4 (1978)

1969. Protein, Corrinoid

SOURCE: Clostridium thermoaceticum
SOLVENT: a. 0.05 M Tris, pH 7.6
b. pH 10-11
COMMENTS: 1. Spectral changes due to cyanide, and cyanide and heat (solvent b)

2. Effect of ammonium sulfate on spectrum in solvent a

REFERENCE: L. G. Ljungdahl, J. LeGall, and J. P. Lee, Biochemistry, 12, 1802 (1973)

1970. Protein, Cortex

SOURCE: Anthocidaris
SOLVENT: 0.6 M KCl-10 mM Phosphate buffer, pH 7.2
COMMENTS: Spectra made before and after DEAE-cellulose treatment
REFERENCE: I. Mabuchi and H. Saka, Devel. Growth and Different., 14, 247 (1972)

1971. Protein, Electron-transferring Flavoprotein (ETF)

SOURCE: Peptostreptococcus elsdenii
SOLVENT:
a. 0.1 M Sodium acetate buffer, pH 5.5, containing 0.0003 M EDTA
b. 0.1 M Potassium phosphate buffer, pH 6.0, containing sodium dithionite at 10°C
c. 0.04 M Sodium pyrophosphate, pH 8.5, containing 0.05 M EDTA

COMMENTS:
1. Spectra of ETF and ETF plus FAD in solvent a
2. Spectra of ETF in solvent b and after reoxidation by ferricyanide
3. Spectral changes due to photoirradiation (0-980 minutes) in solvent c
4. Spectral changes due to the addition of NADH in solvent a

REFERENCE: C. O. Whitfield, and S. G. Mayhew, J. Biol. Chem., 249, 2801 (1974)

1972. Protein, Envelope

SOURCE: Escherichia coli
SOLVENT:
a. 1 mM Sodium bicarbonate
b. 6 M Guanidinium chloride
c. 0.1 N NaOH

REFERENCE: J. R. Rosenbusch, J. Biol. Chem., 249, 8019 (1974)

1973. Protein, apoflavo-

SOURCE: Egg yolk
SOLVENT:
a. 0.5 M Bicarbonate buffer, pH 8.45, containing 10% dioxane
b. 0.1 M Acetate buffer, pH 4.0

COMMENTS:
1. Spectral changes due to the addition of hydrogen peroxide, solvent a

2. Spectral changes due to the addition of N-bromosuccinimide, solvent b

REFERENCE: J. Steckzko and W. Ostrowski, Biochim. Biophys. Acta, 393, 253 (1975)

1974. Protein, Flavohemo-

SOURCE: Alcaligenes eutrophus

SOLVENT: 50 mM Tris-HCl buffer, pH 7.5

COMMENTS: 1. Spectra of oxidized, reduced (dithionite) and CO-reduced (dithionite)

2. Spectral changes due to addition of NADH

REFERENCE: I. Probst, G. Wolf, and H. G. Schlegel, Biochim. Biophys. Acta, 576, 471 (1979)

1975. Protein, Folch-Pi Apo-

SOURCE: Beef brain white matter

SOLVENT: Water

REFERENCE: B. De Foresta, T. Nguyen Le, C. Nicot, and A. Alfsen, Biochimie, 61, 523 (1979)

1976. Protein, Fructose-reactive

SOURCE: Nurse shark (Ginglymostoma cirratum) serum

SOLVENT: a. 0.10 N NaOH

b. 0.10 N HCl

c. 0.01 M Phosphate buffered saline

d. 10% fructose solution

REFERENCE: V. Harisdangkul, E. A. Kabta, R. J. McDonough, and M. M. Siegel, J. Immunol., 108, 1259 (1972)

1977. Protein, Gel

SOURCE: Horseshoe crab (Limulus polyphenus) blood cells

SOLVENT: a. 0.189 N HCl

b. 0.188 N NaOH

c. pH 7.28

REFERENCE: N. O. Solum, Thromb. Diathesis Haem., 23, 170 (1970); J. Levin and F. B. Bang, Thromb. Diathesis Haem., 19, 186 (1968)

1978. Protein, Green Hemo-

SOURCE: Beef erythrocytes

SOLVENT: 0.05 M Potassium phosphate buffer, pH 7.2

COMMENTS: 1. Pyridine hemochrome spectra of forms I and II-

2. Spectra of oxidized, reduced (dithionite), CO-reduced (dithionite), and oxidized-cyanide forms of Forms I and II

REFERENCE: L. J. Defilippi and D. E. Hultquist, J. Biol. Chem., 253, 2946 (1978); L. J. Defilippi and D. E. Hultquist, J. Biol. Chem., 253, 2954 (1978)

1979. Protein, Green Fluorescent

SOURCE: Aequorea
SOLVENT: 10 mM Sodium phosphate buffer, pH 7.0
REFERENCE: H. Morise, O. Shimomura, F. H. Johnson, and J. Winant, Biochemistry, 13, 2656 (1974)

1980. Protein, High Potential Iron

SOURCE: Chromatium
SOLVENT: a. 20 mM Tris-HCl, pH 9.0
b. Solvent a in 80% dimethylsulfoxide
COMMENTS: Spectra of oxidized and reduced (dithionite) forms
REFERENCE: R. Cammack, Biochem. Biophys. Res. Comm., 54, 548 (1973); P. J. Stephens, A. J. Thomson, J. B. R. Dunn, T. A. Keiderling, J. Rawlings, K. K. Rao, and D. O. Hall, Biochemistry, 17, 4770 (1978)

1981. Protein, Histidine-rich Glycoprotein, 3.8S

SOURCE: Human serum
SOLVENT: 0.1 M Sodium phosphate, pH 7.1
REFERENCE: W. T. Morgan, Biochim. Biophys. Acta, 535, 319 (1978)

1982. Protein, Hook

SOURCE: Salmonella 5J25
SOLVENT: 0.15 M NaCl - 0.01 M Tris-HCl buffer, pH 8.2
REFERENCE: H. Kagawa, K. Owaribe, S. Asakura, and N. Takahashi, J. Bacter., 125, 68 (1978)

1983. Protein, Inactivator of Uridine Nucleosidase

SOURCE: Baker's yeast
REFERENCE: P. Natalini, I. Santarelli, S. Ruggieri, and A. Vita, in Limited Proteolysis in Microorganisms (G. N. Cohen and H. Holzer, eds.), 87 (1978), DHEW Publication No. (NIH), 79-1591

1984. Protein, Two-Iron

SOURCE: Desulfovibrio gigas
SOLVENT: 0.01 M Tris-HCl buffer, pH 7.6
REFERENCE: I. Moura, A. V. Xavier, R. Cammack, M. Bruschi, and J. LeGall, Biochim. Biophys. Acta, 533, 156 (1978)

1985. Protein, Iron

SOURCE: Clostridium pasteurianum
SOLVENT: 50 mM Tris-HCl, pH 8.0
COMMENTS: Spectra of oxidized and reduced (dithionite) forms
REFERENCE: T. Ljones, Biochim. Biophys. Acta, 321, 103 (1973)

1986. Protein, Molybdenum-Iron

SOURCE: Soybean bacteroids
SOLVENT: 0.02 M N-Tris-(hydroxymethyl) methyl-2-aminoethane sulfonic acid, pH 7.5, containing 0.25 M NaCl, and 1 mM dithiothreitol
COMMENTS: Spectra of oxidized and reduced (dithionite) forms
REFERENCE: D. W. Israel, R. L. Howard, H. J. Evans, and S. A. Russell, J. Biol. Chem., 249, 500 (1974)

1987. Protein, Iron-chlorin

SOURCE: Aspergillus niger
SOLVENT: 0.1 M Phosphate buffer, pH 7.0
COMMENTS: Spectra of ferric form in the presence and absence of cyanide
REFERENCE: S. Horie, T. Watanabe, and S. Nakamura, J. Biochem. Tokyo, 80, 579 (1976)

1988. Protein, Iron-sulfur

SOURCE: Pseudomonas species
SOLVENT: 0.01 M Tris-HCl buffer, pH 7.5
REFERENCE: T. Matsumoto, J. Tobari, K. Suzuki, K. Kimura, and T. T. Chen, J. Biochem. Tokyo, 79, 937 (1976)

1989. Protein, Iron-sulfur

SOURCE: Desulfovibrio gigas
REFERENCE: J. J. G. Moura, A. V. Xavier, M. Buschi, J. LeGall, D. O. Hall, and R. Cammack, Biochem. Biophys. Res. Comm., 72, 782 (1976)

1990. Protein, Iron-sulfur II

SOURCE: Azotobacter vinelandii
COMMENTS: Spectra of oxidized and reduced (dithionite) forms
REFERENCE: G. Scherings, H. Haaker, and C. Veeger, Eur. J. Biochem., 77, 621 (1977)

1991. Protein, Iron-sulfur

SOURCE: Chromatium, Bacillus stearothermophilus, and Clostridium pasteurianum
REFERENCE: P. J. Stephens, A. J. Thomson, T. A. Keiderling, J. Rawlings, K. K. Rao, and D. O. Hall, Proc. Nat. Acad. Sci., 75, 5273 (1978)

1992. Protein, Iron-sulfur

SOURCE: Beef kidney cortex outer membrane
COMMENTS: Spectra of oxidized and reduced (dithionite) forms
REFERENCE: D. Backstrom, M. Lorusso, K. Anderson, and A. Ehrenberg, Biochim. Biophys. Acta, 502, 276 (1978)

1993. Protein, Molybdenum-iron-sulfur

SOURCE: Desulfvibrio afrikanus
SOLVENT: 50 mM Phosphate buffer, pH 7.6
REFERENCE: E. C. Hatchikian and M. Bruschi, Biochem. Biophys. Res. Comm., 86, 725 (1979)

1994. Protein

SOURCE: Human lens nuclei
SOLVENT: 1% Ammonium bicarbonate
COMMENTS: Normal and cataractous lens were used
REFERENCE: K. J. Dilley and A. Pirie, Exp. Eye Res., 19, 59 (1974)

1995. Protein

SOURCE: Human lens
SOLVENT: 0.01 M Phosphate buffer, pH 7.4
COMMENTS: Effect of exposure to sun in presence and absence of fluorescent compounds on spectrum
REFERENCE: R. Van Heyningen, Exp. Eye Res., 17, 137 (1973)

1996. Protein, Luciferin-binding

SOURCE: Renilla reniformis
SOLVENT: 10 mM Tris, 0.1 M KCl, 1 mM EDTA, pH 7.2
COMMENTS: Effect of calcium ion on spectrum
REFERENCE: H. Charbonneau and M. J. Cormier, J. Biol. Chem., 254, 769 (1979); J. M. Anderson, H. Charbonneau, and M. J. Cormier, Biochemistry, 13, 1195 (1974)

1997. Protein, Major

SOURCE: Escherichia coli
COMMENTS: This protein is released by osmotic shock
REFERENCE: G. R. Jacobson, B. J. Takacs, and J. R. Rosenbusch, Biochemistry, 15, 2297 (1976)

1998. Protein, Maltose-binding

SOURCE: Escherichia coli K12
SOLVENT: 10 mM Tris-HCl, pH 7.1, 2 mM $MgCl_2$ and 0.25 mM 2-mercaptoethanol
REFERENCE: O. Kellermann and S. Szmelcman, Eur. J. Biochem., 47, 139 (1974)

1999. Protein, SDS-binding

SOURCE: Chromatophores (Rhodospirillum rubrum)
SOLVENT: 0.1 M Sodium phosphate buffer, pH 7.0, containing 0.1% SDS
REFERENCE: J. Miyake, S. Ochiai-Yanagi, T. Kasumi, and T. Takagi, J. Biochem. Tokyo, 83, 1679 (1978)

2000. Protein, Membrane

SOURCE: Human erythrocyte membrane
SOLVENT: a. 3% Sodium laurylsulfate
b. Hexafluoroacetone
REFERENCE: R. L. Juliano, Biochim. Biophys. Acta, 266, 301 (1972)

2001. Protein, Mercury-binding

SOURCE: Eel liver
SOLVENT: 0.05 M Ammonium bicarbonate
COMMENTS: Spectra in presence and absence of mercury
REFERENCE: J. M. Bouquegneau, C. Gerday, and A. Distreche, FEBS Lett., 55, 173 (1975)

2002. Protein, Modulator

SOURCE: Beef brain
SOLVENT: 0.10 M Imidazole·HCl and 0.001 M EDTA, pH 7.0
REFERENCE: D. M. Watterson, W. G. Harrelson, P. M. Keller, F. Sharief, and T. C. Vanaman, J. Biol. Chem., 251, 4501 (1976)

2003. Protein, Modulator (17K)

SOURCE: Chicken gizzard
SOLVENT: 10 mM Tris-HCl, pH 7.6
REFERENCE: R. Dabrowska, J. M. F. Sherry, D. K. Aromatorio, and D. J. Hartshorne, Biochemistry, 17, 253 (1978)

2004. Protein, Modulator

SOURCE: Rabbit skeletal muscle
SOLVENT: 0.1 M NaCl, 10 mM $MgCl_2$, 20 mM histidine-KOH, pH 6.8, containing 0.3 mM ethylene glycol-bis (2-aminoethylether)-N,N,N',N'-tetraacetic acid
COMMENTS: Spectra in presence and absence of added calcium ion
REFERENCE: M. Yazawa, H. Kuwayama, and K. Yagi, J. Biochem. Tokyo, 84, 1253 (1978)

2005. Protein

SOURCE: Chicken skeletal muscle
SOLVENT: a. pH 7.3
b. pH 13
REFERENCE: H. Finck, Biochim. Biophys. Acta, 111, 208 (1965)

2006. Protein, Nicotinic Acid Receptor

SOURCE: Electric eel (Electrophorus electricus)
SOLVENT: 10 mM Sodium phosphate buffer, pH 7.1-50 mM NaCl and 0.02% Sodium azide
COMMENTS: Spectrum of detergent-free protein
REFERENCE: J. C. Meunier, R. Sealock, R. Olsen, and J. P. Changeux, Eur. J. Biochem., 45, 371 (1974)

2007. Protein, Nucleo-, II

SOURCE: Pig eye
SOLVENT: 0.01 N HCl
REFERENCE: D. Glasser, E. Weber, J. Frenzel, and H. Hanson, FEBS Lett., 56, 124 (1975)

2008. Protein, Paramagnetic

SOURCE: Clostridium pasteurianum
SOLVENT: a. 0.5 M Phosphate buffer, pH 7.2
b. Tris-HCl buffer, pH 8.1
c. Phosphate buffer, pH 6.5
COMMENTS: 1. Spectra of oxidized and reduced (hydrogenase-ferredoxin-hydrogen) forms in solvent a

2. Spectral changes due to reduction by enzyme systems and dithionite in solvent b
3. Spectral changes due to pyruvate reduction of protein in solvent c

REFERENCE: J. Cardenas, L. E. Mortenson, and O. C. Yoch, Biochim. Biophys. Acta, 434, 244 (1976)

2009. Proteins, Phenylalanine-rich Acidic

SOURCE: Beef brain
SOLVENT: 0.1 M Potassium phosphate buffer, pH 7.1, containing 1 mM EDTA and 0.4 M NaCl
COMMENTS: Spectra of three different proteins
REFERENCE: T. Isobe, T. Nakajima, and T. Okuyama, Biochim. Biophys. Acta, 494, 222 (1977)

2010. Protein, Phosphoryl-group Carrying

SOURCE: Escherichia coli
SOLVENT: 50 mM KCl, pH 7
REFERENCE: E. F. Rossien, G. Dooijewaard and G. T. Robillard, Biochemistry, 18, 5793 (1979)

2011. Protein, Phospho-

SOURCE: Rat liver and spleen
REFERENCE: I. F. Paskevich, I. V. Stolyarov, and V. S. Likhoded, Biokhimiya (Biochemistry), 38, 669 (1974)

2012. Protein, Phospho-

SOURCE: Rat incisor
COMMENTS: Effect of ATP on spectrum
REFERENCE: W. T. Butler, W. T. Hall, and W. S. Richardson, Biochim. Biophys. Acta, 427, 262 (1976)

2013. Protein, Photo-, (Mnemiopsin-2)

SOURCE: Mnemiopsis species
SOLVENT: 0.01 M Tris, pH 8.0 containing 0.001 M EDTA and 0.01 M NaCl
COMMENTS: Spectra in presence and absence of calcium ion
REFERENCE: W. W. Ward and H. H. Seliger, Biochemistry, 13, 1500 (1974)

2014. Protein, Photochlorophyll

SOURCE: Cucumber (Cucumus sativus cv. 'long vert de chine') seedlings
COMMENTS: Low temperature (-196°C) spectra

REFERENCE: B. Elhamouri and C. Sironva, FEBS Lett., 103, 345 (1979)

2015. Protein-Pterin Complex (Phosphodoxin)

COMMENTS: Spectra of freshly isolated and aged complexes
REFERENCE: A. O. Makarov and L. F. Stakhov, Biokhimiya (Biochemistry), 38, 656 (1974)

2016. Protein

SOURCE: Azotobacter chroococcum
SOLVENT: 25 mM N-2-Hydroxyethyl piperazine-N'-2-ethanesulfonic acid buffer, pH 7.8
COMMENTS: Spectra of oxidized and dithionite-reduced forms
REFERENCE: R. L. Robson, Biochem. J., 181, 569 (1979)

2017. Protein, Low Molecular Weight (LMP)

SOURCE: Chicken white breast muscle
COMMENTS: Spectra of LMP, LMP-I, LMP-II, and LMP-III
REFERENCE: S. L. Gaffin and S. Watanabe, J. Mechanochem. Cell Motility, 1, 139 (1972)

2018. Protein, Red

SOURCE: Silkworm (Bombyx mori L.)
SOLVENT: 0.1 M NaCl-0.01 M phosphate buffer, pH 7.0
REFERENCE: J. I. Mukai, J. Inouye, and S. Akune, Agric. Biol. Chem., 33, 125 (1969)

2019. Protein, 9-cis-Retinal-binding

SOURCE: Human plasma
SOLVENT: a. 2 mM Tris, pH 9.0
b. 2 mM Tris, pH 9.0-0.15 M NaCl
COMMENTS: Spectral changes due to prealbumin
REFERENCE: J. Horowitz and J. Heller, J. Biol. Chem., 248, 6317 (1973)

2020. Protein, 11-cis-Retinal-binding

SOURCE: Beef retina
REFERENCE: G. W. Stubbs, J. C. Saari, and S. Futterman, J. Biol. Chem., 254, 8529 (1979)

2021. Protein, Retinoic Acid-binding

SOURCE: Rat testes and beef retina
SOLVENT: 0.05 M Tris-HCl, pH 8.3

REFERENCE: D. E. Ong and F. Chytil, J. Biol. Chem., 253, 4551 (1978); J. C. Saari, S. Futterman and L. Bredberg, J. Biol. Chem., 253, 6432 (1978)

2022. Protein, Retinol-binding

SOURCE: a. Rat plasma
b. Rat testes, cytosol
c. Rat liver
SOLVENT: 0.05 M Tris buffer, pH 7.5 (source c)
REFERENCE: Y. Muto and D. S. Goodman, J. Biol. Chem., 247, 2533 (1972); A. C. Ross, Y. I. Takahashi, and D. S. Goodman, J. Biol. Chem., 253, 6591 (1978); D. E. Ong and F. Chytil, J. Biol. Chem., 253, 828 (1978)

2023. Protein, Retinol-binding

SOURCE: Beef retina
REFERENCE: J. Heller, J. Biol. Chem., 251, 2952 (1976); J. C. Saari, S. Futterman, and L. Bredberg, J. Biol. Chem., 253, 6432 (1978)

2024. Protein, Retinol-binding (RPB)

SOURCE: Human plasma
SOLVENT: a. 0.0033 M Phosphate buffer, pH 7.0, 0.1 M NaCl
b. 2 mM Tris-HCl, pH 9
c. 50 mM Sodium acetate buffer, pH 5.5
d. 1 mM Tris-HCl, pH 7.5
e. 0.05 M Tris, 1 M KCl, pH 7.44
f. pH 11.45
g. pH 12.45
COMMENTS: 1. Spectra of RPB and Apo-RPB (solvent a)
2. Spectra in solvent b, solvents d-g
3. Spectra of native RPB, hydroxynitrobenzyl-modified RPB and hydroxynitrobenzyl-modified Apo-RBP in solvent c
REFERENCE: J. Heller and J. Horowitz, J. Biol. Chem., 248, 6308 (1973); J. Horowitz and J. Heller, J. Biol. Chem., 249, 4712 and 7181 (1974); J. Heller, J. Biol. Chem., 250, 6549 (1975); J. Heller and J. Horowitz, J. Biol. Chem., 250, 3019 (1975)

2025. Protein, trans-3-Dehydroretinol-binding

SOURCE: Human plasma
SOLVENT: 0.066 M Phosphate buffer, pH 7
REFERENCE: J. Horowitz and J. Heller, J. Biol. Chem., 249, 4712 (1974)

2026. Protein, Selenium-containing

SOURCE: Lamb muscle
COMMENTS: Spectral changes due to dithionite
REFERENCE: P. O. Whanger, N. O. Pederson, and P. H. Weswig, Biochem. Biophys. Res. Comm., 53, 1031 (1973)

2027. Protein, Temperature Sensitive

SOURCE: Dog milk
SOLVENT: 0.01 M Phosphate buffer, pH 7.0
REFERENCE: M. J. Swanson and B. E. Sanders, Prep. Biochem., 4, 523 (1974)

2028. Protein, Polyhedron

SOURCE: Cocoons
REFERENCE: E. A. Kozlov, V. M. Sogulyaeva, T. L. Levitina, V. Sereshak, and S. Serebryanyi, Biokhimiya (Biochemistry), 34, 543 (1969)

2029. Protein

SOURCE: Bacteriophage ϕX174
SOLVENT: 0.1 M Sodium borate, pH 8.5
REFERENCE: E. A. Carusi and R. L. Sinsheimer, J. Mol. Biol., 7, 388 (1963)

2030. Protein

SOURCE: Bacteriophage fd
SOLVENT: a. Phosphate buffer, pH 8.0
b. 0.1 N NaOH
COMMENTS: Spectrum of DNA-free protein in solvent b; spectra of phage, succinylated-protein, and tetrafluoro-succinylated-protein
REFERENCE: F. Asbeck, K. Beyreuther, H. Kohler, G. V. Wettstein, and G. Braunitzer, Hoppe-Seyler's Z. Physiol. Chem., 350, 1047 (1969); R. W. Williams and A. K. Dunker, J. Biol. Chem., 252, 6253 (1977)

2031. Protein

SOURCE: a. Tobacco streak virus
b. Cow pea chlorotic mottle virus
c. Bromemosaic virus
d. Citrus variegation virus
e. Citrus leaf rugose virus
f. Alfalfa mosaic virus
REFERENCE: D. Gonsalves and S. M. Garnsey, Virology, 67, 319 (1975)

2032. Protein

SOURCE: Adenovirus type 2
SOLVENT: 6 M Guanidine HCl, 0.02 M Phosphate buffer, pH 7.0
REFERENCE: L. Prager and U. Pettersson, Virology, 45, 364 (1971)

2033. Protein

SOURCE: Cowpea chlorotic mottle virus
SOLVENT: a. pH 7.5 plus 1 M NaCl
b. pH 5
COMMENTS: Protein spectrum in solvent a and nucleo-protein in solvent b
REFERENCE: J. Kruse, B. J. M. Verduin, and A. J. W. G. Visser, Eur. J. Biochem., 95, 21 (1979)

2034. Protein

SOURCE: a. Tobacco mosaic virus
b. Green tomato atypical mosaic virus
c. Cucumber
REFERENCE: A. Magne, D. Gerard, L. Hirth, and G. Laustriat, Biochim. Biophys. Acta, 495, 189 (1977)

2035. Protein Complex, Vitamin B_{12} -

SOURCE: Human saliva
SOLVENT: 0.05 M Potassium phosphate, pH 7.5, containing 0.75 M NaCl
REFERENCE: R. Grasbeck, K. Visuri, and U. H. Stenman, Biochim. Biophys. Acta, 263, 721 (1972); R. L. Burger and R. Allen, J. Biol. Chem., 249, 7270 (1974)

2036. Protein, Vitamin B_{12}-binding

SOURCE: Human milk
SOLVENT: a. 0.05 M Potassium phosphate buffer, pH 7.5, containing 0.75 M NaCl
b. 0.005 M Potassium phosphate, pH 7.5
REFERENCE: R. L. Burger and R. H. Allen, J. Biol. Chem., 249, 7270 (1974); R. L. Burger, S. Waxman, H. S. Gilbert, C. S. Mehlman, and R. H. Allen, J. Clin. Inves., 56, 1262 (1975)

2037. Protein, Vitamin B_{12}-binding

SOURCE: Human hepatoma
SOLVENT: 0.005 M Potassium phosphate, pH 7.5

REFERENCE: R. L. Burger, S. Waxman, H. S. Gilbert, C. S. Mehlman, and R. H. Allen, J. Clin. Inves., 56, 1262 (1975)

2038. Protein, Vitamin B_{12}-binding

SOURCE: Human granulocyte
SOLVENT: 0.05 M Potassium phosphate, pH 7.5, containing 0.75 M NaCl
REFERENCE: R. H. Allen and P. W. Majerus, J. Biol. Chem., 247, 7202 (1972)

2039. Proteinase, Ac_1-

SOURCE: Snake (Agkistrodon acutus)
REFERENCE: T. Nakai, H. Sugihara, and T. Tanaka, Yakugaku Zasshi, 97, 507 (1977)

2040. Proteinase

SOURCE: Rhizopus niveus
SOLVENT: a. 0.5 M Acetate buffer, pH 5.5
b. 0.1 N NaOH
REFERENCE: Y. Kurono, M. Chidimatsu, K. Horikoshi, and Y. Ikeda, Agric. Biol. Chem., 35, 1668 (1971)

2041. Proteinase

SOURCE: Lotus seed
SOLVENT: 50 mM Acetate buffer, pH 4.0
REFERENCE: S. Shinano and K. Fukushima, Agric. Biol. Chem., 35, 1488 (1971)

2042. Proteinase

SOURCE: Pseudomonas maltophilia
SOLVENT: 10 mM Cacodylate-5 mM $CaCl_2$ buffer, pH 6.0
REFERENCE: R. S. Boethling, J. Bacteriol., 121, 933 (1975)

2043. Proteinase

SOURCE: Serratia strain E-15
SOLVENT: 0.01 M Phosphate buffer, pH 7.0
REFERENCE: K. Miyata, K. Maejima, K. Tomoda, and M. Isono, Agric. Biol. Chem., 34, 310 (1970)

2044. Proteinase (Awamorin)

SOURCE: Aspergillus awamori

COMMENTS: Spectra of native enzyme and enzyme inhibited by diazoacetyl-DNP-ethylenediamine
REFERENCE: L. S. Lobareva, G. G. Kovalava, M. P. Shimanskaya, and V. N. Stepanov, Biokhimiya (Biochemistry), 37, 164 (1973)

2045. Proteinase I and Proteinase II

SOURCE: Snake (Vipera russelli)
SOLVENT: 0.1 M Tris-HCl buffer, pH 8.6
REFERENCE: G. D. Dimitrov, Toxicon, 9, 33 (1971)

2046. Proteinase I and Proteinase II

SOURCE: Aspergillus sojae
SOLVENT: a. 50 mM Tris-HCl buffer, pH 7.3
b. 0.1 N NaOH
REFERENCE: H. Sekine, Agric. Biol. Chem., 36, 198 (1972)

2047. Proteinase, Neutral

SOURCE: Streptomyces naraensis
SOLVENT: a. 0.005 M Tris-HCl buffer, pH 7.5
b. 0.1 N NaOH
REFERENCE: A. Hiramatsu and T. Ouchi, J. Biochem. Tokyo, 71, 767 (1972)

2048. Proteinase, Neutral

SOURCE: Bacillus subtilis var. amylosaccharíticus
SOLVENT: a. 0.002 M Calcium acetate, pH 6.7
b. 0.1 N NaOH
REFERENCE: D. Tsuru, T. Yamamoto, and J. Fukumoto, Agric. Biol. Chem., 30, 651 (1966)

2049. Proteinase

SOURCE: Micrococcus caseolyticus
SOLVENT: 0.005 M Tris-maleate plus 0.0015 M $CaCl_2$, pH 7.0
REFERENCE: M. Desmazeaud and J. Hermier, Ann. Biol. Anim. Bioch. Biophys., 8, 565 (1968)

2050. Proteinase

SOURCE: Bacillus R-4
SOLVENT: a. 0.2 M Tris-HCl buffer, pH 7.6
b. 0.1 N NaOH
REFERENCE: Y. Tominaga and Y. Tsujisaka, Agric. Biol. Chem., 40, 1937 (1978)

2051. Proteinase, Blood-coagulating

SOURCE: Cephalosporium species
SOLVENT: 50 mM Tris-HCl buffer, containing 0.2 M NaCl and 1 mM Ca^{2+}
REFERENCE: T. Satoh, T. Beppu, and K. Arima, Agric. Biol. Chem., 41, 293 (1978)

2052. Proteinase, Alkaline

SOURCE: Bacillus species
SOLVENT: pH 12.5
REFERENCE: Y. Morikawa, M. Shimizu, and K. Kimura, Agric. Chem., 48, 106 (1974)

2053. Proteinase, Alkaline

SOURCE: Bacillus No. 221
SOLVENT: a. 0.1 M Tris-HCl buffer, pH 7.0
b. 0.2 M KCl-NaOH buffer, pH 12
REFERENCE: K. Horikoshi, Agric. Biol. Chem., 35, 1407 (1971)

2054. Proteinase, Alkaline

SOURCE: Bacillus subtilis var. amylosaccharitícus
SOLVENT: a. 0.01 M Tris-HCl buffer, pH 7.3
b. 0.1 M NaOH
REFERENCE: D. Tsuru, H. Kira, T. Yamamoto, and J. Fukumoto, Agric. Biol. Chem., 31, 330 (1967)

2055. Proteinase, Alkaline

SOURCE: Aspergillus oryzae
SOLVENT: a. 0.125 M Acetate buffer, pH 5.0
b. 0.1 N NaOH
REFERENCE: T. Nakadai, S. Nasuno, and N. Iguchi, Agric. Biol. Chem., 37, 2685 (1973)

2056. Proteinase, Alkaline

SOURCE: Aspergillus sojae
SOLVENT: a. 0.2 M Acetate buffer, pH 5.1
b. 0.1 N HCl
c. 0.1 N NaOH
REFERENCE: K. Hayashi, D. Fukushima, and K. Mogi, Agric. Biol. Chem., 31, 1171 (1967)

2057. Proteinase, Alkalophilic

SOURCE: Streptomyces species
SOLVENT: a. 0.01 M Phosphate buffer, pH 7.0
b. 0.1 N NaOH
REFERENCE: T. Nakanichi, Y. Matsumura, N. Minamura, and T. Yamamoto, Agric. Biol. Chem., 38, 37 (1974)

2058. Proteinase, Acid

SOURCE: Paecilomyces varioti BAINIER Tpr-220
SOLVENT: a. Water
b. 0.1 N NaOH
REFERENCE: J. Sawada, Agric. Biol. Chem., 30, 393 (1966)

2059. Proteinase B, Acid

SOURCE: Scytalidium lignicolum
SOLVENT: 0.01 M Acetate buffer, pH 4.0
REFERENCE: K. Oda, S. Murao, T. Oka, and K. Morihara, Agric. Biol. Chem., 39, 477 (1975)

2060. Proteinase, Acid

SOURCE: Cladosporium species No. 45-2' and 45-2
SOLVENT: 0.01 M Sodium acetate buffer, pH 4.5
REFERENCE: S. Murao, S. Funakoshi, and K. Oda, Agric. Biol. Chem., 36, 1327 (1972); K. Oda, S. Funakoshi, and S. Murao, Agric. Biol. Chem., 37, 1723 (1973)

2061. Protein Kinase, Catalytic Subunit

SOURCE: Rabbit skeletal muscle
SOLVENT: 5 mM 2-(N-Morpholino) ethane sulfonic acid buffer, pH 6.5, plus 0.1 N NaCl, 15 mM 2-mercaptoethanol, and 0.1 mM EDTA
REFERENCE: P. J. Bechtel, J. A. Beavo, and E. G. Krebs, J. Biol. Chem., 252, 2691 (1977)

2062. Proteinoid

SOURCE: Synthetic
COMMENTS: 1. Spectra of oxidized and reduced (dithionite) proteinoid
2. Spectrum of proteinoid polymerized in presence in hemin
REFERENCE: J. L. Fox and C. F. Muska, Experientia, 28, 143 (1972)

2063. Proteoglycan

SOURCE: Beef
SOLVENT: a. Phosphate buffer
b. 3 M Guanidine HCl
REFERENCE: W. Strider, S. Pal, and L. Rosenberg, Biochim. Biophys. Acta, 379, 271 (1975)

2064. Proteoglycan

SOURCE: Pig larynx cartilage
SOLVENT: 0.1 M Sodium acetate, pH 4.0
COMMENTS: Spectral changes due to N-bromosuccinimide
REFERENCE: T. E. Hardingham, R. J. F. Ewins, and H. Muir, Biochem. J., 157, 127 (1976)

2065. Protocatechuate 3,4-Dioxygenase

SOURCE: Acinetobacter calcoaceticus and Pseudomonas aeruginosa
SOLVENT: 0.05 M Tris-HCl, pH 8.5
REFERENCE: C. T. Hou, M. O. Lillard, and R. D. Schwartz, Biochemistry, 15, 582 (1976)

2066. Protocatechuate 3,4-Dioxygenase

SOURCE: Azotobacter vinelandii
SOLVENT: Tris-HCl, pH 8.5
COMMENTS: Spectra of oxidized and reduced (protocatechuate) forms
REFERENCE: D. R. Durham, L. A. Stirling, L. N. Orston, and J. J. Perry, Biochemistry, 19, 149 (1980)

2067. Protocatechuate 3,4-Dioxygenase

SOURCE: Pseudomonas putida
SOLVENT: pH 6.0
COMMENTS: Spectra of free enzyme, enzyme-protocatechuate complex, and enzyme-β-carboxy-cis,cis-muconate complex
REFERENCE: D. Ballou and C. Bull in Biochemical and Clinical Aspects of Oxygen (W. S. Caughey, ed.), 573 (1979), Academic Press, New York

2068. Protocatechuate 3,4-Dioxygenase

SOURCE: Pseudomonas aeruginosa
COMMENTS: Spectra of native enzyme, enzyme-substrate complex and oxygenated intermediate

REFERENCE: O. Hayaishi, Y. Ishimura, H. Fujisawa, and M. Nozaki, in International Symposium on Oxidases and Related Redox Systems (T. E. King, H. S. Mason, and M. Morrison, eds.), Vol. 1, 126 (1971), University Park Press, Baltimore

2069. Protocatechuate 3,4-Dioxygenase

SOURCE: Pseudomonas aeruginosa ATCC 23975
SOLVENT: a. 50 mM Tris-HCl buffer, pH 8.5
b. Potassium phosphate buffer, pH 7.5
COMMENTS: 1. Spectra of enzyme before and after addition of protocatechuic acid (solvent a)
2. Spectral change upon addition of protocatechualdehyde to enzyme-protocatechuic acid complex (solvent a)
3. Spectra of native enzyme, reduced (dithionite) and reoxidized (ferricyanide) forms (solvent a)
4. Spectra of enzyme-3,4-dihydroxyphenylacetate complex and 3,4-dihydroxyphenylpropionate complex (solvent b)
REFERENCE: H. Fujisawa, M. Uyeda, Y. Kojima, M. Nozaki, and O. Hayaishi, J. Biol. Chem., 247, 4414 and 4422 (1972)

2070. Protocatechuate 3,4-Dioxygenase

SOURCE: Pseudomonas aeruginosa
SOLVENT: 50 mM Tris-Acetate buffer, pH 8.5
COMMENTS: Spectra of native, apo-, and reconstituted enzymes
REFERENCE: M. Fujiwara and M. Nozaki, Biochim. Biophys. Acta, 327, 306 (1973)

2071. Protocatechuate 3,4-Dioxygenase

SOURCE: Pseudomonas aeruginosa
SOLVENT: 50 mM Tris-HCl buffer, pH 8.5
COMMENTS: Spectra of cobalt-free and cobalt-containing enzyme
REFERENCE: H. Fujisawa and M. Uyeda, Eur. J. Biochem., 45, 223 (1974)

2072. Protocatechuate 3,4-Dioxygenase

SOURCE: Pseudomonas aeruginosa
SOLVENT: 50 mM Tris-HCl buffer, pH 8.5
COMMENTS: Spectra of native enzyme, enzyme-protocatechuic acid complex, enzyme-protocatechualdehyde complex; all spectra made in absence in air
REFERENCE: O. R. Zabrowsky, C. T. Hou, and J. Oglefree, Biochem. Biophys. Acta, 386, 18 (1975)

2073. Pullulanase

SOURCE: Aerobacter aerogenes
REFERENCE: S. Ueda and R. Ohba, Agric. Biol. Chem., 36, 2381 (1972)

2074. Putidamonooxin

SOURCE: Pseudomonas putida
SOLVENT: 50 mM Potassium phosphate buffer, pH 8.0, containing 30% glycerol
COMMENTS:
1. Spectra of oxidized and reduced (dithionite) forms
2. Spectra of oxidized protein in presence and absence of 4-methoxybenzoate at 80°K
3. Spectral changes exhibited by oxidized protein in presence of glucose oxidase system upon addition of increasing amounts of 4-methoxybenzoate at 10°C

REFERENCE: F. H. Bernhardt, E. Heymann, and P. S. Traylor, Eur. J. Biochem., 92, 209 (1978)

2075. Putidaredoxin

SOURCE: Pseudomonas putida strain PpG786
SOLVENT: 50 mM Tris-HCl buffer, pH 7.4
COMMENTS: Spectra of protein and des-TRY-GLN-putidaredoxin formed by action of carboxypeptidase A on either reduced or oxidized protein
REFERENCE: S. G. Sligar, P. G. Debrunner, J. D. Lipscomb, M. J. Namtvedt, and I. C. Gunsalus, Proc. Nat. Acad. Sci., 71, 3906 (1974)

2076. Putidaredoxin and Putidaredoxin Trimer

SOURCE: Pseudomonas putida
SOLVENT: 50 mM Tris-HCl buffer, pH 7.4
REFERENCE: K. Dus, Adv. Exp. Med. Biol., 58, 287 (1975)

2077. Putrescine Oxidase

SOURCE: Micrococcus rubens
SOLVENT: 0.03 M Phosphate buffer, pH 7.0
COMMENTS: Spectra of oxidized, reduced I (putrescine), and reduced II (dithionite) forms
REFERENCE: O. Adachi, H. Yamada, and K. Ogata, Agric. Biol. Chem., 30, 1202 (1966)

2078. Putrescine Oxidase

SOURCE: Micrococcus rubens
SOLVENT: a. 1.5 M KBr
b. 0.025 M Phosphate buffer
c. 0.01 M Phosphate, pH 7, 0.02 M EDTA
COMMENTS: 1. Spectral changes upon addition of putrescine under anaerobic conditions and then admitting air (solvent b)
2. Spectral changes upon irradiation under anaerobic conditions (solvent c)
REFERENCE: R. J. De Sa, J. Biol. Chem., 247, 5527 (1972)

2079. Putrescine Oxidase

SOURCE: Micrococcus rubens
COMMENTS: Spectral changes due to 1,10-diaminodecane
REFERENCE: W. F. Swain and R. J. De Sa, Biochim. Biophys. Acta, 429, 331 (1976)

2080. Putrescine-peptide Conjugate

SOURCE: Human plasma
REFERENCE: T. W. Seale, W. Y. Chan, J. Shulka, and O. M. Rennert, Arch. Biochem. Biophys., 198, 164 (1979)

2081. Pyocin R1, Core Particle

SOURCE: Pseudomonas aeruginosa P15
SOLVENT: 0.01 M Tris-HCl buffer, pH 7.5, containing 0.1 M NaCl
REFERENCE: T. Hasegawa and S. Ishii, J. Biochem. Tokyo, 85, 403 (1979)

2082. Pyocin R

SOURCE: Pseudomonas aeruginosa
SOLVENT: a. 0.01 M Tris-HCl, 0.01 M EDTA, pH 9.0
b. 0.01 M Tris-KOH, 0.01 M EDTA, pH 12.5
COMMENTS: 1. Spectra of pyocin R and sheaths in solvent a
2. Spectrum of basal-rod fraction in solvent b
REFERENCE: P. P. Goryaev and B. F. Poglazov, Biokhimiya (Biochemistry), 34, 469 (1969)

2083. Pyocin S2

SOURCE: Pseudomonas aeruginosa M47
SOLVENT: a. 0.01 M Sodium phosphate buffer, pH 7.0, containing 0.2 M NaCl
b. 0.1 N NaOH

REFERENCE: I. Ohkawa, M. Kageyama, and F. Egami, J. Biochem. Tokyo, 73, 281 (1973)

2084. Pyrazon Dioxygenase, Components A_1, A_2 and B

SOURCE: Pyrazon degrading bacteria, strain E
SOLVENT: a. 0.1 M Potassium phosphate buffer, pH 7.0, 5 mM in dithiothreitol
b. pH 6.8 (Water)
COMMENTS: 1. Spectra of A_1: Oxidized and reduced (dithionite) in solvent a
2. Spectrum of A_2 in solvent a
3. Spectra of B: Oxidized and reduced (dithionite) in solvent b
REFERENCE: K. Sauber, C. Frohner, G. Rosenberg, J. Eberspacher, and F. Lingens, Eur. J. Biochem., 74, 89 (1977)

2085. Pyridine Nucleotide Dehydrogenase

SOURCE: Beef erythrocytes
SOLVENT: 0.1 M Sodium phosphate buffer, pH 7.0
REFERENCE: K. Adachi and T. Okuyama, Biochim. Biophys. Acta, 268, 629 (1972)

2086. Pyridine Nucleotide Reductase

SOURCE: Spinach
REFERENCE: K. T. Fry and A. San Pietro in Photosynthetic Mechanisms of Green Plants, 252 (1963), Publication 1145, Nat. Acad. Sci. Nat. Res. Council

2087. Pyridine Nucleotide Transhydrogenase

SOURCE: Yellowtail (Seriola quinqueradiata)
REFERENCE: T. Sakai, C. Shimizu, and F. Matsuura, Bull. Jap. Soc. Sci. Fish., 41, 753 (1975)

2088. Pyridine Nucleotide Transhydrogenase

SOURCE: Azotobacter vinelandii
SOLVENT: 0.05 M Potassium phosphate buffer, pH 6.8
COMMENTS: Spectra of holo- and apo-enzymes
REFERENCE: L. E. Middleditch, R. W. Atchison, and A. E. Chung, J. Biol. Chem., 247, 6802 (1972)

2089. Pyridoxamine 5-Phosphate Oxidase

SOURCE: Rabbit liver
SOLVENT: 0.02 M Potassium phosphate, pH 7
COMMENTS: 1. Spectra of holo- and apo-enzymes

2. Spectral changes of holoenzyme upon addition of dithionite under anaerobic conditions

REFERENCE: M. N. Kazarinoff and O. B. McCormick, J. Biol. Chem., 250, 3436 (1975)

2090. Pyridoxine 5-Phosphate Oxidase

SOURCE: Pig brain
SOLVENT: 0.1 M Potassium phosphate, pH 7
COMMENTS: Spectra of holo- and apo-enzymes
REFERENCE: F. Kwok and J. E. Churchich, J. Biol. Chem., 255, 882 (1980)

2091. Pyrocatechase and Metapyrocatechase

SOURCE: Pseudomonas
SOLVENT:
a. 0.05 M Tris buffer, pH 8.0
b. 0.05 M Phosphate buffer, pH 7.5
c. 0.05 M Tris buffer, pH 9.0
d. 0.05 M Phosphate buffer, pH 6.0
e. NaOH solution, pH 12

COMMENTS:
1. Spectra of pyrocatechase in solvents a, c, d, and e
2. Spectrum of metapyrocatechase in solvent b

REFERENCE: O. Hayaishi, in Oxidases and Related Redox Systems (T. E. King, H. S. Mason, M. Morrison, eds.), 286 (1965), J. Wiley, New York

2092. Pyrocatechase

SOURCE: Brevibacterium fuscum p-13
SOLVENT: 0.1 M Tris-HCl, pH 7.5
COMMENTS: Spectral changes due to $HgCl_2$
REFERENCE: K. Nagami, Biochem. Biophys. Res. Comm., 51, 357 (1973); K. Nagami and S. Senoh, J. Biochem. Tokyo, 75, 389 (1974)

2093. Pyrocatechase

SOURCE: Brevibacterium fuscum p-13
COMMENTS:
1. Spectra of native, apo-enzyme, and reconstituted enzyme
2. Spectral changes due to the addition of ethylprotocatechuate

REFERENCE: K. Nagami and Y. Mizake, Biochem. Biophys. Res. Comm., 46, 198 (1972)

2094. Pyrocatechase I

SOURCE: Pseudomonas species B13
SOLVENT: 20 mM Tris-HCl buffer, pH 8.0, containing 1 mM 2-mercaptoethanol
REFERENCE: E. Dorn and H. J. Knackmuss, Biochem. J., 174, 73 (1978)

2095. Pyrophosphatase

SOURCE: Baker's yeast
SOLVENT: a. Water
b. 0.1 M HCl
c. 0.1 M NaOH
REFERENCE: M. Kunitz, J. Gen. Physiol., 35, 423 (1952)

2096. Pyrophosphatase, inorganic

SOURCE: Yeast
SOLVENT: 5 M Guanidine HCl
COMMENTS: Spectra of native and denatured protein
REFERENCE: P. Heitmann, C. Scholz, and G. Hansen, Studia Biophysica, 33, 99 (1972)

2097. Pyrophosphate Phosphohydrolase

SOURCE: Lathyrus sativa
SOLVENT: 8 M Urea
COMMENTS: Spectra of native and denatured proteins
REFERENCE: S. Niculescu, I, Dumitru, and D. Iordachescu, Acta Vitaminol. Enzymol., 26, 13 (1972)

2098. Pyruvate Decarboxylase

SOURCE: Yeast
SOLVENT: pH 6.8
COMMENTS: Spectra of holo- and apo-enzymes
REFERENCE: J. Ullrich and A. Wollmer, Hoppe-Seyler's Z. Physiol. Chem., 352, 1635 (1971)

2099. Pyruvate Decarboxylase

SOURCE: Yeast
SOLVENT: pH 6.5-pH 12.5
REFERENCE: I. I. Stepuro and Y. M. Ustrovsky, Bioorganischeskaya Chemiya, 1, 821 (1975)

2100. Pyruvate Dehydrogenase

SOURCE: Azotobacter vinelandii
SOLVENT: 0.05 M Potassium phosphate buffer, pH 7.0
COMMENTS: Spectra of oxidized and reduced (dithionite) enzyme
REFERENCE: T. W. Bresters, R. A. De Abreu, A. De Kok, J. Visser, and C. Veeger, Eur. J. Biochem., 59, 335 (1975)

2101. Pyruvate Dehydrogenase Complex

SOURCE: Escherichia coli
REFERENCE: D. C. Speckhard and P. A. Frey, Biochem. Biophys. Res. Comm., 62, 614 (1975)

2102. Pyruvate Kinase, trinitrophenyl-

SOURCE: Rabbit skeletal muscle
SOLVENT: 0.05 M Tris buffer, pH 7.5, 0.2 M KCl, 0.008 M $MgCl_2$, 0.002 M phosphoenolpyruvate and 0.025 M potassium acetate
REFERENCE: M. Flashner, P. F. Hollenerg, and M. J. Coon, J. Biol. Chem., 247, 8114 (1972)

2103. Pyruvate Kinase, Co (II)-

SOURCE: Rabbit skeletal muscle
SOLVENT: pH 7.5
COMMENTS: Spectra of Co(II)-enzyme and of its complexes with ATP, phosphoenolpyruvate, and pyruvate
REFERENCE: C. Y. Kwan, K. Erhard, and R. C. Davis, J. Biol. Chem., 250, 5951 (1975)

2104. Pyruvate Kinase

SOURCE: Pig liver
SOLVENT: a. 0.1 M Potassium phosphate buffer, pH 7.0
b. 0.1 N NaOH
REFERENCE: C. Kutzbach, H. Bischofberger, B. Hess, and H. Zimmermann-Telschow, Hoppe-Seyler's Z. Physiol. Chem., 354, 1473 (1973)

2105. Pyruvate Kinase

SOURCE: Beef skeletal muscle
SOLVENT: 0.05 M Phosphate and 10 mM 2-mercaptoethanol, pH 7.0
COMMENTS: Spectra of native protein and trinitrophenyl derivative of protein

REFERENCE: D. R. Hubbard and J. M. Cardenas, J. Biol. Chem., 250, 4931 (1975)

2106. Pyruvate Oxidase

SOURCE: Escherichia coli
SOLVENT: 0.1 M Sodium phosphate buffer, pH 5.7
COMMENTS: Spectra of unlabeled and of 1-azidonaphthalene-5-sulfonate labeled protein
REFERENCE: D. J. Stevens and R. B. Gennis, J. Biol. Chem., 255, 379 (1980)

2107. Rachis, S-carboxymethyl- and Rachis Keratose

SOURCE: Feathers of white leghorn hens
REFERENCE: B. S. Harrap and E. F. Woods, Biochem. J., 92, 8 (1964)

2108. Reaction Centers

SOURCE: Rhodopseudomonas spheroides, strain 2.4.1/R-26
REFERENCE: R. K. Clayton and R. T. Wang, Meth. Enz., 23, 696 (1971)

2109. Reaction Centers

SOURCE: Rhodospirillium spheroides R-26 and Rhodospirillium rubrum G-9
REFERENCE: M. Y. Okamura, L. A. Steiner, and G. Feher, Biochemistry, 13, 1394 (1974)

2110. Reaction Centers

SOURCE: Rhodospirillium rubrum
SOLVENT: 50 mM Sodium carbonate, pH 9.35
COMMENTS: Spectrum of oxidized (ferricyanide) form, and spectrum of reduced form (ascorbate)
REFERENCE: H. Noel, M. Van der Rest, and G. Gingras, Biochim. Biophys. Acta, 275, 219 (1975)

2111. Reaction Centers

SOURCE: Rhodopseudomonas
SOLVENT:
a. 50 mM Tris buffer, pH 8.0
b. 25.5 mM Tris buffer, pH 8, 55% glycerol

COMMENTS:
1. Spectra of isolated and photooxidized preparations (solvent a)
2. Spectra of photooxidized and reduced (dithionite) forms at 77°K

3. Spectra of reduced (dithionite in the dark) form and form obtained after exposure to strong white light (solvent a)

REFERENCE: T. L. Trosper, D. L. Benson, and J. P. Thornber, Biochim. Biophys. Acta, 460, 318 (1977)

2112. Reaction Centers

SOURCE: Rhodopseudomonas virdis
COMMENTS: Spectrum of form in weak light and spectrum of form in strong white light
REFERENCE: R. K. Clayton and B. J. Clayton, Biochim. Biophys. Acta, 501, 478 (1978)

2113. Reaction Centers

SOURCE: Rhodopseudomonas gelatinosa, strain TG-9/EM-1
COMMENTS: Spectrum of centers in weak light and spectrum of centers in strong white light
REFERENCE: B. J. Clayton and R. K. Clayton, Biochim. Biophys. Acta, 501, 470 (1978)

2114. Reaction Centers

SOURCE: Swiss chard chloroplasts
REFERENCE: C. Bengis and N. Nelson, J. Biol. Chem., 250, 2783 (1975)

2115. Regulator, Ca^{2+}-dependent

SOURCE: Beef brain
SOLVENT: 10 mM Tris-Cl, pH 7.4
REFERENCE: D. J. Wolff, P. G. Poirier, C. O. Brostrom, and M. A. Brostrom, J. Biol. Chem., 252, 4108 (1977)

2116. Relaxin

SOURCE: Pig ovary
SOLVENT: 0.2 M Acetate buffer, pH 4.7
COMMENTS: Spectral changes upon addition of N-bromosuccinimide
REFERENCE: C. Schwave and S. A. Braddon, Biochem. Biophys. Res. Comm., 68, 1126 (1976)

2117. Relaxin, A and B Chains

SOURCE: Pig ovary
REFERENCE: C. Schwabe, J. K. McDonald, and B. G. Steinetz, Biochem. Biophys. Res. Comm., 70, 397 (1976)

2118. Rennin

SOURCE: Fungus (Endothia parasitica)
REFERENCE: J. L. Sardinas, Appl. Microbiol., 16, 248 (1968)

2119. Rennin

SOURCE: Beef, commercial rennet
SOLVENT: 0.01 M Acetic acid
COMMENTS: Spectra of rennin and 1,2-epoxy-3-(nitrophenoxy) propane-modified protein
REFERENCE: W. J. Chang and K. Takahashi, J. Biochem. Tokyo, 74, 231 (1973)

2120. Resistance-enhancing Compound

SOURCE: Mycobacterium tuberculosis
SOLVENT: 0.1 M Phosphate buffer, pH 8.0
REFERENCE: L. W. Hedgecock, Amer. Rev. Resp. Dis., 112, 633 (1975)

2121. Resorcinol Hydroxylase

SOURCE: Pseudomonas putida ORC
SOLVENT: 20 mM Phosphate buffer, pH 6.8, containing 50 mM EDTA
COMMENTS:
1. Spectral changes upon reduction of enzyme by light
2. Spectral changes during reduction by NADH
3. Spectra in presence and absence of resorcinol

REFERENCE: Y. Ohta and D. W. Ribbons, Eur. J. Biochem., 61, 259 (1976)

2122. Reticulin

SOURCE: Pig liver and kidney
SOLVENT: Water
REFERENCE: M. Pras and L. E. Glyn, Brit. J. Exp. Pathol., 54, 449 (1973)

2123. Retina

SOURCE: Frog
SOLVENT: 0.1 M Hydroxylamine, pH 7.1
COMMENTS: Frog retina fixed in glutaraldehyde; spectra made with polarized and unpolarized light
REFERENCE: P. K. Brown, Nature New Biology, 236, 35 (1972)

2124. Rhapidosomes, Sheath and Core

SOURCE: Saprospira grandis
REFERENCE: A. S. Delk and C. A. Dekker, J. Mol. Biol., 64, 287 (1972)

2125. Rhodanese

SOURCE: Beef liver
SOLVENT: 0.1 M Phosphate buffer, pH 7.6
COMMENTS: Spectra in presence and absence of cyanide and thiosulfate
REFERENCE: A. Finazzi-Agro, G. Frederici, C. Giovagnoli, C. Cannella, and D. Cavallini, Eur. J. Biochem., 28 (1972), 89

2126. Rhodanese

SOURCE: Beef kidney
SOLVENT: 0.05 M Sodium acetate
COMMENTS: Spectra in presence and absence of cyanide, cyanide and thiosulfate, and after treatment with ammonium sulfate
REFERENCE: C. Cannella, L. Pecci, B. Pensa, M. Costa, and D. Cavallini, FEBS Lett., 22 (1974)

2127. Rhodanese

SOURCE: Beef liver and kidney
SOLVENT: 50 mM Tris-acetate, 1 mM EDTA, pH 8.0
COMMENTS: Spectra in presence of dithiothreitol and dithiothreitol and cyanide
REFERENCE: L. Pecci, B. Pensa, M. Costa, P. L. Cignini, and C. Cannella, Biochim. Biophys. Acta, 445, 104 (1976)

2128. Rhodanese

SOURCE: Beef liver
SOLVENT: 50 mM Phosphate buffer, pH 8
COMMENTS:
1. Spectrum of enzyme
2. Spectrum of 5,5'-dithio-bis-(2-nitrobenzoic acid) modified enzyme
3. Spectrum of product from 2 after reduction with 2-mercaptoethanol

REFERENCE: B. Pensa, M. Costa, L. Pecci, C. Cannella, and D. Cavallini, Biochim. Biophys. Acta, 484, 368 (1977)

2129. Rhodopsin

SOURCE: Beef
SOLVENT: a. 1.0 M Acetic acid
b. 1.0 M Monochloroacetic acid
c. 1.0 M Dichloroacetic acid
d. 1.0 M Trichloroacetic acid
e. 1.0 M Tribromoacetic acid
f. 1.0 M Trifluoroacetic acid
g. Non-buffered 1% digitonin
h. 1.0 M Sodium trichloroacetate
COMMENTS: 1. Spectra of denatured protein at liquid nitrogen temperature (solvents a-f)
2. Spectra of native and denatured protein at 10° (solvents g and h)
REFERENCE: T. Suzuki and Y. Kito, Ann. Rept. of Biol. Works Fac. Science, Osaka Univ., 18, 71 (1970)

2130. Rhodopsin

SOURCE: Beef rods, outer segments
SOLVENT: a. 1% Ammonyx LO
b. 1.5% Ammonyx LO
c. 0.05 M Octyl glucoside, 0.1 M potassium phosphate buffer, pH 7
REFERENCE: T. G. Ebrey and B. Honig, Proc. Nat. Acad. Sci., 69, 1897 (1972); B. Honig, P. Kahn, and T. G. Ebrey, Biochemistry, 12, 1637 (1973); G. W. Stubbs, H. G. Smith, and B. J. Litman, Biochim. Biophys. Acta, 425, 46 (1976); S. A. Shukolyukov, Biokhimiya (Biochemistry), 37, 764 (1972)

2131. Rhodopsin

SOURCE: Beef retina and rod outer segments
SOLVENT: a. 0.01 M Imidazole-0.1 mM dithiothreitol, pH 7
b. 0.05 M Tris-HCl, pH 7.0, containing 0.3% Ammonyx LO plus 0.05 M dodecyldimethylamine oxide
COMMENTS: Spectra before and after illumination (photo bleaching)
REFERENCE: M. L. Applebury, D. M. Zuckerman, A. A. Lamola, and T. M. Jovin, Biochemistry, 13, 3448 (1974); J. J. Plantnar and E. L. Kean, J. Biol. Chem., 251, 1548 (1976); S. L. Bonting, Protides Biol. Fluids, 21, 97 (1973)

2132. Rhodopsin

SOURCE: Beef opsin plus bacteria retinal
COMMENTS: Spectra before and after irradiation with white light
REFERENCE: J. J. Wolken and C. S. Nakagawa, Biochem. Biophys. Res. Comm., 54, 1262 (1973)

2133. Rhodopsin

SOURCE: Beef
SOLVENT: 0.10 M Potassium phosphate buffer, pH 7.0, containing 1 mM dithiothreitol
COMMENTS: Spectra in presence and absence of sodium cholate
REFERENCE: R. A. Henselman and M. A. Cusanovich, Biochemistry, 13, 5199 (1974)

2134. Rhodopsin

SOURCE: Beef retinas
SOLVENT: pH 4.80-pH 8.93
COMMENTS: Spectral changes due to pH and irradiation
REFERENCE: S. E. Ostroy, Arch. Biochem. Biophys., 164, 275 (1974)

2135. Rhodopsin

SOURCE: a. Beef retinas
b. Grass frog (Rana pipiens) retinas
c. Pigeon (Columbia livia) retinas
SOLVENT: 2% Digitonin
REFERENCE: J. H. Porkes, J. H. Rockey, and P. A. Liebman, Biochim. Biophys. Acta, 428, 1 (1976)

2136. Rhodopsin and Rod Outer Segments and Rod Outer Segment Membranes

SOURCE: Beef retinas
SOLVENT: a. 100 mM Tridecyltrimethylammonium bromide-5 mM N-2-hydroxyethylpiperazine-N'-2-ethanosulfonic acid, pH 6.6
b. 1% Emulphogene BC-720, pH 7.0
REFERENCE: D. F. Obrien, L. F. Cosita, and R. A. Ott, Biochemistry, 16, 1295 (1977); C. N. Rafferty, Biophys. Str. Mech., 3, 123 (1977)

2137. Rhodopsin

SOURCE: Beef retinas
SOLVENT: Glycerol-water (1:1)

COMMENTS: Spectral changes at different temperatures (-196°C-0°C), before and after illumination of rhodopsin-asolectin recombinant
REFERENCE: D. F. O'Brien, L. C. Costa, and R. A. Ott, Biochemistry, 16, 1295 (1977)

2138. Rhodopsin

SOURCE: Beef retinas
SOLVENT: 200 mM Dodecyltrimethylammonium bromide in 66 mM phosphate buffer, pH 6.8
COMMENTS: Spectral changes after 3 minutes illumination at 23°C (0-40 minutes)
REFERENCE: E. O. Plante, M. B. Maude, and R. E. Anderson, Exp. Eye Res., 24, 377 (1977)

2139. Rhodopsin

SOURCE: Beef rod outer segments
SOLVENT: 0.5% Digitonin in 1 mM potassium phosphate, pH 7.2
COMMENTS: Spectra of isochromic forms
REFERENCE: H. Shichi, S. Kawamura, C. G. Muellenberg, and T. Yoshizawa, Biochemistry, 16, 5376 (1977)

2140. Rhodopsin, modified with N-diazoacetyl-N'-(2,4-dinitrophenyl)-ethylenediamine

REFERENCE: E. M. Slobodyanskaya, Bioorganicheskaya Chemie, 4, 1054 (1978)

2141. Rhodopsin, Meta I Derivatives

SOURCE: Beef rod outer segments
SOLVENT: 2% (W/V) Digitonin containing 0.067 M potassium phosphate buffer, pH 6.8
COMMENTS: Spectra before and after illumination at 0°C
REFERENCE: G. J. Sale, P. Towner, and M. Akhtar, Biochemistry, 16, 5641 (1977)

2142. Rhodopsin and Metarhodopsin (Acid and Alkaline)

SOURCE: Octopus (Octopus variabilis and Paroctopus defleini) eyes
SOLVENT: a. pH 5.5-5.6
b. pH 10.5
COMMENTS: Spectra before and after irradiation with blue and/or red light
REFERENCE: T. Suzuki, M. Sugahara, K. Azuma, M. Azuma, Y. Saimi, and Y. Kito, Biochim. Biophys. Acta, 333, 149 (1973); M. Tsuda, Biochim. Biophys. Acta, 578, 372 (1979)

2143. Rhodopsin

SOURCE: Squid (Ommastrephes sloani pacificus) eyes
SOLVENT: a. Trichloroacetic acid, 0.05 M-3 M
b. Trichloroacetic acid, 5 M
c. 0.1 M HCl
COMMENTS: Spectra made at liquid nitrogen temperature (solvent a) and at 5°C (solvents b and c)
REFERENCE: T. Suzuki and Y. Kito, Photochem. Photobiol., 15, 275 (1972)

2144. Rhodopsin

SOURCE: Squid (Watasenia scintillans) eyes
SOLVENT: 0.1% L-1690
COMMENTS: Spectra of rhodopsin, and trans-retinal-oxime and opsin
REFERENCE: K. Nashima, M. Mitsudo, and Y. Kito, Biochim. Biophys. Acta, 579, 155 (1979)

2145. Rhodopsin and N-Retinylopsin

SOURCE: Squid (Loligo pealii)
SOLVENT: 1.6% Digitonin in pH 6.4 phosphate buffer
COMMENTS: Spectral changes due to heat and light in presence of hydroxylamine
REFERENCE: F. M. Hagins, J. Biol. Chem., 248, 3298 (1973)

2146. Rhodopsin

SOURCE: Squid (Todarodes pacificus) eyes
SOLVENT: 1% Digitonin and 0.1 M borate-NaOH buffer, pH 10.5
COMMENTS: Spectra before and after irradiation with light at 6°C
REFERENCE: T. Suzuki, M. Sugahara, K. Azuma, M. Azuma, Y. Saimi, and Y. Kito, Biochim. Biophys. Acta, 333, 149 (1974)

2147. Rhodopsin

SOURCE: Squid (Todarodes pacificus) eyes
SOLVENT: a. 0.1 M Phosphate buffer, pH 5.6
b. 0.1 M Sodium carbonate buffer, pH 10.1
COMMENTS: Spectral changes during irradiation
REFERENCE: F. Tokunaga, Y. Shichichida, and T. Yoshisawa, FEBS Lett., 55, 229 (1975)

2148. Rhodopsin

SOURCE: Squid (Todarodes pacificus) eyes
SOLVENT: a. 2% LM (Lauryl ester of sucrose)
b. Glycerol pH 10.5-10.8
c. Glycerol pH 6.8
COMMENTS: Spectral changes due to illumination at various low temperatures
REFERENCE: K. Nashima, M. Mitsudo, and Y. Kito, Biochim. Biophys. Acta, 536, 78 (1978); Y. Shichida, F. Tokunaga, and T. Yoshizawa, Biochim. Biophys. Acta, 504, 413 (1978)

2149. Rhodopsin and Metarhodopsin

SOURCE: Drosophia melanogaster eyes
SOLVENT: a. 1% Triton X-100
b. 2% Digitonin extract
COMMENTS: 1. Rhodopsin in solvent a
2. Metarhodopsin in solvent b
REFERENCE: S. E. Ostroy, M. Wilson, and W. L. Pak, Biochem. Biophys. Res. Comm., 59, 960 (1974)

2150. Ribonuclease T_1, Modified

SOURCE: Commercial, Takadiastase powder
SOLVENT: 0.01 M Ammonium acetate
COMMENTS: Spectra of untreated, citraconylated, aconitylated, maleylated, and succinylated forms
REFERENCE: K. Takahashi, J. Biochem. Tokyo, 81, 641 (1977)

2151. Ribonuclease T_1

SOURCE: Commercial, Takadiastase powder
SOLVENT: a. Water, pH 5.6
b. 0.01 M Tris-HCl buffer, pH 7.5 containing 0.7 M hydroxylamine
COMMENTS: Spectra of native enzyme (solvent a), acetylated enzyme (solvent a) and deacetylated enzyme (solvent b)
REFERENCE: H. Kasai, K. Takahashi, and T. Ando, J. Biochem. Tokyo, 81, 1751 (1977)

2152. Ribonuclease T_1

SOURCE: Takadiastase powder
SOLVENT: a. Water
b. 0.05 M Tris-HCl buffer containing 0.1 M NaCl, pH 7.5
c. Buffer b containing 6 M guanidine-HCl

COMMENTS: 1. Spectral changes due to ozonation (solvent a)
2. Spectral changes due to deformylation of N'-formyl kynurenine-ribonuclease at position 59
3. Spectra of ribonuclease T_1 with tryptophan-59 replaced by kynurenine (solvents b and c)

REFERENCE: H. Tamaoki, F. Sakiyama, and K. Narita, J. Biochem. Tokyo, 83, 771 (1978)

2153. Ribonuclease Ms and Trinitrophenyl Derivatives

SOURCE: Aspergillus saitoi
SOLVENT: 0.1 M Acetate buffer, pH 5.0
REFERENCE: K. Ohgi and M. Irie, J. Biochem. Tokyo, 77, 1085 (1975); K. Ohgi, M. Kiryu, Y. Torano, Y. Hori, H. Watanabe, M. Iwama, and M. Irie, J. Biochem. Tokyo, 82, 939 (1977)

2154. Ribonuclease

SOURCE: Human serum
SOLVENT: Water
REFERENCE: K. K. Reddi, Biochem. Biophys. Res. Comm., 67, 110 (1975)

2155. Ribonuclease

SOURCE: Yeast (Rhodotorula glutinis)
SOLVENT: 0.01 M Phosphate buffer, pH 7.0
REFERENCE: K. Ogata, S. Song, and H. Yamada, Agric. Biol. Chem., 35, 58 (1971)

2156. Ribonuclease, Ribonuclease C_2 and Ribonuclease T_1

SOURCE: a. Penicillium chrysogenum 152A (RNA)
b. Aspergillus clavatus (RNA-C_2)
c. Aspergillus oryzae (RNA-T_1)

REFERENCE: S. I. Bezboradova, V. M. Grishchenko, and N. Y. Markelova, Biokhimiya (Biochemistry), 38, 275 (1973)

2157. Ribonuclease BS-1

SOURCE: Beef seminal fluid
COMMENTS: Spectra of ribonuclease BS-1 and pyridoxal-phosphate-modified ribonuclease BS-1
REFERENCE: G. Sannia and G. D'Alessio, Bull. Soc. Ital. Biol. Sperim., 48, 1 (1973)

2158. Ribonuclease BS-1

SOURCE: Bull seminal plasma
SOLVENT: a. Water
b. 6 M Guanidine HCl, 0.02 M phosphate buffer, pH 6.5
REFERENCE: G. D'Alessio, A. Floridi, R. De Prisco, A. Pignero, and E. Leone, Eur. J. Biochem., 26, 153 (1972)

2159. Ribonuclease U_1 (RU_1), Ribonuclease N_1 (RN_1), and Ribonuclease T_1 (RT_1)
SOURCE: Ustilago sphaerogena
SOLVENT: a. 0.01 M Acetate buffer, pH 6.0
b. 0.1 M Phosphate buffer, pH 8.0
c. Water
COMMENTS: Spectra of RU_1 in solvent a, RN_1 in solvent b, and RT_1 in solvent c
REFERENCE: J. Hashimoto, T. Uchida, and F. Egami, J. Biochem. Tokyo, 70, 903 (1971)

2160. Ribonuclease U_2

SOURCE: Ustilago sphaerogena
COMMENTS: Spectra of enzyme and ethoxyformylated enzyme
REFERENCE: S. Sato and T. Uchida, J. Biochem. Tokyo, 77, 795 (1975)

2161. Ribonuclease (R) and Asparaginase (A)

SOURCE: Beef pancreas (R), Erwinia carotova (A), and Escherichia coli (A)
SOLVENT: μ = 0.1 Carbonate buffer, pH 8.5
COMMENTS: Effect of Fremy's salt on spectra of ribonuclease and asparaginase
REFERENCE: R. B. Homer, S. R. Allsop, and J. E. Arrieta, FEBS Lett., 59, 173 (1975)

2162. Ribonuclease and Dimethylaminonaphthylsulfonyl-ribonuclease

SOURCE: Beef pancreas
SOLVENT: a. 0.1 M Sodium acetate buffer, pH 5.0
b. 0.1 M Acetic acid, pH 3.4
c. 0.1 M Sodium acetate, pH 7.3
COMMENTS: Spectra of native enzyme (solvent a) and dimethylaminonaphthylsulfonyl enzyme (solvents a, b and c)
REFERENCE: F. H. White, Jr., J. Biol. Chem., 239, 1032 (1964)

2163. Ribonuclease

SOURCE: Beef pancreas
SOLVENT: Water
COMMENTS: Spectrum of ribonuclease and 2,4-dinitrophenylene-1,5-ribonuclease
REFERENCE: P. S. Marfey, H. Nowack, M. Uziel, and D. Yphantis, J. Biol. Chem., 240, 3264 (1965)

2164. Ribonuclease

SOURCE: Beef pancreas
SOLVENT: pH 8
COMMENTS: Spectral changes due to reaction with 1.5% phenylglyoxal (0-60 minutes)
REFERENCE: K. Takahashi, J. Biol. Chem., 243, 6171 (1968)

2165. Ribonuclease, 4-chloro-3,5-dinitrophenacyl-

SOURCE: Beef pancreas
SOLVENT: Water
REFERENCE: J. Diopoh and M. Olomucki, Biochim. Biophys. Acta, 263, 220 (1972)

2166. Ribonuclease

SOURCE: Beef pancreas
COMMENTS: Spectra of O-acylated and unacylated protein
REFERENCE: A. Previero, L. G. Barry, and M. A. Coletti-Previero, Biochim. Biophys. Acta, 263, 7 (1972)

2167. Ribonuclease

SOURCE: Beef pancreas
COMMENTS: Spectral changes due to U.V. irradiation (0-260 minutes)
REFERENCE: S. Aktipis and A. J. Iammartino, Biochim. Biophys. Acta, 278, 239 (1972)

2168. Ribonuclease A, Carboxymethyl-

COMMENTS: Spectra before and after reaction with malonic aldehyde
REFERENCE: A. V. Muranov and N. N. Modjanov, Bioorganiskaya Chimiya, 5, 210 (1979)

2169. Ribonuclease

SOURCE: Beef pancreas
SOLVENT: pH 7

COMMENTS:	Spectral changes due to the addition of dimethylsulfoxide (0-80%)
REFERENCE:	A. Steinschneider and K. Druck, Biochim. Biophys. Acta, 287, 77-89 (1972)

2170. Ribonuclease A, Pyridoxamine Phosphate Derivatives

SOURCE:	Beef pancreas
SOLVENT:	pH 1.3-pH 11.0 in 0.2 M NaCl
COMMENTS:	Pyridoxamine phosphate is on the ε-amino groups of lysine-7 and lysine-41
REFERENCE:	S. M. Dudkin, L. V. Karabachyan, S. N. Borisova, S. V. Shlyapnikov, M. Y. Karpeisky, and T. G. Geidarov, Biochim. Biophys. Acta, 386, 275 (1975)

2171. Ribonuclease A

SOURCE:	Beef pancreas
SOLVENT:	pH 8.0
COMMENTS:	Spectral changes due to reaction with either phenylglyoxal, glyoxal, or methylglyoxal
REFERENCE:	J. I. Kuriaski, and K. Yamauchi, J. Biochem. Tokyo, 81, 443 (1977)

2172. Ribonuclease A, 5'-(4-azophenylphosphoryl)-uridine 2'(3')-phosphate-derivative

SOURCE:	Beef pancreas
SOLVENT:	a. pH 6 b. pH 11
REFERENCE:	M. Gorecki and M. Wilchek, Biochim. Biophys. Acta, 532, 81 (1978)

2173. Ribonuclease

SOURCE:	Beef pancreas
SOLVENT:	0.1 M Phosphate buffer, pH 7.0
REFERENCE:	T. Ichikawa and H. Terada, Biochim. Biophys. Acta, 580, 120 (1978)

2174. Ribonuclease, Nitrated

SOURCE:	Beef pancreas
SOLVENT:	0.2 M Tris-HCl, pH 8.0
REFERENCE:	R. Van Dar Zee, F. J. Duisterwinkel, and G. W. Welling, Eur. J. Biochem., 77, 125 (1977)

2175. Ribonuclease S, Deamidated

SOURCE:	Ribonuclease A

SOLVENT: 0.2 M Sodium phosphate buffer, pH 6.02
COMMENTS: Spectra of ribonucleases and deamidated-ribonuclease S
REFERENCE: B. N. Manjula, A. S. Acharya, and P. J. Vithayathil, Biochem. J., 165, 337 (1977)

2176. Ribonucleoproteins, 15S and 20S and Ribosomes

SOURCE: Rabbit reticulocytes, cytoplasm
REFERENCE: H. V. Tan and G. Shapira, Eur. J. Biochem., 85, 271 (1978)

2177. Ribonucleotide Reductase, Protein B1

SOURCE: Escherichia coli
SOLVENT: 0.1 M Tris-HCl, pH 7.6
COMMENTS: Spectra of 2'-deoxy-2'-chlorocytidine diphosphate-inactivated protein and native protein
REFERENCE: L. Thelender, B. Larsson, J. Hobbs, and F. Eckstein, J. Biol. Chem., 251, 1398 (1976)

2178. Ribonucleotide Reductase, Protein B2

SOURCE: Escherichia coli
REFERENCE: B. M. Sjoberg and A. Graslund, in Symposium on Molecular Interactions and Activity in Proteins, CIBA Foundation Symposium 60 (New Series), 187 (1978), Excerpta Medica, New York; A. Ehrenberg and P. Reichard, J. Biol. Chem., 247, 3485 (1972)

2179. Ribonucleotide Reductase

SOURCE: Induced in Escherichia coli by infection with bacteriophage T4
SOLVENT: 0.01 M N-2-Hydroxyethylpiperazine-N'-2-ethane sulfonic acid, pH 7.6-0.1 M KCl
REFERENCE: O. Berglund, J. Biol. Chem., 247, 7270 (1972)

2180. Ribosomal Particles

SOURCE: Escherichia coli
SOLVENT: 20 mM Tris, 20 mM $MgCl_2$, and 360 mM KCl, pH 7.6
COMMENTS: Spectra of 70S Ribosome, 50S subunit, and 30S subunit
REFERENCE: S. H. Allen and K. P. Wong, Arch. Biochem. Biophys., 195, 112 (1979)

2181. Ribulose Bisphosphate Carboxylase

SOURCE: Rhodospirillum rubrum
SOLVENT: 50 mM Potassium phosphate, pH 7.5-1 mM EDTA
COMMENTS: Spectra of native enzyme and enzyme modified with pyridoxal phosphate and sodium borohydride
REFERENCE: W. B. Whitman and F. R. Tabita, Biochemistry, 17, 1282 (1978)

2182. Ribulose 1,5-Bisphosphate Carboxylase-Oxygenase

SOURCE: Tobacco
SOLVENT: 0.025 M Tris-HCl, pH 7.4, 0.10 M NaCl, and 0.5 mM EDTA
REFERENCE: R. Chollet, L. L. Anderson, and L. D. Hovespain, Biochem. Biophys. Res. Comm., 64, 97 (1975)

2183. Ribulose 1,5-Diphosphate Oxygenase

SOURCE: Parsley
SOLVENT: pH 8.3
REFERENCE: R. Branden, Biochem. Biophys. Res. Comm., 81, 539 (1978)

2184. Ricin D

SOURCE: Castor bean
SOLVENT: 0.005 M Tris-HCl, pH 8.5
REFERENCE: M. Funatsu, K. Hara, M. Ishiguro, and G. Funatsu, Proc. Jap. Acad., 49, 754 (1973)

2185. tRNA Nucleotidyltransferase

SOURCE: Rabbit liver
SOLVENT: 0.01 M Potassium phosphate, pH 7.5
REFERENCE: M. P. Deutscher, J. Biol. Chem., 247, 450 (1972)

2186. RNA Polymerase, σ Subunit

SOURCE: Escherichia coli
SOLVENT: 0.1 M K_2SO_4, 5 mM Potassium phosphate, pH 7.0
REFERENCE: P. A. Lowe, D. A. Hager, and R. P. Burgess, Biochemistry, 18, 1344 (1979)

2187. Rubredoxin and Ferredoxin

SOURCE: Spirochaeta stenostrepta, Treponema denticola, Spirochaeta aurantia, and Clostridium pasteurianum
SOLVENT: a. 0.05 M Tris-HCl, pH 7.3
b. 0.15 M Tris-HCl, pH 7.3

COMMENTS: 1. Spectra of oxidized and reduced (sodium borohydride) rubredoxin in solvent a
2. Spectra of ferredoxins in solvent b
REFERENCE: P. W. Johnson and E. Canale-Parola, Arch. Microbiol. 89, 341 (1973)

2188. Rubredoxin

SOURCE: Pseudomonas oleovorans
COMMENTS: Spectra of 1 Fe- and 2 Fe-rubredoxins
REFERENCE: M. J. Coon, A. P. Autor, R. F. Boyer, E. T. Lode, and H. W. Strobel, in International Symposium on Oxidases and Related Redox Systems (T. E. King, H. S. Mason, and M. Morrison, eds.), 529 (1971), University Park Press, Baltimore

2189. Rubredoxin

SOURCE: Pseudomonas oleovorans
SOLVENT: 0.1 M Tris-Cl buffer, pH 7.3
COMMENTS: Spectra of soluble, immobilized, oxidized, and reduced (NADPH or dithionite) rubredoxin
REFERENCE: S. W. May and J. Y. Kuo, J. Biol. Chem., 252, 2390 (1977)

2190. Rubredoxin

SOURCE: Acinetobacter calcoaceticus
COMMENTS: Spectra of oxidized and reduced (dithionite) forms
REFERENCE: H. Aurich, D. Sorger, and O. Asperger, Acta Biol. Med. Germ., 35, 443 (1976)

2191. Rubredoxin

SOURCE: Ch. ethylica
COMMENTS: Spectra of 1 Fe-rubredoxin in the oxidized and reduced forms
REFERENCE: D. O. Hall, R. Cammack, and K. K. Rao, in Iron in Biochemistry and Medicine (A. Jacobs and M. Worwood, edts.), 279 (1974), Academic Press, New York

2192. Rubredoxin, Cobalt

SOURCE: Rubredoxin in pseudomonas oleovorans
SOLVENT: 0.05 M Tris, pH 7.3
REFERENCE: S. W. Ma and J. Y. Kuo, Biochemistry, 17, 3333 (1978)

2193. Rubredoxin Reductase, DPNH-dependent

SOURCE: Pseudomonas oleovorans
SOLVENT: 0.02 M Tris buffer, pH 7.3, containing 10% glycerol and 0.0005 M EDTA
REFERENCE: T. Ueda, E. T. Lode, and M. J. Coon, J. Biol. Chem., 247, 2109 (1972)

2194. Rubredoxin Reductase, DPNH-dependent

SOURCE: Pseudomonas oleovorans
SOLVENT: 0.02 M Tris-Cl buffer, pH 7.3, containing 10% (V/V) glycerol and 0.0005 M EDTA
COMMENTS:
1. Spectral changes due to urea
2. Spectral changes due to DPNH
3. Spectral changes due to DPN on EDTA-light-reduced enzyme (anaerobic)

REFERENCE: T. Ueda and M. J. Coon, J. Biol. Chem., 247, 5010 (1972)

2195. Rusticyanin

SOURCE: Thiobacillus ferro-oxidans
SOLVENT: 10 mM H_2SO_4, pH 2.0
COMMENTS: Spectra of oxidized and reduced (10 mM $FeSO_4$) forms
REFERENCE: J. C. Cox and D. H. Boxer, Biochem. J., 174, 497 (1978)

2196. Saccharopine Dehydrogenase

SOURCE: Baker's yeast
SOLVENT: 0.1 M Potassium phosphate buffer, pH 6.9
COMMENTS: Spectral change due to diethylpyrocarbonate
REFERENCE: M. Fujioka, Y. Takata, H. Ogawa, and M. Okamoto, J. Biol. Chem., 255, 937 (1980)

2197. Salicylate Hydroxylase

SOURCE: Pseudomonas putida
SOLVENT: 20 mM Tris-HCl, pH 8.0
COMMENTS:
1. Spectral changes due to salicylate
2. Spectral changes during the anaerobic addition of NADH in presence of salicylate

REFERENCE: M. Katagiri, S. Takemori, M. Nakamura, and T. Nakamura, in International Symposium on Oxidases and Related Redox Systems (T. E. King, H. S. Mason, and M. Morrison, eds.), 165 (1977) University Park Press, Baltimore

2198. Salicylate Hydroxylase

SOURCE: Microbial
SOLVENT: 90 mM Phosphate buffer, pH 7.62
COMMENTS:
1. Spectral changes during anaerobic addition of DPNH to enzyme, enzyme and benzoate, and enzyme plus salicylate
2. Spectral changes during anaerobic illumination in the presence of EDTA of enzyme, enzyme plus benzoate, and enzyme plus salicylate
3. Spectral changes during anaerobic addition of sodium sulfite to enzyme and enzyme plus benzoate

REFERENCE: R. H. White-Stevens, H. Kamin, and Q. H. Gibson, J. Biol. Chem., 247, 2371 (1972)

2199. Salicylate Hydroxylase

SOURCE: Microbial
SOLVENT: 0.0 M Potassium phosphate
COMMENTS: Spectra of reduced (dithionite) and oxidized forms
REFERENCE: R. H. White-Stevens, and H. Kamin, J. Biol. Chem., 247, 2351 (1972)

2200. Salmine, dinitrophenyl-

SOURCE: Salmon (Onchorhynchus keta) mature milts
SOLVENT: a. 1% Sodium bicarbonate
b. 1 N HCl
REFERENCE: T. Ando, M. Yamasaki, E. Abukumagawa, S. Ishi, and Y. Nagai, J. Biochem. Tokyo, 45, 429 (1958)

2201. Salmine

SOURCE: Keta salmon
SOLVENT: 0.1 N HCl
COMMENTS: Spectral changes due to reaction with fluorodinitrobenzene
REFERENCE: D. M. P. Phillips, Biochem. J., 60, 403 (1955)

2202. Sarcoplasmic Reticulum, Solubilized

SOURCE: Rabbit skeletal muscle
SOLVENT: 10 mM Tris-N-Tris-(hydroxymethyl)-methyl-2-aminoethane sulfonic acid, 0.1 mM $CaCl_2$, 0.35 M KCl, 0.25 M sucrose, pH 8.0
REFERENCE: T. Nakamura, Y. Tonomura, and B. Hagihara, J. Biochem. Tokyo, 86, 443 (1979)

2203. Sarcosine Dehydrogenase

SOURCE: Pseudomonas putida
SOLVENT: 50 mM Tris-HCl buffer, pH 7.2
COMMENTS: Spectral changes due to addition of sarcosine and sarcosine plus sodium sulfite
REFERENCE: J. Oka, T. Yoshimoto, K. Rikitake, S. Ogushi, and D. Tsuru, Agric. Biol. Chem., 43, 1197 (1979)

2204. Sarcosine Dehydrogenase

SOURCE: Rat liver mitochondria
SOLVENT: 20 mM Potassium phosphate buffer, pH 7.2
REFERENCE: M. Sato, N. Ohishi, and K. Yagi, Biochem. Biophys. Res. Comm., 87, 706 (1979)

2205. D-Serine Dehydratase

SOURCE: Escherichia coli
SOLVENT: 100 mM Potassium phosphate buffer, pH 7.8
COMMENTS:
1. Spectral changes due to reaction with D-serine
2. Spectral changes due to reaction with 4-(2-methyl-3-hydroxy-5-(phosphooxy-methyl)-4-pyridinyl)-2-oxo-3-butenoic acid

REFERENCE: K. D. Schnackerz, J. H. Ehrlich, W. Giesemann, and T. A. Reed, Biochemistry, 18, 3557 (1979)

2206. Serine Sulfhydrase

SOURCE: Chicken liver
SOLVENT: 0.1 M Tris buffer, pH 8.2
COMMENTS:
1. Spectrum of enzyme
2. Spectrum of enzyme plus serine
3. Spectrum of enzyme plus cysteine

REFERENCE: N. D. Lac, Biokhimiya (Biochemistry), 34, 696 (1969)

2207. Serine Sulfhydrase

SOURCE: Chicken liver
SOLVENT: 0.1 M Tris buffer, pH 8.5
COMMENTS:
1. Spectrum of enzyme
2. Spectrum of enzyme plus L-serine
3. Spectrum of enzyme plus DL-homocysteine
4. Spectrum of enzyme plus β-mercaptoethanol
5. Spectrum of enzyme plus glutathione
6. Spectrum of enzyme plus L-cysteine

REFERENCE: L. L. Efremova and E. V. Goryachenkova, Biokhimiya (Biochemistry), 40, 422 (1975)

2208. Serine-threonine Dehydratase

SOURCE:	Sheep liver
SOLVENT:	0.1 M Phosphate buffer, pH 7.2
COMMENTS:	Spectra of enzyme and of phenylhydrazone of enzyme
REFERENCE:	G. Kapke and L. Davis, Biochemistry, 14, 4273 (1975)

2209. Serine Transhydroxymethylase-glycine Complex

SOURCE:	Rabbit liver
SOLVENT:	0.01 M Potassium phosphate, pH 5.8-pH 7.9
COMMENTS:	Spectral changes of glycine-enzyme complex with pH and temperature
REFERENCE:	L. Schirch and A. Diller, J. Biol. Chem., 246, 3961 (1971); L. Schirch, J. Biol. Chem., 250, 1939 (1975)

2210. Spectrin-heme Complex

SOURCE:	Human erythrocytes
SOLVENT:	50 mM Phosphate buffer, pH 6.6
COMMENTS:	Spectra of oxidized and reduced (dithionite) forms
REFERENCE:	R. Cassoly, FEBS Lett., 85, 357 (1978)

2211. Spermidine Dehydrogenase

SOURCE:	Serratia marcescene
SOLVENT:	0.033 M Potassium phosphate buffer, pH 7.2
COMMENTS:	Spectra of enzyme and enzyme plus spermidine
REFERENCE:	C. W. Tabor and P. D. Kellog, J. Biol. Chem., 245, 5424 (1970)

2212. Stearylcoenzyme A Desaturase

SOURCE:	Rat liver microsomes
REFERENCE:	P. Strittmatter, L. Spatz, D. Corcoran, M. J. Rogers, B. Setlow, and R. Redune, Proc. Nat. Acad., 71, 4565 (1974)

2213. Stearylcoenzyme A Desaturase

SOURCE:	Rat liver microsomes
SOLVENT:	Tris buffer containing 0.2% dodecylsulfate
COMMENTS:	Spectra of native and acetylated enzyme, and of acetylated plus hydroxylamine
REFERENCE:	H. G. Enoch and P. Strittmatter, Biochemistry, 17, 4927 (1978)

2214. Stellacyanin

SOURCE: Lacquer tree (Rhus vernicifera)
COMMENTS: Spectra of native and reconstituted protein
REFERENCE: L. Morpurgo, A. Finazzi-Agro, G. Rotilio, and B. Mondovi, Biochim. Biophys. Acta, 271, 292 (1972)

2215. Stellacyanin

SOURCE: Rhus vernicifera
SOLVENT: 0.1 M Sodium phosphate, pH 7.5
REFERENCE: K. E. Falk and B. Reinhammar, Biochim. Biophys. Acta, 285, 84 (1972)

2216. Stellacyanin, Cobalt (II)

SOURCE: Rhus vernicifera
SOLVENT: Tris buffer, pH 9.5
REFERENCE: D. R. McMillin, O. R. A. Holwerda, and H. B. Gray, Proc. Nat. Acad. Sci., 71, 1339 (1974)

2217. Stellacyanin

SOURCE: Rhus vernicifera
SOLVENT: 0.67 M Carbonate buffer, pH 9, pH 10, pH 11
REFERENCE: L. Murpurgo, G. Rotilio and A. Finazzi-Agro, B. Mondovi, Eur. J. Biochem., 64, 453 (1976); B. Mondovi, L. Morpurgo, G. Rotilio, and A. Finazzi-Agro, Adv. Exp. Med. Biol., 74, 424 (1976)

2218. Stellacyanin

SOURCE: Rhus vernicifera
COMMENTS: Spectra of films at 270°K and 35°K
REFERENCE: E. I. Solomon, J. W. Hare, and H. B. Gray, Proc. Nat. Acad. Sci., 73, 1389 (1976)

2219. Stroma Suspension

SOURCE: Human erythrocytes, Hb S
SOLVENT: a. Water
b. 5 mM Phosphate, 0.5 mM EDTA, pH 7.5
c. 10 mM Phosphate, pH 7.5
d. 20 mM Phosphate, pH 7.5
COMMENTS: Spectra of suspensions of stroma prepared by four different methods
REFERENCE: T. Asakura, K. Minakata, K. Adachi, M. Russell, and E. Schwartz, J. Clin. Invest., 59, 641 (1977)

2220. Subtilopeptidase and Derivatives

SOURCE: Bacillus subtilis var. amylosaccharitus
SOLVENT: pH 8.0
COMMENTS: Spectra of native, nitrotyrosyl-, and aminotyrosyl-protéins
REFERENCE: D. Tsuru and T. Hirose, J. Biochem. Tokyo, 71, 555 (1972)

2221. Subtilisin, Diazoactyl-

SOURCE: Commercial subtilisin
COMMENTS: Spectral changes during photolysis
REFERENCE: Y. Stefanovsky and F. Westheimer, Proc. Nat. Acad. Sci., 70, 1132 (1973)

2222. Subtilisin Carlsberg, Phenylmethylsulfonyl-

SOURCE: Commercial source
SOLVENT: 0.1 M Phosphate buffer, pH 6.8
REFERENCE: J. Schlessinger, R. S. Roche, and I. Z. Steinberg, Biochemistry, 14, 255 (1975)

2223. Succinic Dehydrogenase

SOURCE: Rhodospirillum rubrum
SOLVENT: 50 mM Tris-acetate, pH 8.0
COMMENTS: Spectra of oxidized enzyme, enzyme plus succinate, and after adding dithionite to succinate-treated enzyme
REFERENCE: Y. Hatefi, K. A. Davis, H. Baltscheffsky, M. Baltscheffsky, and B. C. Johansson, Arch. Biochem. Biophys., 152, 613 (1972)

2224. Succinate Dehydrogenase

SOURCE: Heart mitochondria
SOLVENT: 0.1 M Carbonate-bicarbonate buffer, pH 10
COMMENTS:
1. Spectral decay of enzyme
2. Spectrum of dithionite-reduced enzyme
3. Spectral decay during cyanolysis of enzyme

REFERENCE: T. E. King, T. Ohnishi, D. B. Winter, and J. T. Wu, Adv. Exp. Med. Biol., 74, 182 (1976)

2225. Succinyl-CoA Synthetase

SOURCE: Escherichia coli
SOLVENT: 0.1 M Sodium phosphate buffer, pH 6.35
COMMENTS: Spectra before and after reaction with ethoxyformic anhydride

REFERENCE: G. E. Collier and J. S. Nishimura, J. Biol. Chem., 254, 10925 (1979)

2226. Sulfite Oxidase

SOURCE: Chicken liver
SOLVENT: a. 0.05 M Potassium phosphate, pH 7.8
b. 0.1 M Phosphate buffer, pH 7.4
c. 0.10 M Tris-HCl, pH 8.5
COMMENTS: 1. Spectra before and after addition of sodium sulfite (solvent a)
2. Spectra of dithionite-reduced enzyme and core (solvent b)
3. Spectra of ferricyanide-treated and untreated enzyme (solvent c)
REFERENCE: D. L. Kessler and K. V. Rajagopalan, J. Biol. Chem., 247, 6566 (1972); B. Guiard and F. Lederer, Eur. J. Biochem., 74, 81 (1977); D. L. Kessler, and K. V. Rajagopalan, Biochim. Biophys. Acta, 370, 399 (1974)

2227. Sulfite Oxidase

SOURCE: Rat liver
SOLVENT: a. 0.05 M Potassium phosphate, pH 7.8 and pH 7.0
b. 0.1 M Tris-HCl, pH 7.0 and pH 9.0
COMMENTS: 1. Spectra of protein and molybdenum-free protein
2. Spectra of enzyme and molybdenum-free enzyme in presence of 1 mM sodium sulfite or dithionite
3. Spectra of oxidized and sulfite-reduced fragments containing molybdenum
REFERENCE: J. L. Johnson, H. J. Cohen, and K. V. Rajagopalan, J. Biol. Chem., 249, 5046 (1974); J. L. Johnson and K. V. Rajagopalan, J. Biol. Chem., 252, 2017 (1977)

2228. Sulfite Oxidase, Reduced

SOURCE: Rat liver
SOLVENT: 0.05 M Phosphate, pH 7.5
COMMENTS: Spectra taken at liquid nitrogen temperature
REFERENCE: H. J. Cohen, S. Betcher-Lange, D. L. Kessler, and K. V. Rajagopalan, J. Biol. Chem., 247, 7759 (1972)

2229. Sulfite Oxidase

SOURCE: Rat liver
SOLVENT: 0.05 M Potassium phosphate

COMMENTS: 1. Spectra of oxidized, reduced (sulfite or cytochrome b_5) heat inactivated enzyme
2. Spectra of molybdenum domain, heat-inactivated and untreated
REFERENCE: W. B. Southerland and K. V. Rajagopalan, J. Biol. Chem., 253, 8753 (1978)

2230. Sulfite Oxidase

SOURCE: a. Rat liver
b. Human liver
SOLVENT: 0.05 M Potassium phosphate, pH 7.8
REFERENCE: J. L. Johnson and K. V. Rajagopalan, J. Clin. Invest., 58, 543 (1976)

2231. Sulfite Reductase

SOURCE: Red alga (Porphyra yezoensis)
SOLVENT: 0.05 M Potassium phosphate buffer, pH 7.5
REFERENCE: E. Saito and G. Tamura, Agric. Biol. Chem., 35, 491 (1971)

2232. Sulfite Reductase

SOURCE: Desulfovibrio vulgaris
SOLVENT: 0.01 M Tris buffer, pH 7.0
REFERENCE: K. Kobayashi, E. Takahashi, and M. Ishimoto, J. Biochem. Tokyo, 72, 879 (1972)

2233. Sulfite Reductase and Desulfoviridin

SOURCE: a. Escherichia coli
b. Desulfotomaculum nigrificans
c. Desulfovibrio gigas
SOLVENT: 0.05 M Potassium phosphate buffer, pH 7.7, containing 0.1 mM EDTA
REFERENCE: M. J. Murphy and L. M. Siegal, J. Biol. Chem., 248, 6911 (1973)

2234. Sulfite Reductase

SOURCE: Escherichia coli
SOLVENT: a. 0.1 M Potassium phosphate, pH 7.7
b. 0.05 M Potassium phosphate, pH 7.7
COMMENTS: 1. Spectra of enzyme, p-chloromercuryphenylsulfonate enzyme (PCMPS-E), PCMPS-E plus NADPH, and PCMPS-E plus NADPH plus dithionite in solvent a
2. Spectra of oxidized enzyme (OE), OE plus sulfite, OE plus NADPH, and OE plus NADPH-sulfite

3. Spectra of enzyme, enzyme plus sodium arsenite, enzyme plus NADPH, and enzyme plus arsenite plus NADPH

REFERENCE: L. M. Siegel, P. S. Davis, and H. Kamin, J. Biol. Chem., 249, 1572 (1974)

2235. Sulfite Reductase, Desulforubidin, and Desulfoviridin

SOURCE: Escherichia coli, Desulfovibrio desulfuricans, Norway and Desulfovibrio gigas

COMMENTS: Spectra of sulfite reductase, desulforubidin, their CO-complexes, and desulfoviridin

REFERENCE: M. J. Murphy, L. M. Siegel, H. Kamin, D. V. Der Vartanian, J. P. Lee, J. LeGall and H. D. Peck, Biochem. Biophys. Res. Comm., 54, 82 (1973)

2236. Sulfite Reductase, NADPH-

SOURCE: Escherichia coli B

SOLVENT:
a. 0.05 M Potassium phosphate, pH 7.7, containing 0,1 mM EDTA
b. 0.1 N NaOH
c. 0.02 N NaOH

COMMENTS:
1. Spectral changes due to NADPH and to dithionite
2. Spectra of enzyme-pyridine complex, and enzyme-cyanide complex, both in presence of dithionite in solvent b
3. Spectrum of reduced enzyme (dithionite) in solvent c
4. Spectral changes due to CO
5. Spectral changes due to cyanide

REFERENCE: L. M. Siegel, M. J. Murphy, and H. Kamin, J. Biol. Chem., 248, 251 (1973)

2237. Sulfite Reductase Hemoprotein and CO-Complex

SOURCE: Escherichia coli

SOLVENT: 0.5 M Potassium phosphate, pH 7.7, containing 0.1 mM EDTA

REFERENCE: M. J. Murphy, L. M. Siegel, and H. Kamin, J. Biol. Chem., 249, 1610 (1974); L. M. Siegel and P. S. Davis, J. Biol. Chem., 249, 1587 (1974)

2238. Sulfite Reductase Flavoprotein

SOURCE: Escherichia coli

SOLVENT: 0.05 M Potassium phosphate-0.1 mM EDTA, pH 7.7

REFERENCE: L. M. Siegel and P. S. Davis, J. Biol. Chem., 249, 1587 (1974)

2239. Sulfurylase, ATP-

SOURCE: Penicillium chrysogenum
SOLVENT: 0.001 M Potassium phosphate
REFERENCE: J. W. Tweedie and I. H. Segal, Prep. Biochem., 1, 91 (1971)

2240. Sulphatase A Derivatives

SOURCE: Sulphatase of beef liver
SOLVENT: 0.1 M NaOH
COMMENTS: Spectra of sulphanilazosulphatase and trimethylaminobenzeneazosulphatase
REFERENCE: A. Jerfy and A. B. Roy, Biochim. Biophys. Acta, 371 (1974)

2241. Superoxide Dismutase

SOURCE: Human liver
REFERENCE: J. M. McCord, J. A. Boyle, E. D. Day, Jr., L. J. Rizzolo, and M. L. Salin, in Superoxide and Superoxide Dismutases (A. M. Michelson, J. M. McCord, and I. Fridovich, eds.), 129 (1977), Academic Press, New York; J. M. McCord, Adv. Exp. Med. Biol., 74, 540 (1973)

2242. Superoxide Dismutase (SOD)

SOURCE: Human erythrocytes
SOLVENT: a. Water
b. pH 1.9
COMMENTS: Spectra of SOD I, SOD II, Apo-SOD I, and Apo SOD II
REFERENCE: W. H. Bannister, A. Anastasi, and J. V. Bannister, in Superoxide and Superoxide Dismutases (A. M. Michelson, J. M. McCord and I. Fridovich, eds.), 107 (1977), Academic Press, New York

2243. Superoxide Dismutase

SOURCE: Human erythrocytes
SOLVENT: 10 mM Sodium phosphate, pH 7.5
REFERENCE: S. Marklund, G. Beckman, and T. Stigbrand, Eur. J. Biochem., 65, 415 (1976)

2244. Superoxide Dismutase

SOURCE: Pig liver and pig erythrocytes
REFERENCE: A. Bartkowiak, W. Leyko, and R. Fried, Comp. Biochem. Physiol., 62B, 61 (1978)

2245. Superoxide Dismutase (SOD)

SOURCE: Chicken liver (cytoplasm and mitochondrial), Streptococcus mutans, and Escherichia coli
SOLVENT: 0.05 M Potassium phosphate, 0.0001 M EDTA, pH 7.8
COMMENTS: Spectra of oxidized and reduced (dithionite) forms of cytoplasmic SOD
REFERENCE: R. A. Weisiger and I. Fridovich, J. Biol. Chem., 248, 5382 (1973)

2246. Superoxide Dismutase

SOURCE: Swordfish (Xiphias gladius L.) liver and beef liver
SOLVENT: Water
REFERENCE: J. V. Bannister, A. Anastasi, and W. H. Bannister, Comp. Biochem. Physiol., 56B, 235 (1977)

2247. Superoxide Dismutase

SOURCE: Beef
SOLVENT: a. 0.05 M Phosphate buffer, pH 7.2
b. Water
COMMENTS: 1. Spectral changes due to cyanide in solvent a
2. Spectral changes due to azide in solvent b
3. Spectral changes due to fluoride in solvent b
4. Spectra of type I and type II apo-enzymes to which copper and zinc ions were added
REFERENCE: G. Rotilio, L. Morpurgo, C. Giovagnoli, L. Calabrese, and B. Mondovi, Biochemistry, 11, 2187 (1972)

2248. Superoxide Dismutase (Cuprein)

SOURCE: Beef liver and yeast (Saccharomyces cerevisiae)
SOLVENT: a. 5 mM Potassium phosphate buffer, pH 7.2
b. 50 mM Potassium phosphate buffer, pH 7.2
REFERENCE: U. Weser, R. Prinz, A. Schallies, A. Fretzdorff, P. Krauss, W. Voelker, and W. Voetsch, Hoppe-Seyler's Z. Physiol. Chem., 353, 1821 (1972);
U. Weser, A. Fretzdorff, and R. Prinz, FEBS Lett., 27, 267 (1972)

2249. Superoxide Dismutase

SOURCE: Beef erythrocytes
COMMENTS: Spectral changes due to reaction with ferrocyanide
REFERENCE: G. Rotilio, L. Morpurgo, L. Calabrese, and B. Mondovi, Biochim. Biophys. Acta, 302, 229 (1973)

2250. Superoxide Dismutase

SOURCE: Beef erythrocytes
SOLVENT: a. Water
b. 0.05 M Phosphate buffer, pH 7.8
COMMENTS: 1. Spectra of native and apo-enzyme in solvent a
2. Spectra of reconstituted enzyme: apo-enzyme plus the following metal ions: copper plus cadmium, mercury plus copper, copper plus zinc, copper plus cobalt, and copper alone (solvents a and b)
REFERENCE: K. M. Beem, W. E. Rich, and K. V. Rajagopalan, J. Biol. Chem., 249, 7298 (1974)

2251. Superoxide Dismutase

SOURCE: Beef erythrocytes
SOLVENT: Water
COMMENTS: Spectra of enzyme and cobalt (II) enzyme
REFERENCE: G. Rotilio, L. Calabrese, B. Mondovi, and W. E. Blumberg, J. Biol. Chem., 249, 3159 (1974)

2252. Superoxide Dismutase

SOURCE: Beef erythrocytes
SOLVENT: a. 0.05 M Sodium carbonate-0.0001 M EDTA, pH 10.2
b. 0.02 M Sodium pyrophosphate, pH 9.0
c. 0.05 M Potassium phosphate-0.0001 M EDTA, pH 7.8
COMMENTS: 1. Spectral changes due to azide (solvent a)
2. Spectral changes due to hydrogen peroxide (solvent b)
3. Spectral changes due to hydrogen peroxide in the presence of cyanide (solvent c)
REFERENCE: E. K. Hodgson and I. Fridovich, Biochemistry, 14, 5294 (1975)

2253. Superoxide Dismutase

SOURCE: Beef erythrocytes
SOLVENT: a. pH 2.99-pH 5.56
b. pH 2.96
COMMENTS: 1. Spectrum of Co (II)-Cu (II) protein in solvent a
2. Spectrum of Zn (II)-Cu (II) protein in solvent b
REFERENCE: L. Calabrese, D. Cocco, L. Morpurgo, B. Mondovi, and G. Rotilio, FEBS Lett., 59, 29 (1975)

2254. Superoxide Dismutase

SOURCE: Beef erythrocytes
SOLVENT: pH 3.2-pH 4.2
REFERENCE: J. A. Fee and W. D. Phillips, Biochim. Biophys. Acta, 412, 26 (1975); J. A. Fee, in Superoxide and Superoxide Dismutase (A. M. Michelson, J. M. McCord and I. Fridovich, eds.), 173 (1977), Academic Press, New York

2255. Superoxide Dismutase, Co (II)-, and Co (II)-Cu (II) (SOD-CuCo)

SOURCE: Beef erythrocytes
SOLVENT: a. 0.1 M Borate buffer, pH 8.9
b. Water
c. pH 3.5
d. pH 11.5
e. pH 3.0
COMMENTS: 1. Spectral changes due to cyanide (solvent a)
2. Spectra in solvents b-d
3. Spectra in solvent a and solvent b (SOD-CuCo)
REFERENCE: L. Calabrese, D. Cocco, L. Morpurgo, B. Mondovi, and G. Rotilio, Eur. J. Biochem., 64, 465 (1976)

2256. Superoxide Dismutase

SOURCE: Beef liver, kidneys, and adrenals
REFERENCE: N. A. Grigoryan, G. V. Gyulkhandanyan, M. A. Simonyan, and R. M. Nalbandyan, Biokhimiya (Biochemistry), 42, 1171 (1977)

2257. Superoxide Dismutase

SOURCE: Beef erythrocytes
SOLVENT: 0.156 M Sodium bicarbonate, pH 8.0
COMMENTS: Spectral changes upon reaction with phenylglyoxal
REFERENCE: D. P. Malinowski and I. Fridovich, Biochemistry, 18, 5909 (1977)

2258. Superoxide Dismutase, Cu-Zn-

SOURCE: Beef erythrocytes
SOLVENT: 50 mM Potassium phosphate, pH 7.4
COMMENTS: Spectral changes upon reaction with diethyldithiocarbamate
REFERENCE: H. P. Misra, J. Biol. Chem., 254, 11623 (1979)

2259. Superoxide Dismutase

SOURCE: Beef erythrocytes
SOLVENT: 0.56 M Sodium borate, pH 8.0
COMMENTS: Spectra of native enzyme and enzyme modified with diazonium 1H-tetrazole
REFERENCE: D. P. Malinowski and I. Fridovich, Biochemistry, 18, 237 (1979)

2260. Superoxide Dismutase

SOURCE: Bacillus megaterium
SOLVENT: pH 7.8
COMMENTS: Spectra of holoprotein and apoprotein
REFERENCE: A. Anastasi, J. V. Bannister, and W. H. Bannister, Int. J. Biochem., 7, 541 (1976)

2261. Superoxide Dismutase

SOURCE: Chlorobium thiosulfatophilum
SOLVENT: 10 mM Potassium phosphate, pH 7.8
REFERENCE: S. Kanematsu and K. Asada, FEBS Lett., 91, 94 (1978)

2262. Superoxide Dismutase

SOURCE: Chromatium vinosum
SOLVENT: a. 10 mM Potassium phosphate, pH 7.8
b. 50% glycerol (V/V)-0.1 M Tris-HCl, pH 7.5
COMMENTS: 1. Spectrum in solvent a at 23°C
2. Spectrum in solvent b at 77°C
REFERENCE: K. Asada, S. Kanematsu, M. A. Takahashi, and Y. Kona, Adv. Exp. Med., 74, 551 (1976); S. Kanematsu and K. Asada, Arch. Biochem. Biophys., 185, 473 (1978)

2263. Superoxide Dismutase

SOURCE: Desulfovibrio desulfuricans (Norway 4)
SOLVENT: 10 mM Potassium phosphate, pH 7
REFERENCE: E. C. Hatchikian and Y. A. Henry, Biochimie, 59, 153 (1977); E. C. Hatchikan, J. E. LeGall, and G. R. Bell, in Superoxide and Superoxide Dismutases (A. M. Michelson, J. M. McCord, and I. Fridovich, eds.), 159 (1977), Academic Press, New York

2264. Superoxide Dismutase

SOURCE: Escherichia coli
SOLVENT: 0.05 M Phosphate buffer, pH 7.8

REFERENCE: F. J. Yost and I. Fridovich, J. Biol. Chem., 248, 4905 (1973)

2265. Superoxide Dismutase

SOURCE: Escherichia coli B
SOLVENT: 50 mM Potassium phosphate, pH 7.8
REFERENCE: H. W. Dougherty, S. J. Sadowski, and E. E. Baker, J. Biol. Chem., 253, 5220 (1978)

2266. Superoxide Dismutase

SOURCE: Escherichia coli
SOLVENT: 0.05 M Phosphate buffer, pH 7.8
COMMENTS:
1. Spectral changes upon addition of hydrogen peroxide; anaerobically and after admission of air
2. Spectral changes upon addition of ferrocyanide; anaerobically and after admission of air

REFERENCE: B. B. Klee, G. Giovagnoli and G. Rotilio, Physiol. Chem. Physics, 7, 1 (1975)

2267. Superoxide Dismutase

SOURCE: Escherichia coli
COMMENTS: Spectra of manganese-containing protein and cobalt-substituted derivative
REFERENCE: D. E. Ose and I. Fridovich, Arch. Biochem. Biophys., 194, 360 (1979)

2268. Superoxide Dismutase

SOURCE: Green alga (Euglena gracilis)
REFERENCE: E. Lengfelder and E. F. Elstner, Z. Naturforsch., 34C, 374 (1979)

2269. Superoxide Dismutase I and II

SOURCE: Green alga (Euglena gracilis)
SOLVENT: 10 mM Potassium phosphate, pH 7.8
REFERENCE: S. Kanematsu and K. Asada, Arch. Biochem. Biophys., 195, 535 (1979)

2270. Superoxide Dismutase

SOURCE: Green pea (Pisum sativum)
SOLVENT:
a. 10 mM Potassium phosphate, pH 7.0
b. pH 7.8
c. pH 2.8

REFERENCE: Y. Sawada, T. Ohyama, and I. Yomazaki, Biochim. Biophys. Acta, 268, 305 (1972); Y. Sawada, T. Ohyama, and I. Yamazaki, in International Symposium on Oxidases and Related Redox Systems (T. E. King, H. S. Mason, and M. Morrison, eds.), 82 (1971), University Park Press, Baltimore; E. A. Permyakov, E. A. Burstein, Y. Sawada, and I. Yamazaki, Biochim. Biophys. Acta, 491, 149 (1977)

2271. Superoxide Dismutase

SOURCE: Mycobacterium lepraermurium
SOLVENT: 50 mM Potassium phosphate buffer, pH 7.0
REFERENCE: K. Ichihara, E. Kusunose, M. Kusunose, and T. Mori, J. Biochem. Tokyo, 81, 1427 (1977)

2272. Superoxide Dismutase

SOURCE: Mycobacterium tuberculosis
SOLVENT: 50 mM Potassium phosphate buffer, pH 7.0
REFERENCE: E. Kusunose, K. Ichihara, Y. Noda and M. Kusunose, J. Biochem. Tokyo, 80, 1343 (1976)

2273. Superoxide Dismutase

SOURCE: Neurospora crassa
SOLVENT: 0.05 M Potassium phosphate, pH 7.8
REFERENCE: H. P. Misra and I. Fridovich, J. Biol. Chem., 247, 3410 (1972)

2274. Superoxide Dismutase

SOURCE: Photobacterium leignathi
SOLVENT: a. 0.005 M Phosphate, pH 7.4-7.8
b. 0.05 M NaCl-0.05 M Glycine, pH 8.6
c. Buffer b-50% ethanol
COMMENTS: Spectra of native and heat-denatured enzyme (solvent a-c)
REFERENCE: K. Puget and A. M. Michelson, Biochem. Biophys. Res. Comm., 58, 830 (1974); K. Puget and A. M. Michelson, Biochimie, 56, 1255 (1974)

2275. Superoxide Dismutase

SOURCE: Plectonema boryanum
SOLVENT: a. 10 mM Potassium phosphate, pH 7.8
b. Solvent a-50% glycerol (V/V)
COMMENTS: Spectra at room temperature in solvent a and at 77°K in solvent b

REFERENCE: K. Asada, K. Yoshikawa, M. A. Takahashi, Y. Maeda, and K. Enmanji, J. Biol. Chem., 25, 2801 (1975)

2276. Superoxide Dismutase

SOURCE: Pleurotus olearius
SOLVENT: 0.01 M Phosphate, pH 7.0
REFERENCE: F. Lavelle, P. Durosay and A. M. Michelson, Biochimie, 56, 451 (1974)

2277. Superoxide Dismutase

SOURCE: Porphyridium cruentum
SOLVENT: 0.05 M Potassium phosphate, pH 7.8
REFERENCE: H. P. Misra and I. Fridovich, J. Biol. Chem., 252, 6421 (1977)

2278. Superoxide Dismutase

SOURCE: Pseudomonas ovalis
SOLVENT: 50 mM Potassium phosphate buffer, pH 7.8
COMMENTS: 1. Spectra of enzyme, apoenzyme, and reconstituted enzyme
2. Spectrum of reduced (dithionite) enzyme
REFERENCE: F. Yamakura, J. Biochem. Tokyo, 83, 849 (1978); F. Yamakura, Biochim. Biophys. Acta, 422, 280 (1976); F. Yamakura and K. Suzuki, Biochem. Biophys. Res. Comm., 72, 1108 (1976)

2279. Superoxide Dismutase

SOURCE: Rhodopseudomonas spheroides
SOLVENT: 50 mM Tris-HCl, 0.1 M KCl, pH 7.8
COMMENTS: Spectrum taken at 77°K
REFERENCE: J. Lumsden, R. Cammack, and D. O. Hall, Biochim. Biophys. Acta, 438, 380 (1976)

2280. Superoxide Dismutase

SOURCE: Spinach
SOLVENT: 10 mM Potassium phosphate, pH 7.8
REFERENCE: K. Asada, M. Urano, and M. Takahashi, Eur. J. Biochem., 36, 257 (1973)

2281. Superoxide Dismutase

SOURCE: Streptococcus mutans 6715
SOLVENT: 0.05 M Tris-HCl, 0.1 M KCl, pH 7.8
REFERENCE: P. G. Vance, B. B. Keele, Jr., and K. V. Rajagopalan, J. Biol. Chem., 247, 4782 (1972)

2282. Superoxide Dismutase

SOURCE: Thermus aquaticus
SOLVENT: 5 mM Tris-HCl buffer, pH 7.5, containing 0.5 mM EDTA
COMMENTS: Spectra of native and reconstituted enzyme
REFERENCE: S. Sato and J. I. Harris, Eur. J. Biochem., 73, 373 (1977)

2283. Superoxide Dismutase

SOURCE: Thermus thermophilus HB8
SOLVENT: 10 mM Tris-Cl buffer, pH 7.5
REFERENCE: S. Sato and K. Nakazawa, J. Biochem. Tokyo, 83, 1165 (1978)

2284. Superoxide Dismutase, Isozymes I and II

SOURCE: Wheat germ
SOLVENT: 0.05 M Potassium phosphate, pH 7.8
REFERENCE: C. O. Beauchamp and I. Fridovich, Biochem. Biophys. Acta, 317, 50 (1973)

2285. Superoxide Dismutase

SOURCE: Baker's yeast (Saccharomyces cerevisiae)
SOLVENT: 0.0025 M Potassium phosphate, pH 7.8
REFERENCE: S. A. Goscin and I. Fridovich, Biochim. Biophys. Acta, 289, 276 (1972)

2286. Superoxide Dismutase

SOURCE: Baker's yeast mitochondria (Saccharomyces cerevisiae)
SOLVENT: a. 0.05 M Tris-HCl, pH 7.4
b. 0.05 M Potassium phosphate, pH 7.8
COMMENTS: Spectra taken before and after addition of ascorbate (solvent a)
REFERENCE: S. Ravindranath and I. Fridovich, J. Biol. Chem., 250, 6107 (1975)

2287. Synechococcus 6312

COMMENT: Spectra at room temperature and at -196°C
REFERENCE: C. Lemasson, M. Tandeau de Marsae, and G. Cohen-Bazire, Proc. Nat. Acad. Sci., 70, 3130 (1973)

2288. Tannase

SOURCE: Aspergillus flavus
SOLVENT: 0.01 M Acetate buffer, pH 5.0
REFERENCE: O. Adachi, M. Watanabe, and H. Yamada, Agric. Biol. Chem., 32, 1079 (1968)

2289. Tau

SOURCE: Pig brain
SOLVENT: Purification buffer
REFERENCE: D. W. Cleveland, S. Y. Hwo, and M. W. Kirschner, J. Mol. Biol., 116, 227 (1977)

2290. Tetrahydrofolate Dehydrogenase

SOURCE: Subline of L1210 lymphoma
REFERENCE: J. Gauldie and B. L. Hillcoat, Biochem. Biophys. Acta, 268, 35 (1972)

2291. Thermolysin

SOURCE: Bacillus thermoproteolyticus Rokko
COMMENTS: Spectra of native enzyme and acetylimidazole-acetylated (pH 7.5) enzyme
REFERENCE: Y. Ohta, H. Nakamura, and T. Samejima, J. Biochem. Tokyo, 72, 521 (1972)

2292. Thermolysin

SOURCE: Commercial (Bacillus thermoproteolyticus)
COMMENTS: Spectra of thermolysin and ethoxyformylthermolysin
REFERENCE: S. Blumberg, B. Holmquist, and B. L. Vallee, Biochem. Biophys. Res. Comm., 51, 987 (1973)

2293. Thiamine Dehydrogenase

SOURCE: Soil bacterium ATCC No. 25589
SOLVENT: 50 mM Tris-phosphate, pH 7.8
COMMENTS:
1. Spectral changes upon formation of a sulfite adduct
2. Spectra of oxidized, semiquinone, reduced (substrate) and reduced (dithionite) forms

REFERENCE: D. E. Edmondson and C. Gomez-Moreno, in International Symposium on Mechanisms of Oxidizing Enzymes (T. P. Singer and R. N. Ondarza, eds.), 45 (1977), Elsevier, Amsterdam

2294. Thiamin Dehydrogenase

SOURCE: Soil bacterium ATCC No. 25589
SOLVENT: a. 25 mM Tris-phosphate, 10 mM EDTA, pH 8.0
b. 50 mM Tris-phosphate, 1 mM EDTA, pH 7.8
COMMENTS: 1. Spectral changes due to light irradiation (anaerobically) for 0-10 minutes
2. Spectral changes due to anaerobic addition of thiamin
REFERENCE: C. Gomez-Moreno, M. Choy, and D. E. Edmundson, J. Biol. Chem., 254, 7630 (1979)

2295. Thionein, Cd, Zn

SOURCE: Rat and chicken liver
SOLVENT: pH 6.6
COMMENTS: Spectra of protein and two different preparations of apoproteins
REFERENCE: U. Weser, H. Rupp, F. Donay, F. Lineman, W. Voetter, W. Voetsch, and G. Jung, Eur. J. Biochem., 39, 127 (1973)

2296. Thionein, Cu-

SOURCE: Yeast (Saccharomyces cerevisiae) and chicken liver
SOLVENT: a. 5 mM Potassium phosphate buffer, pH 7.5
b. 8.3 mM Potassium phosphate buffer, pH 7.0
COMMENTS: 1. Spectra of copper-containing protein in solvents a and b
2. Spectrum of metal-free protein from yeast
REFERENCE: R. Prinz and U. Weser, Hoppe-Seyler's Z. Physiol. Chem., 356, 767 (1975)

2297. Thionein, Cu-

SOURCE: Beef fetal liver
SOLVENT: 10 mM Tris-HCl buffer, pH 8.6
REFERENCE: H. J. Hartmann and U. Weser, Biochim. Biophys. Acta, 491, 211 (1977)

2298. Thionein, Cd, Zn-; Zn-; Apo-

SOURCE: Rat liver
SOLVENT: a. 25 mM Tris buffer, pH 8.6
b. Water, pH 6.0
COMME c. 20 mM Sodium phosphate buffer, pH 6.8
COMMENTS: 1. Spectrum of Cd, Zn-thionein in solvent a
2. Spectrum of Zn-thionein in solvent b
3. Spectrum of Zn-thionein after reduction with dithiothreitol, solvent c

REFERENCE: G. Sokolowski and U. Weser, Hoppe-Seyler's Z. Physiol. Chem., 356, 1715 (1975)

2299. Thioredoxin-S_2

SOURCE: Escherichia coli
SOLVENT: 0.10 M Sodium acetate buffer, pH 4.0
COMMENTS: Spectral changes due to reaction with N-bromosuccinimide
REFERENCE: A. Holmgren, J. Biol. Chem., 248, 4106 (1973)

2300. Thioredoxin

SOURCE: Escherichia coli B
SOLVENT: 8 M Urea-0.05 M Sodium acetate, pH 4.4
COMMENTS: Spectra of thioredoxin, 1-formyltryptophanyl thioredoxin and deformylated thioredoxin
REFERENCE: A. Holmgren, Eur. J. Biochem., 26, 528 (1972)

2301. Thioredoxin

SOURCE: Calf liver
SOLVENT: a. 10 mM Tris-Cl, pH 7.5-0.20 M NaCl-1 mM EDTA
b. 0.1 M NaOH
REFERENCE: N. E. Engstrom, A. Holmgren, A. Larsson, and S. Soderhall, J. Biol. Chem., 249, 205 (1974)

2302. Thioredoxin Reductase

SOURCE: Baker's yeast
SOLVENT: 0.1 M Phosphate buffer, pH 7.0
REFERENCE: M. L. Speranza, S. Ronchi, and L. Minchiotti, Biochim. Biophys. Acta, 327, 274 (1973)

2303. Thioredoxin Reductase

SOURCE: Escherichia coli
SOLVENT: 50 mM Phosphate buffer, pH 7.6, 3 mM EDTA
REFERENCE: V. P. Piriet and R. R. Conley, J. Biol. Chem., 252, 6367 (1977)

2304. Thread-like Particles

SOURCE: Leaves and bark of citrus trees with Citrus tristeza disease
REFERENCE: M. Bar-Joseph, G. Loebenstein, and J. Cohen, Virology, 50, 821 (1972)

2305. Threonine Aldolase

SOURCE: Candida humicola
SOLVENT: 0.03 M Potassium phosphate buffer, pH 6.4, containing 0.001 M 2-mercaptoethanol and 0.001 M EDTA
COMMENTS: Spectral changes due to addition of cysteine (0-13.6 m moles)
REFERENCE: H. Kumagai, T. Nagata, H. Yoshida, and H. Yamada, Biochim. Biophys. Acta, 258, 779 (1972)

2306. Threonine Deaminase

SOURCE: Escherichia coli
SOLVENT: a. 0.1 M Potassium phosphate, pH 7.4, containing 1 mM isoleucine, 1 mM EDTA, and 1 mM mercaptoethanol
b. 0.05 M Tris-HCl, pH 8.5
COMMENTS: Spectra of native enzyme (solvent a) and apoenzyme (solvent b)
REFERENCE: K. Koerner, I. Rahimi-Laridjani, and H. Grimminger, Biochim. Biophys. Acta, 397, 220 (1975)

2307. Threonine Deaminase

SOURCE: Escherichia coli B mutant 1P7
SOLVENT: 0.1 M Potassium phosphate, pH 7.4, containing 1 mM mercaptoethanol and 1 mM EDTA
REFERENCE: J. Feldner and H. Grimminger, J. Bacteriol., 126, 100 (1976)

2308. Threonine Dehydratase

SOURCE: Escherichia coli
SOLVENT: 100 mM Potassium phosphate buffer, pH 8.0, plus 3 mM ATP
COMMENTS: 1. Spectral changes due to glyoxylate and pyruvate in presence of L-threonine
2. Spectral changes due to glyoxylate in absence of L-threonine
REFERENCE: L. S. Park and P. Datta, J. Biol. Chem., 254, 7927 (1979)

2309. Threonine Dehydratase

SOURCE: Pseudomonas putida and Clostridium tetanomorphum
SOLVENT: 0.05 M Potassium phosphate, pH 7.2, containing 1 mM dithiothreitol
REFERENCE: M. S. Cohn and A. T. Phillips, Biochemistry, 13, 1208 (1974)

2310. Threonyl-tRNA Synthetase

SOURCE: Saccharomyces carlsbergensis
REFERENCE: H. Yamada, J. Biochem. Tokyo, 83, 1583 (1978)

2311. Thrombin and Nitrothrombin

SOURCE: Beef
REFERENCE: I. P. Baskova, V. L. Glotova, M. Strukova, and V. Y. Chernyak, Biokhimiya (Biochemistry), 38, 327 (1973)

2312. Thymidylate Synthetase

SOURCE: Bacteriophage T_2 and Lactobacillus casei (methotrexate-resistant)
SOLVENT: 0.1 M Potassium phosphate, pH 7.0, containing 0.02 M 2-mercaptoethanol
REFERENCE: J. Galivan, G. F. Maley, and E. Maley, Biochemistry, 13, 2282 (1974)

2313. Thymidylate Synthetase

SOURCE: Lactobacillus casei
SOLVENT: 50 mM Potassium phosphate, pH 6.8, containing 25 mM 2-mercaptoethanol
COMMENTS: Spectral changes due to reaction with N-bromosuccinimide at pH 6.5
REFERENCE: J. A. Lyon, A. L. Pollard, R. B. Loeble, and R. B. Dunlap, Cancer, Biochem. Biophys., 1, 121 (1975)

2314. Thymidylate Synthetase

SOURCE: Lactobacillus casei (amethopterin-resistant)
SOLVENT: 50 mM Potassium phosphate, pH 6.8
COMMENTS:
1. Spectra of forms II and III in buffer in presence and absence of 4 M guanidinium chloride
2. Spectra of enzyme, FdUMP, and 0-2.0 fold excess of (+)-5, 10-methylene tetrahydrofolate in buffer plus 25 mM-2-mercaptoethanol

REFERENCE: H. Donato, J. L. Aull, J. A. Lyon, J. W. Reinsch, and R. B. Dunlap, J. Biol. Chem., 251, 1303 (1976)

2315. Thyroalbumin

SOURCE: Human serum
REFERENCE: E. E. Gussakovsky and T. Saatov, Biophyzika, 20 (1975); T. Saatov and T. K. Turakulov, Biokhimiya (Biochemistry), 38, 712 (1975)

2316. Toxin I

SOURCE: Anemonia sulcata
COMMENTS: 1. Spectra of mono- and di-iodinated toxin
2. Spectrum of 4-azido-2-nitrophenyl-modified iodinated toxin
REFERENCE: F. Hucho, L. Beress, and S. Stengelin, in International Symposium on Function and Molecular Aspects of Biomembrane Transport (E. Quagliariello, F. Palmieri, S. Papa, and M. Klinkenberg, eds.), 43 (1979), Elsevier, Amsterdam

2317. Toxin (Bungarotoxin)

SOURCE: Bungarus multicinctus
COMMENTS: Spectra of native and diiodinated toxins
REFERENCE: V. A. Eterovic, R. G. Aune, and E. L. Bennett, Anal. Biochem., 68, 394 (1975)

2318. Toxin

SOURCE: Bacillus cereus
COMMENTS: Spectral changes due to the separate reactions with 0.1 M $FeSO_4$, 6 M urea, 0.05 M EDTA, and 0.8% formalin
REFERENCE: F. S. Fluer and Y. V. Ezepchuk, Biokhimiya (Biochemistry), 38, 113 (1974)

2319. Toxin

SOURCE: Bacillus pertussis
SOLVENT: 0.85% saline
REFERENCE: K. Onoue, M. Kitagawa, and Y. Yamamoto, J. Bacteriol., 86, 648 (1963)

2320. Toxin e, Hemorrhagic

SOURCE: Western diamondback rattlesnake (Crotalus atrox) venom
COMMENTS: Spectra of holotoxin e, apotoxin e, and zinc-regenerated toxin e
REFERENCE: J. B. Bjarnason and A. T. Tu, Biochemistry, 17, 3395 (1978)

2321. Toxin, Wilting (Cerato-Ulmin)

SOURCE: Ceratocystis ulmi
SOLVENT: 60% aqueous ethanol
REFERENCE: S. Taka and W. C. Richards, Phytopath. Z., 91, 129 (1978)

2322. Toxin (Necrotoxin)

SOURCE: Tarantula (Dugsiella hentzi) venom
SOLVENT: 0.1 M Sodium acetate buffer, pH 4.0
COMMENTS: Spectra of native and N-bromosuccinimide-modified toxins
REFERENCE: C. K. Lee, T. K. Chan, B. C. Ward, D. E. Howell, and G. V. Odell, Arch. Biochem. Biophys., 164, 341 (1974)

2323. Toxin I and Toxin II

SOURCE: Central Asian cobra (Naja oxiana Eichwald) venom
REFERENCE: Y. K. Turakulov, V. M. Sorokin, S. A. Nishankhodzhaeva, and L. Y. Yukelson, Biokhimiya (Biochemistry), 36, 1073 (1971); Y. K. Turakulov, V. M. Sorokin, and S. A. Nishankhodzhaeva, Biokhimiya (Biochemistry), 37, 102 (1972)

2324. Toxin B

SOURCE: Indian cobra (Naja naja) venom
SOLVENT: Formic acid
COMMENTS: Spectral changes during ozonization
REFERENCE: M. Ohta and K. Hayashi, Biochem. Biophys. Res. Comm., 57, 973 (1974)

2325. Toxin A and Toxin B (Cytotoxins)

SOURCE: Indian cobra (Naja naja) venom
SOLVENT: pH 7.0
REFERENCE: A. G. Szabó, K. R. Lynn, D. T. Krajcarski, and D. M. Rayner, FEBS Lett., 94, 249 (1978)

2326. α-Toxin

SOURCE: Staphylococcus aureus
SOLVENT:
a. 0.1 N HCl in 0.05 M Sodium phosphate buffer, pH 1.12
b. 0.05 M Sodium phosphate buffer, pH 7.0
c. 0.1 N NaOH in 0.05 M Sodium phosphate buffer, pH 12.3

REFERENCE: M. Watanabe and I. Kato, Jap. J. Exp. Med., 44, 165 (1974)

2327. Toxin, Tetanus

SOLVENT:
a. pH 7.0
b. 0.1 N NaOH

COMMENTS: Spectra of toxin T_6, aged toxin T_6, toxin T_7 and anatoxin T_6
REFERENCE: B. Bizzini, A. Turpin, and M. Raynaud, Ann. Inst. Pasteur., Paris, 116, 686 (1969)

2328. Transcobalamin II

SOURCE: Human serum
SOLVENT: 0.05 M Potassium phosphate, pH 7.5, containing 0.75 M NaCl
COMMENTS: Spectrum of transcobalamin II-Vitamin B_{12} complex
REFERENCE: R. H. Allen and P. W. Majerus, J. Biol. Chem., 247, 7709 (1972)

2329. Transferrin

SOURCE: Beef blood
SOLVENT: 0.01 M Tris-0.01% $NaHCO_3$, pH 7.8
COMMENTS: Spectra of protein and tryptic fragments
REFERENCE: J. H. Brock and F. R. Arzabe, FEBS Lett., 69, 63 (1976)

2330. Transferrin

SOURCE: Beef blood
SOLVENT: a. 0.01 M Sodium phosphate buffer, pH 7.4
b. 0.1 M Sodium bicarbonate
REFERENCE: G. D. Efremov, L. L. Smith, B. P. Barton and T. H. J. Huisman, Anim. Bld. Grps. Biochem. Genet., 2, 159 (1971)

2331. Transferrin and Gut Iron Binding Protein

SOURCE: Guinea pig serum and intestinal mucosa
REFERENCE: S. Pollack and F. D. Lasky, Biochem. Biophys. Res. Comm., 70, 533 (1976)

2332. Transferrin

SOURCE: Hagfish
SOLVENT: 0.005 M Tris-HCl buffer, pH 7.0
REFERENCE: R. Aisen, A. Leibman, and C. L. Sia, Biochemistry, 11, 3460 (1972)

2333. Transferrin and carboxyl-modified Protein

SOURCE: Human serum
COMMENTS: Spectra of native and 16 carboxyl groups-modified transferrins

REFERENCE: A. Bezkorovainy and D. Grohlich, Biochim. Biophys. Acta, 214, 37 (1970)

2334. Transferrin

SOURCE: Human serum
COMMENTS: 1. Spectral changes due to reaction of apotransferrin with different iron salts
2. Effect of hydrogen peroxide
REFERENCE: G. W. Bates, E. F. Workman, Jr., and M. R. Schlabach, Biochem. Biophys. Res. Comm., 50, 84 (1973)

2335. Transferrin

SOURCE: Human serum
SOLVENT: pH 7.4
COMMENTS: 1. Spectra of transferrin and Fe^{3+} plus each of the following: 0.1 M sodium bicarbonate, 0.1 M citrate; Fe^{3+} and nitrilotriacetic acid in 0.1 M sodium bicarbonate; serum albumin
2. Spectra of Fe^{3+}-transferrin-ion complexes; the ions are nitrilotriacetate, carbonate, salicylate, thioglycolate, glycine, glyoxylate, and glycolate
REFERENCE: G. W. Bates, and M. R. Schlabach, J. Biol. Chem., 250, 2177 (1975); M. R. Schlabach and G. W. Bates, J. Biol. Chem., 250, 2182 (1977); G. W. Bates and M. R. Schlabach, in Proteins of Iron Storage and Transport in Biochemistry and Medicine (R. R. Crighton, ed.), 51 (1975), North Holland Publ. Co., Amsterdam

2336. Transferrin, Fe(III)-

SOURCE: Commercial (human and chicken)
SOLVENT: pH 7
REFERENCE: Y. Tominatsu, S. Kint, and J. R. Scherer, Biochem. Biophys. Res. Comm., 54, 1067 (1973)

2337. Transpeptidase, DD-Carboxypeptidase-

SOURCE: Streptomyces
SOLVENT: 1 mM Tris-HCl buffer, pH 7.4
REFERENCE: M. Nieto, H. R. Perkins, J. M. Frere, and J. M. Ghuysen, Biochem. J., 135, 493 (1973)

2338. Tremorogen A-10

SOURCE: Tremella mesenterica Fr_1
SOLVENT: 80% methanol:acetic acid (99:1)

REFERENCE: Y. Sakagami, A. Isogai, A. Suzuki, S. Tamura, E. Tsuchiya, and S. Fukui, Agric. Biol. Chem., 42, 1093 (1978)

2339. Trimethylamine Dehydrogenase

SOURCE: Bacterium W3A1
SOLVENT: 75 mM Sodium pyrophosphate, pH 7.7
COMMENTS: Spectral changes due to each of the following: trimethylamine, sodium dithionite, sodium borohydride, 8 M urea, diethylamine, and phenylhydrazine (oxidized enzyme used)
REFERENCE: D. J. Steenkamp and T. P. Singer, Biochem. Biophys. Res. Comm., 71, 1289 (1976); D. J. Steenkamp and J. Mallinson, Biochim. Biophys. Acta, 429, 705 (1979); J. Nagy, W. C. Kenney, and T. P. Singer, J. Biol. Chem., 254, 2684 (1979)

2340. Trimethylamine Dehydrogenase

SOURCE: Bacterium W3A1
SOLVENT: 0.1 M Sodium pyrophosphate, pH 8.5
COMMENTS:
1. Spectral changes during anaerobic titration with dithionite
2. Spectral changes during anaerobic titration with diethylamine
3. Spectral changes during anaerobic titration with dithionite in presence of tetramethyl ammonium chloride
4. Spectra of oxidized enzyme, after addition of trimethylamine, and after addition of tetramethyl ammonium chloride

REFERENCE: D. J. Steenkamp, H. Beinert, W. McIntire, and T. P. Singer, in International Symposium on Mechanisms of Oxidizing Enzymes (T. P. Singer and R. N. Ondarza, eds.), 127 (1977), Elsevier, Amsterdam

2341. Trimethylamine Dehydrogenase

SOURCE: Bacterium W3A1
SOLVENT: 0.1 M Sodium pyrophosphate, pH 7.7
COMMENTS:
1. Spectral changes during anaerobic titration with dithionite
2. Spectral changes during anaerobic titration with trimethylamine

REFERENCE: D. J. Steenkamp, T. P. Singer, and H. J. Beinert, Biochem. J., 169, 361 (1978)

2342. Trimethylamine Dehydrogenase Electron Acceptor Flavoprotein

SOURCE: Bacterium W3A1
COMMENTS: 1. Spectra of reduced and oxidized forms
2. Spectral changes during anaerobic dithionite titration
REFERENCE: D. J. Steenkamp and M. Gallup, J. Biol. Chem., 253, 4026 (1978)

2343. Tropomyosin and fluoresceinisothiocarbamyl- derivative

SOURCE: Rabbit skeletal muscle
SOLVENT: 10 mM Phosphate buffer, pH 7.3, containing 10 mM KCl
REFERENCE: A. Satoh and K. Mihashi, J. Biochem. Tokyo, 71, 597 (1972); K. Mihashi, J. Biochem. Tokyo, 71, 607 (1972)

2344. Tropomysin

SOURCE: Rabbit skeletal muscle
SOLVENT: 50 mM Tris-H_2SO_4 buffer, pH 7.0
COMMENTS: Spectra of the following derivatives of tropomyosin: N-(1-anilinonaphthyl-4) maleimide and N-(3-pyrene) maleimide
REFERENCE: T. Ohyashiki, Y. Kanaoka, and T. Sekine, Biochim. Biophys. Acta, 420, 27 (1976)

2345. Tropomysin

SOURCE: Carp
SOLVENT: 0.1 M KCl-20 mM Phosphate, pH 7.0
REFERENCE: N. Seki and K. Konno, Bull. Jap. Soc. Sci. Fish., 41, 1135 (1975)

2346. Troponin Components

SOURCE: Rabbit skeletal muscles, chicken skeletal muscles, and beef cardiac muscles
SOLVENT: 2 mM Potassium bicarbonate and 0.2 mM EGTA, pH 7.2
REFERENCE: I. Staprans, H. Takahashi, M. P. Russel, and S. Watanabe, J. Biochem. Tokyo, 72, 223 (1972)

2347. Troponin A

SOURCE: Rabbit skeletal muscles
SOLVENT: 50 mM Tris, pH 7.6
REFERENCE: A. C. Murray and C. M. Kay, Biochemistry, 11, 2622 (1972)

2348. Troponin C

SOURCE: Frog (Rana esculenta) skeletal muscle
REFERENCE: J. P. Van Eerd, J. P. Capony, C. Ferraz, and J. F. Pechere, Eur. J. Biochem., 91, 231 (1978)

2349. Troponin C

SOURCE: Chicken, rabbit, and lamprey skeletal muscles
SOLVENT: 0.3 mM $NaHCO_3$, 0.1 M $CaCl_2$
REFERENCE: T. Hirabayashi and S. V. Perry, Biochim. Biophys. Acta, 351, 273 (1974)

2350. Troponin-C, Troponin-I, and Troponin-T

SOURCE: Rabbit muscle
SOLVENT: a. 0.3 mM Bicarbonate
b. 0.3 mM Bicarbonate-0.3 M KCl
c. 1 M KCl and 5 mM Tris, pH 7.5
REFERENCE: J. P. Van Eerd and Y. Kawasaki, Biochemistry, 12, 4972 (1973); M. L. Greaser and J. Gergely, J. Biol. Chem., 248, 2125 (1973)

2351. Troponin-C

SOURCE: Rabbit white skeletal muscle
SOLVENT: 0.1 M KCl-20 mM Tris-15 mM HCl (pH 7.6)-0.1 mM $CaCl_2$
REFERENCE: J. F. Head and S. V. Perry, Biochem. J., 137, 145 (1974)

2352. Troponin-C

SOURCE: Dogfish (Squalis acanthias)
SOLVENT: 2 mM Tris-HCl, pH 8.0
REFERENCE: D. A. Malencik, G. W. Heizmann, and E. H. Fischer, Biochemistry, 14, 715 (1975)

2353. Troponin-C and nitrotyrosyl-Troponin-C

SOURCE: Beef cardiac muscle
SOLVENT: 0.15 M KCl, 50 mM Tris-HCl, 1 mM EDTA, pH 8 and pH 9
REFERENCE: L. D. Burtnick, W. D. McCubbin, and C. M. Kay, Can. J. Biochem., 53, 15 (1975); W. C. McCubbin and C. M. Kay, FEBS Lett., 55, 183 (1975)

2354. Troponin-C

SOURCE: Rabbit muscle
SOLVENT: 20 mM Tris-1 mM magnesium acetate, pH 7.5
REFERENCE: F. C. Stevens, M. Walsh, H. C. Cho. T. S. Teo, and J. H. Wang, J. Biol. Chem., 251, 4495 (1976)

2355. Troponin-I

SOURCE: Beef cardiac muscle
SOLVENT: 50 mM Tris-HCl, pH 7.5, 0.5 M KCl, and 1 mM EGTA
REFERENCE: L. D. Burtnick, W. D. McCubbin, and C. M. Kay, Can. J. Biochem., 53, 1207 (1975)

2356. Troponin, Basic Components

SOURCE: Rabbit skeletal muscle
SOLVENT: 0.3 M KCl, 50 mM Tris-HCl, pH 7.4
COMMENTS: Spectra of fractions MB1, MB2 and MB3
REFERENCE: G. Berson, and P. Bouveret, Biochimie, 57, 1237 (1975)

2357. Troponin-C-like Protein

SOURCE: Beef adrenal medulla
SOLVENT: 50 mM Tris-Cl, pH 7.6
REFERENCE: I. C. Y. Kuo and C. J. Coffee, J. Biol. Chem., 251, 1603 (1976)

2358. Trypsin

SOURCE: Commercial
SOLVENT: 50% Ethanol, pH 8
COMMENTS: Spectrum of mixture of trypsin and chloranil
REFERENCE: J. B. Birks and M. A. Slifkin, Nature, 197, 42 (1963)

2359. Trypsin Derivative

SOURCE: Commercial
SOLVENT: 0.05 M Sodium phosphate buffer, pH 7.0
COMMENTS: Spectrum of N-α-dansyl-L-lysyl-trypsino (histidine-46) methane
REFERENCE: G. Schoellman, Int. J. Pep. Prot. Res., 4, 221 (1972)

2360. Trypsin

SOURCE: Beef

COMMENTS: Spectra of trypsin plus toluidine blue, and trypsin, λ-carrageenan, plus toluidine blue
REFERENCE: I. L. Gatfield and R. Stute, FEBS Lett., 28, 29 (1972)

2361. Trypsin

SOURCE: Commercial
SOLVENT: a. pH 7.3-7.6 solution saturated with N_2O
b. pH 10.8-11.9 solution saturated with N_2O
COMMENTS: 1. Spectra after pulse radiolysis in solvents a and b
2. Spectra after pulse radiolysis in solvents a and b plus each of the additional salts: potassium thiocyanate, potassium bromide, and potassium iodide
REFERENCE: G. E. Adams, R. H. Bisby, R. B. Cundal, and J. L. Redpath, J. Chem. Soc. Farad. Trans. I, 69, 1608 (1973)

2362. Trypsin-DNP-trypsin Inhibitor

SOURCE: Beef (trypsin) and soybeans (inhibitor)
REFERENCE: M. Wilchek and M. Gorecki, FEBS Lett., 31, 149 (1975)

2363. Trypsin

SOURCE: Streptomyces griseus
COMMENTS: Spectra of native and acetylated enzyme
REFERENCE: W. M. Awad and M. S. Ochoa, Biochem. Biophys. Res. Comm., 59, 527 (1974)

2364. Trypsin

SOURCE: Commercial
SOLVENT: 0.001 M HCl
COMMENTS: Spectra of soluble protein and insolubilized (attached to Sepharose 6B) protein
REFERENCE: R. Koelsch, J. Lasch, I. Marquardt, and H. Hanson, Analyt. Biochem., 66, 556 (1975)

2365. β-Trypsin

SOURCE: Commercial, bovine
SOLVENT: pH 7.0-pH 12.0
REFERENCE: N. Ramachandran and C. A. Ghiron, Biochim. Biophys. Acta, 532, 286 (1978)

2366. Trypsin-like Enzyme Derivative

SOURCE: Streptomyces erythreus
SOLVENT: pH 3
COMMENTS: Spectrum of L-1-chloro-3-(2,4-dinitroanilino)-7-aminoheptan-2-one modified protein
REFERENCE: N. Yoshida, A. Sasaki, and K. Inouye, Biochim. Biophys. Acta, 321, 615 (1973)

2367. Trypsinogen

SOURCE: Commercial
SOLVENT: pH 8.0 and pH 12.1, both solutions saturated with N_2O
COMMENTS: Spectra of protein after pulse radiolysis
REFERENCE: G. E. Adams, R. H. Bisby, R. B. Cundall and J. L. Redpath, J. Chem. Soc. Farad. Trans. I, 69, 1608 (1975)

2368. Tryptophanase

SOURCE: Escherichia coli, Escherichia aurescens, Proteus morganii, Proteus vulgaris, and Shigella alkalescens
SOLVENT: 0.1 M Potassium phosphate, pH 6.8 containing DTT
COMMENTS:
1. Spectra of holoenzymes and apoenzymes
2. Spectral changes due to pH: pH 6.2-7.8
3. Spectra of holoenzymes plus alanine

REFERENCE: C. Simard, A. Mardini, and L. M. Bordeleau, Can. J. Microbiol., 21, 834 (1975)

2369. Tryptophanase

SOURCE: Bacillus alvei
SOLVENT: 0.05 M Potassium phosphate, pH 7.7, containing 0.0001 M pyridoxal phosphate
COMMENTS:
1. Spectra of high and low pH forms
2. Spectra of holoenzyme and reconstituted enzyme
3. Spectra of holoenzyme in presence of each of the following: anthranilic acid, anthranilic acid and L-α,β-diaminopropionic acid, L-tryptophan, D-tryptophan, L-alanine, L-phenylalanine, and D-phenylalanine

REFERENCE: H. Isom and R. D. Demoss, Biochemistry, 14, 4291 (1975); J. D. Fenske and R. O. Demoss, J. Biol. Chem., 250, 7554 (1975)

2370. Tryptophanase

SOURCE: Proteus rettgeri
SOLVENT: 0.1 M Potassium phosphate buffer, pH 8.0, containing 10 mM mercaptoethanol

REFERENCE: H. Yoshida, H. Kumagai, and H. Yamada, FEBS Lett., 48, 56 (1975); H. Yoshida, H. Kumagai, H. Yamada, and H. Matsubara, Biochim. Biophys. Acta, 391, 494 (1975)

2371. Tryptophanase

SOURCE: Escherichia coli B/1t 7-A
SOLVENT: 0.1 M Potassium phosphate buffer, pH 8.0, pyridoxal-phosphate and DTT
COMMENTS: Spectra before and after photoinactivation
REFERENCE: T. Nihira, T. Toraya, and S. Fukui, Eur. J. Biochem., 101, 341 (1979)

2372. Tryptophanase

SOURCE: Escherichia coli
SOLVENT: 0.1 M Potassium phosphate buffer, pH 7.7
COMMENTS: Spectral changes due to ozonization (0-6 minutes)
REFERENCE: M. Tokushige, Y. Fukuda, and Y. Watanabe, Biochem. Biophys. Res. Comm., 86, 976 (1979)

2373. Tryptophanase

SOURCE: Escherichia coli
SOLVENT: 0.025 M Tetramethylamine-N-2-hydroxyethyl piperazine-N'-2-ethanesulfonic acid, pH 8.0
COMMENTS: Spectra of holotryptophanase, and plus each of the following: 0.1 N KCl, 0.1 M KCl plus 10 mM ethionine, and 10 mM ethionine
REFERENCE: C. H. Suelter and E. E. Snell, J. Biol. Chem., 252, 1852 (1977)

2374. L-Tryptophan 2,3-Dioxygenase

SOURCE: Rat liver
SOLVENT: 50 mM Tris-HCl buffer, pH 7.2
COMMENTS:
1. Spectra of ferriheme form and ferroheme (dithionite) form
2. Spectra of ferroheme form plus bathocuproine sulfonate, and plus copper and bathocuproine sulfonate

REFERENCE: R. Makino and Y. Ishimura, J. Biol. Chem., 251, 7722 (1976)

2375. Tryptophan 2,3-Dioxygenase

SOURCE: Pseudomonas fluorescens (ATCC 11299)
SOLVENT: 50 mM Tris-acetate buffer, pH 7
COMMENTS: Spectral changes due to following: native enzyme (ferric form) to which was added successively sodium nitrite, dithionite, and L-tryptophan
REFERENCE: Y. Henry, Y. Ishimura, and J. Peisach, J. Biol. Chem., 251, 1578 (1976)

2376. Tryptophan 2,3-Dioxygenase

SOURCE: Pseudomonas acidovorans and rat liver
SOLVENT: 0.1 M Potassium phosphate, pH 7.0
COMMENTS:
1. Spectrum of half-reduced (Fe^{3+}-Cu^{+}) enzyme plus L-tryptophan and oxygen
2. Spectrum of fully reduced (Fe^{2+}-Cu^{+}) enzyme plus L-tryptophan, anaerobic
3. Spectrum of fully reduced (Fe^{2+}-Cu^{+}) enzyme plus L-tryptophan and oxygen
4. Spectra of fully reduced (Fe^{2+}-Cu^{+}) enzyme plus O_2 plus each of the following: DL-5-fluorotryptophan, DL-α-methyltryptophan, DL-5-fluorotryptophan and DL-α-methyltryptophan together

REFERENCE: F. O. Brady and P. Feigelson, J. Biol. Chem., 250, 5041 (1975)

2377. Tryptophan 2,3-Dioxygenase

SOURCE: Rat liver
SOLVENT: 0.1 M Sodium phosphate, pH 7.0
COMMENTS: Spectra of ferri- and ferro- (dithionite) forms of enzyme in presence and absence of bathocuproinesulfonate
REFERENCE: F. O. Brady, M. Monaco, H. J. Forman, G. Schutz, and P. Feigelson, J. Biol. Chem., 247, 7915 (1972)

2378. Tryptophan 2,3-Dioxygenase

SOURCE: Pseudomonad
REFERENCE: O. Hayaishi, Y. Ishimura, H. Fujisawa, and M. Nozaki, in International Symposium on Oxidases and Related Redox Systems, 1971 (T. E. King, H. S. Mason, and M. Morrison, eds.), 125 (1973), University Park Press, Baltimore

2379. Tryptophan Side Chain Oxidase: Dehydrogenase Component Components

SOURCE: Pseudomonas ATCC 29574
SOLVENT: 50 mM Potassium phosphate buffer, pH 6.0
COMMENTS: Spectra of oxidized and dithionite-reduced forms
REFERENCE: H. Ushiro, K. Takai, S. Narumiya, S. Ito, and O. Hayaishi, J. Biol. Chem., 254, 11794 (1979)

2380. Tryptophan Synthase

SOURCE: Neurospora crassa strain td201
SOLVENT: 0.05 M Potassium phosphate buffer, pH 7.2
COMMENTS: 1. Spectra of apo- and holoenzyme
2. Spectra of aged holoenzyme and after sodium borohydride reduction
REFERENCE: H. Tsai and S. R. Suskind, Biochim. Biophys. Acta, 284, 324 (1972)

2381. Tryptophan Synthase Inactivase

SOURCE: Yeast
SOLVENT: 0.1 M Potassium phosphate buffer, pH 7
REFERENCE: E. H. Schott and H. Holzer, Eur. J. Biochem., 42, 61 (1974)

2382. Tryptophan Synthetase

SOURCE: Escherichia coli
SOLVENT: 0.1 M Potassium phosphate, pH 7.0, 0.02 M pyridoxal phosphate, 1 mM DTT, 1 mM phenylmethanesulfonyl fluoride, and 5 mM EDTA
REFERENCE: O. Adachi, L. D. Cohn, and E. W. Miles, J. Biol. Chem., 249, 7756 (1974)

2383. Tryptophan Synthetase

SOURCE: Escherichia coli
SOLVENT: 0.04 mM Potassium phosphate, pH 7.8
COMMENTS: Spectral changes due to addition of trans-L-2-amino-4-methoxy-3-butenoic acid
REFERENCE: E. W. Miles, Biochem. Biophys. Res. Comm., 66, 94 (1975)

2384. Tryptophan Synthetase, B Protein

SOURCE: Pseudomonas putida
SOLVENT: 0.1 M Potassium phosphate buffer, pH 7.0, containing 1 mM 2-mercaptoethanol

COMMENTS: Spectra of holoenzyme, apoenzyme, and reduced (sodium borohydride) form
REFERENCE: R. Maurer, and I. P. Crawford, Arch. Biochem. Biophys., 144, 193 (1971)

2385. Tryptophan Synthetase, β_2-subunit

SOURCE: Escherichia coli
SOLVENT: 0.1 M Potassium phosphate, pH 7.8, with or without EDTA and pyridoxal-phosphate
COMMENTS:
1. Spectra of washed crystals in solvent and after addition of pyridoxal-phosphate and dithiothreitol
2. Spectra of untreated and diethylpyrocarbonate-treated subunit in presence of pyridoxal-phosphate
3. Spectra of β_2-subunit and its complexes with (α-H) serine and (α-D) serine in the presence and absence of ammonium ion

REFERENCE: O. Adachi and E. W. Miles, J. Biol. Chem., 249, 5430 (1974); E. W. Miles and H. Kumagi, J. Biol. Chem., 249, 2483 (1974); E. W. Miles and P. M. McPhie, J. Biol. Chem., 249, 2852 (1974)

2386. Tryptophan Synthetase, β_2

SOLVENT: 0.1 M Phosphate buffer, pH 7.8, containing β-mercaptoethanol
COMMENTS:
1. Spectrum of protein plus 0.1 mM pyridoxal 5'-phosphate isonicotinyl hydrazone and L-serine
2. Spectra of protein plus L-serine, indole, pyridoxal 5'-phosphate, pyridoxal 5'-phosphate isonicotinyl hydrazone, and sodium borohydride

REFERENCE: M. Camier, P. Gonnard, and M. E. Goldberg, Biochimie, 55, 1011 (1973)

2387. Tubulin, denatured

SOURCE: Rat brain
SOLVENT:
a. 0.02 M Sodium phosphate, pH 6.5, containing 6 M guanidine HCl
b. 0.02 M Sodium phosphate, pH 7.2, containing 0.3% sodium dodecylsulfate

REFERENCE: B. A. Eipper, J. Biol. Chem., 249, 1407 (1974)

2388. Tyrosinase

SOURCE: Mushroom
SOLVENT: 35 mM Sodium phosphate buffer, pH 6.8

COMMENTS: Spectra of hydroxylamine-treated and untreated enzyme
REFERENCE: A. J. M. Schoot Uiterkamp, L. H. Evans, R. L. Jolley, and H. S. Mason, Biochim. Biophys. Acta, 453, 200 (1976)

2389. Tyrosinase

SOURCE: Mushroom (Agaricus bisporus)
COMMENTS: Spectra of tyrosinase and tyrosinase treated with hydrogen peroxide
REFERENCE: R. L. Jolley, L. H. Evans, N. Makino, and H. S. Mason, J. Biol. Chem., 249, 335 (1974)

2390. Tyrosinase

SOURCE: Rana pipiens
SOLVENT: 0.02 M Sodium phosphate, pH 7.2
COMMENTS: Spectra of native enzyme and enzyme extracted in 50% acetone
REFERENCE: R. B. Mikkelsen, D. H. Tang, and E. L. Triplett, Biochem. Biophys. Res. Comm., 63, 980 (1975)

2391. Tyrosinase

SOURCE: Neurospora crassa
COMMENTS:
1. Spectrum of enzyme treated aerobically with 2 mM cysteamine or 2-mercaptoethylamine
2. Spectra of enzyme before and after treatment with hydrogen peroxide

REFERENCE: R. Aasa, J. Deinum, K. Lerch, and B. Reinhammar, Biochim. Biophys. Acta, 535, 287 (1978); K. Lerch, FEBS Lett., 69, 157 (1976); J. Deinum, K. Lerch, and B. Reinhammer, FEBS Lett., 69, 161 (1976); S. Gutteridge and S. Robb, Eur. J. Biochem., 54, 107 (1975)

2392. Tyrosinase

SOURCE: Streptomyces glaucescens
SOLVENT: 0.05 M Tris-HCl buffer, pH 7.5
REFERENCE: K. Lerch and L. Ettlinger, Eur. J. Biochem., 31, 427 (1972)

2393. Tyrosine Phenol Lyase

SOURCE: Erwinia herbicola and Erwinia intermedia
SOLVENT: 0.05 M Potassium phosphate buffer, pH 7.0, containing 2-mercaptoethanol

COMMENTS: 1. Spectra in presence and absence of pyridoxal phosphate
2. Spectral changes due to addition of each of the following amino acids to enzyme: tyrosine, phenylalanine, and alanine

REFERENCE: H. Kumaga, N. Kashima, H. Torii, H. Yamada, H. Enei, and S. Okumura, Agric. Biol. Chem., 36, 472 (1972); T. Muro, H. Nakatani, K. Hiromi, H. Kumagain, and H. Yamada, J. Biochem. Tokyo, 84, 633 (1978)

2394. Tyrosine Phenol Lyase

SOURCE: Escherichia intermedia

SOLVENT: a. 0.1 M Potassium phosphate buffer, pH 6.0
b. 0.1 M Potassium phosphate buffer, pH 8.0
c. Solvent b plus 0.1 mM pyridoxal phosphate, 0.5 mM 2-mercaptoethanol and 0.1 mM EDTA

COMMENTS: 1. Spectral changes due to reaction with diethyl pyrocarbonate in solvent a
2. Spectra of untreated and diethylpyrocarbonate-treated enzyme in presence and absence of L-alanine in solvent b
3. Spectra of untreated and diethylpyrocarbonate-treated enzyme in presence of ammonium ion, phenol, and pyruvate in solvent c

REFERENCE: H. Kumagai, T. Utagawa, and H. Yamada, J. Biol. Chem., 250, 1661 (1975)

2395. Tyrosyl-tRNA Synthetase

SOURCE: Baker's yeast

SOLVENT: 0.15 M Tris-HCl, pH 7.6, 0.15 M KCl, 10 mM $MgSO_4$, 1 mM DTT, and 8% (V/V) glycerol

REFERENCE: H. Beikirch, F. Von der Haar, and F. Cramer, Eur. J. Biochem., 26, 182 (1972)

2396. Ubiquinone Reductase

SOURCE: Neurospora crassa mitochondria membrane

COMMENTS: Spectra of air-oxidized and dithionite-reduced forms

REFERENCE: H. Weiss and H. J. Kolb, Eur. J. Biochem., 99, 139 (1979)

2397. Urate Oxidase

SOURCE: Yeast (Candida utilis)

SOLVENT: a. 0.0002 M Sodium phosphate
b. 0.1 N NaOH

COMMENTS: Spectra of enzyme and dithionite-reduced enzyme
REFERENCE: K. Itaya, T. Yamamoto, and J. Fukumoto, Agric. Biol. Chem., 31, 1256 (1967); K. Itaya, J. Fukumoto, and T. Yamamoto, Agric. Biol. Chem., 35, 813 (1971)

2398. Uricase

SOURCE: Beef kidney
SOLVENT: 1% Sodium carbonate
REFERENCE: R. Farina, M. R. F. Mennella, and E. Leone, Ital. J. Biochem., 28, 270 (1979)

2399. Uridine Diphosphate-N-acetylenol-pyruvoylglucosamine Reductase

SOURCE: Escherichia coli
SOLVENT: 50 mM Phosphate buffer, pH 7.5, 3 mM EDTA
REFERENCE: R. A. Anwar and M. Vlaovic, Can. J. Biochem., 57, 188 (1979)

2400. Uridine Diphosphoglucose-4-epimerase

SOURCE: Mung bean and calf liver
REFERENCE: T. N. Bruzhinina, M. A. Novikova, and V. N. Shibaev, Biokhimiya (Biochemistry), 34, 87 (1969)

2401. Urocanase

SOURCE: Pseudomonas putida
SOLVENT: 33 mM Potassium phosphate buffer, pH 7.4
COMMENTS:
1. Spectra of active and inactive enzymes
2. Spectra of active and inactive enzymes in the presence and absence of imidazopropionate

REFERENCE: D. H. Hug, P. S. O'Donnell, and J. K. Hunter, J. Biol. Chem., 253, 7622 (1978)

2402. Uronic Acid Dehydrogenase

SOURCE: Pseudomonas syringae
SOLVENT: 0.02 M Sodium diphosphate buffer, pH 8.0
REFERENCE: G. Wagner and S. Hollman, Eur. J. Biochem., 61, 589 (1976)

2403. Uteroglobulin

SOURCE: Rabbit uterine luminal fluid
SOLVENT: Tris-Saline, pH 7.5
REFERENCE: A. Nieto, H. Ponstingl, and M. Beato, Arch. Biochem. Biophys., 180, 82 (1977)

2404. Virus, Adeno-, Type-2 Hexon

SOLVENT: 0.01 M Sodium phosphate, pH 7.0
REFERENCE: L. A. Day, R. M. Franklin, U. Pettersson, and L. Philipson, Eur. J. Biochem., 29, 537 (1972)

2405. Virus, Citrus variegation

COMMENTS: Spectra of RNAs 1+2+3, RNAs 4+4a, and protein
REFERENCE: D. Gonsalves and S. M. Garnsey, Virology, 67, 311 (1975)

2406. Virus, Corona-, 229E

SOLVENT: a. 0.3 M Phosphate buffer, pH 7.2
b. 0.15 M NaCl/0.005 M EDTA/0.02 M Tris, pH 7.2
REFERENCE: J. C. Hierholzer, Virology, 75, 155 (1976)

2407. Virus, Mosquito Iridescent (MIV)

COMMENTS: Spectra of RMIV and TMIV
REFERENCE: G. W. Wagner, J. D. Paschke, W. R. Campbell, and S. R. Webb, Virology, 52, 72 (1973)

2408. Virus, Potato Aucuba Mosaic

SOLVENT: a. Water
b. 10% Acetic acid
COMMENTS: Spectra of virus in solvent a and virus protein in solvent b
REFERENCE: T. Miki and N. Oshima, Virology, 48, 386 (1972)

2409. Virus X, Potato

SOLVENT: 0.05 M Phosphate buffer, pH 7
COMMENTS:
1. Spectra of virus and pyridoxal-modified virus
2. Spectral changes on reaction with trinitrobenzenesulfonate
3. Spectral changes on reaction with chlorogenoquinone

REFERENCE: W. S. Pierpoint, J. Gen. Virol., 25, 303 (1974); W. S. Pierpoint, R. J. Ireland, and J. M. Carpenter, Phytochemistry, 16, 29 (1977)

2410. Virus, Sugar Beet Yellow

SOLVENT: 1% Sodium dodecylsulfate
REFERENCE: M. Bar-Joseph and R. Hull, Virology, 62, 552 (1974)

2411. Virus, Tobacco Mosaic

SOLVENT: a. 0.01 M Phosphate buffer, pH 7.0
b. Sodium pyrophosphate buffer, pH 7.25
REFERENCE: G. Jonard, H. Guilley, and L. Hirth, Virology, 64, 1 (1975)

2412. Virus, Tobacco Mosaic, Vulgare Type, Dahlemense Type and U2 Type

SOLVENT: a. 1 N NaOH
b. 0.1 N NaOH
c. M/15 Pyrophosphate, pH 7
d. 0.1 N HCl
e. 8 M Urea
REFERENCE: G. Paulsen, Zeit. Naturforsch., 27b, 427 (1972)

2413. Virus, Tobacco Mosaic

REFERENCE: M. Taniguchi, A. Yamaguchi, and T. Taniguchi, Biochim. Biophys. Acta, 251, 164 (1971)

2414. Virus, Tobacco Rattle

SOLVENT: a. 0.1 M NaOH
b. 0.02 M Tris-HCl, pH 7.4
COMMENTS: Spectra of proteins TRV-Y, TRV-Y/Z II, and TRV-Z
REFERENCE: S. A. Ghabrial and R. M. Lister, Virology, 52, 1 (1973)

2415. Virus, Tobacco Streak

SOLVENT: a. 0.001 M EDTA, pH 7.2
b. 0.02 M Tris-HCl buffer, pH 7.4
COMMENTS: Spectra of RNA and protein
REFERENCE: S. A. Ghabrial and R. M. Lister, Virology, 57, 1 (1974)

2416. Virus, Turnip Mosaic

SOLVENT: a. 0.025 M Phosphate, pH 7.4, containing 0.5 M urea
b. 0.02 M Phosphate, pH 6.5, containing 6.0 M guanidine-HCl
c. Water
COMMENTS: 1. RNA in solvent c
2. Protein in solvent b
3. Virus in solvent a
REFERENCE: J. H. Hill and R. J. Shepherd, Virology, 47, 807 (1972)

2417. Vesicles

SOURCE: Torpedo californica and Narcine brasiliensis electric organs
SOLVENT: 0.4 M NaCl, 10 mM N-2-Hydroxyethyl-piperazine-N'-2-ethanesulfonic acid, 10 mM EGTA, pH 7.0
COMMENTS: Spectra before and after lysis with 1% sodium dodecylsulfate
REFERENCE: S. C. Carlson, J. A. Wagner, and R. B. Kelly, Biochemistry, 17, 1188 (1978)

2418. Vitreous

SOURCE: Beef vitreous body
SOLVENT: a. pH 6
b. pH 10
REFERENCE: E. A. Balazs, Am. J. Ophthal., 38, 21 (1954)

2419. Xanthine Dehydrogenase

SOURCE: Pseudomonas synxantha
SOLVENT: 50 mM Pyrophosphate buffer, pH 8.5, containing 0.1 mM EDTA and 2-mercaptoethanol, pH 8.5
REFERENCE: T. Sakai and H. K. Jun, Agric. Biol. Chem., 43, 753 (1979)

2420. Xanthine Dehydrogenase

SOURCE: Pseudomonas acidovorans
SOLVENT: Tris-EDTA buffer
REFERENCE: I. L. Sin, Biochim. Biophys. Acta, 410, 12 (1975)

2421. Xanthine Dehydrogenase

SOURCE: Chicken liver
COMMENTS: 1. Spectra of native enzyme and flavin-free protein from NADH-reduced enzyme
2. Spectral changes during aerobic reduction of xanthine-specific flavin-free enzyme and NADH-specific flavin-free enzyme by xanthine and NADH
REFERENCE: M. Kanda and K. V. Rajagopalan, J. Biol. Chem., 247, 2177 (1972)

2422. Xanthine Dehydrogenase

SOURCE: Chicken liver
COMMENTS: Spectra of native enzyme and flavin-free protein
REFERENCE: M. Kanda, F. O. Brady, K. V. Rajagopalan, and P. Handler, J. Biol. Chem., 247, 765 (1972)

2423. Xanthine Dehydrogenase

SOURCE: Turkey liver
SOLVENT: 0.05 M Phosphate buffer, pH 7.8, containing 0.1 mM EDTA
COMMENTS: Spectra of the following: native enzyme, after reduction with xanthine, NADH, and dithionite under aerobic conditions.
REFERENCE: W. F. Cleere and M. P. Coughlan, Comp. Biochem., 50B, 311 (1975)

2424. Xanthine Dehydrogenase, Types D and O

SOURCE: Rat liver
REFERENCE: W. R. Waud and K. V. Rajagopalan, Arch. Biochem. Biophys., 172, 354 (1976)

2425. Xanthine Dehydrogenase

SOURCE: Streptomyces cyanogenus
COMMENTS: Spectra of oxidized and hydrosulfite-reduced forms
REFERENCE: T. Ohe and Y. Watanabe, J. Biochem. Tokyo, 86, 45 (1979)

2426. Xanthine Dehydrogenase

SOURCE: Neurospora crassa
REFERENCE: E. S. Lyon and R. H. Garrett, J. Biol. Chem., 253, 2604 (1978)

2427. Xanthine Oxidase

SOURCE: Cow milk
COMMENTS: Spectra of oxidized, xanthine-reduced, and dithionite-reduced forms
REFERENCE: K. Garbett, R. D. Gillard, P. F. Knowles, and J. E. Stangroom, Nature, 215, 824 (1967)

2428. Xanthine Oxidase

SOURCE: Cow milk
SOLVENT: 0.05 M Phosphate, pH 7.8
REFERENCE: W. R. Waud, F. O. Brady, R. D. Wiley, and K. V. Rajagopalan, Arch. Biochem. Biophys., 169, 695 (1975)

2429. Xanthine Oxidase

SOURCE: Cow milk
COMMENTS: Spectra of native enzyme and flavin-free protein
REFERENCE: M. Kanda, F. O. Brady, K. V. Rajagopalan, and P. Handler, J. Biol. Chem., 247, 765 (1972)

2430. Xanthine Oxidase

SOURCE: Cow milk
SOLVENT: 0.10 M Potassium phosphate, pH 7.0, 0.005% EDTA
REFERENCE: G. R. Nathan and E. P. Kirbyhade, Biochim. Biophys. Acta, 526, 328 (1978)

2431. Xanthine Oxidase and Xanthine Oxidase Cross-reacting Material

SOURCE: Rat liver
REFERENCE: J. L. Johnson, W. R. Waud, H. J. Cohen, and K. V. Rajagopalan, J. Biol. Chem., 249, 5056 (1974)

2432. Xanthine Oxidase

SOURCE: Arthrobacter S-2
SOLVENT: 0.01 M Phosphate, pH 7.2, plus salts
REFERENCE: C. A. Woolfolk and J. S. Downard, J. Bacteriol., 135, 422 (1978)

2433. Xylanase

SOURCE: Streptomyces strain E-86
SOLVENT: a. McIlvane buffer, pH 6.5
b. 0.1 N NaOH
REFERENCE: I. Kusakabe, M. Kawaguchi, and T. Yasui, Nippon Nogei Kakaku Kaishi, 51, 429 (1977)

2434. B-Xylosidase

SOURCE: Fungus (Malbrancea pulchella var. sulfurea #48)
SOLVENT: 0.1 M Phosphate buffer, pH 6.7
REFERENCE: M. Matsuo, T. Yasui, and T. Kobayashi, Agric. Biol. Chem., 41, 1593 (1977)

NAME INDEX

Numbers refer to page and (protein number)

SOURCE INDEX

ADDITIVE/MODIFIER/MODIFICATION INDEX

CORRECTIONS TO ATLAS OF PROTEIN SPECTRA IN THE ULTRAVIOLET AND VISIBLE REGIONS

Volume 1:

p. 124[top]: Comments: Upper six curves

p. 186[top]: Comments: Insert: Concentration

p. 197[bottom]: Solvent: 1 mM

p. 272: Carboxypeptidase

A, bovine pancreas

A_{α}, bovine pancreas

Casein, 44, 192

p. 273: Gelatin, 44, 101, 102, 192

Volume 2:

p. 216[bottom]: Glutamate Decarboxylase

EXCITED STATES OF BIOPOLYMERS

Edited by Robert F. Steiner
University of Maryland, Catonsville, Maryland

Fluorescence techniques have, over the past ten years, assumed central importance in the field of biochemistry. Many laser techniques have evolved from esoteric research specialties into standard procedures that are applied routinely to biochemical problems. Owing to the many technological advances made, this book represents a new and timely study on selected topics of major importance to this growing field of research.

Presenting the most up-to-date material, the contributors to this volume provide extensive coverage on areas such as laser techniques, radiationless energy transfer, and nanosecond fluorometry. Also covered here in exhaustive detail are topics concerned with fluorescent probes, laser-related research, lipoproteins, and molecular dynamics.

Excited States of Biopolymers will be especially valuable to those working in the fields of photobiology, biochemistry, and organic chemistry. Workers and researchers in molecular biology, biophysics, physical chemistry, and analytical chemistry and applied spectroscopy will also find much interesting material here.

CONTENTS: Some principles governing the luminescence of organic molecules, ***Robin M. Hochstrasser***. Covalent fluorescent probes, ***Richard P. Haugland***. Nanosecond pulse fluorimetry of proteins, ***Tsung-I Lin and Robert M. Dowben***. The use of fluorescence anisotropy decay in the study of biological macromolecules, ***Robert F. Steiner***. Plasma lipoproteins: Fluorescence as a probe of structure and dynamics, ***William W. Mantulin and Henry J. Pownall***. Fluorescent dye-nucleic acid complexes, ***Robert F. Steiner and Yukio Kubota***. Index.

272 pp., illus., 1983, ISBN 0-306-41316-7

MIX
Papier aus verantwortungsvollen Quellen
Paper from responsible sources
FSC® C105338

If you have any concerns about our products,
you can contact us on
ProductSafety@springernature.com

In case Publisher is established outside the EU,
the EU authorized representative is:
Springer Nature Customer Service Center GmbH
Europaplatz 3, 69115 Heidelberg, Germany

Printed by Libri Plureos GmbH
in Hamburg, Germany